W9-CFA-365

# Computational Statistics Handbook with
# MATLAB®

# Computational Statistics Handbook with MATLAB®

Wendy L. Martinez
Angel R. Martinez

CHAPMAN & HALL/CRC

Boca Raton   London   New York   Washington, D.C.

## Library of Congress Cataloging-in-Publication Data

Catalog record is available from the Library of Congress

**Visit the CRC Press Web site at www.crcpress.com**

© 2002 by Chapman & Hall/CRC

No claim to original U.S. Government works
International Standard Book Number 1-58488-229-8
Printed in the United States of America 1 2 3 4 5 6 7 8 9 0
Printed on acid-free paper

*To*

*Edward J. Wegman*

*Teacher, Mentor and Friend*

# Table of Contents

# Chapter 7
## Data Partitioning

# Chapter 8
## Probability Density Estimation

# Chapter 9
## Statistical Pattern Recognition

# Chapter 10
## Nonparametric Regression

# *Preface*

Computational statistics is a fascinating and relatively new field within statistics. While much of classical statistics relies on parameterized functions and related assumptions, the computational statistics approach is to let the data tell the story. The advent of computers with their number-crunching capability, as well as their power to show on the screen two- and three-dimensional structures, has made computational statistics available for any data analyst to use.

Computational statistics has a lot to offer the researcher faced with a file full of numbers. The methods of computational statistics can provide assistance ranging from preliminary exploratory data analysis to sophisticated probability density estimation techniques, Monte Carlo methods, and powerful multi-dimensional visualization. All of this power and novel ways of looking at data are accessible to researchers in their daily data analysis tasks. One purpose of this book is to facilitate the exploration of these methods and approaches and to provide the tools to make of this, not just a theoretical exploration, but a practical one. The two main goals of this book are:

- To make computational statistics techniques available to a wide range of users, including engineers and scientists, and
- To promote the use of MATLAB® by statisticians and other data analysts.

MATLAB and Handle Graphics® are registered trademarks of The MathWorks, Inc.

There are wonderful books that cover many of the techniques in computational statistics and, in the course of this book, references will be made to many of them. However, there are very few books that have endeavored to forgo the theoretical underpinnings to present the methods and techniques in a manner immediately usable to the practitioner. The approach we take in this book is to make computational statistics accessible to a wide range of users and to provide an understanding of statistics from a computational point of view via algorithms applied to real applications.

This book is intended for researchers in engineering, statistics, psychology, biostatistics, data mining and any other discipline that must deal with the analysis of raw data. Students at the senior undergraduate level or beginning graduate level in statistics or engineering can use the book to supplement course material. Exercises are included with each chapter, making it suitable as a textbook for a course in computational statistics and data analysis. Scien-

tists who would like to know more about programming methods for analyzing data in MATLAB would also find it useful.

We assume that the reader has the following background:

- Calculus: Since this book is computational in nature, the reader
  needs only a rudimentary knowledge of calculus. Knowing the
  definition of a derivative and an integral is all that is required.
- Linear Algebra: Since MATLAB is an array-based computing language, we cast several of the algorithms in terms of matrix algebra.
  The reader should have a familiarity with the notation of linear
  algebra, array multiplication, inverses, determinants, an array
  transpose, etc.
- Probability and Statistics: We assume that the reader has had introductory probability and statistics courses. However, we provide a
  brief overview of the relevant topics for those who might need a
  refresher.

We list below some of the major features of the book.

- The focus is on implementation rather than theory, helping the
  reader understand the concepts without being burdened by the
  theory.
- References that explain the theory are provided at the end of each
  chapter. Thus, those readers who need the theoretical underpinnings will know where to find the information.
- Detailed step-by-step algorithms are provided to facilitate implementation in any computer programming language or appropriate
  software. This makes the book appropriate for computer users who
  do not know MATLAB.
- MATLAB code in the form of a Computational Statistics Toolbox
  is provided. These functions are available for download at:

  `http://www.infinityassociates.com`

  `http://lib.stat.cmu.edu`.

  Please review the **readme** file for installation instructions and information on any changes.
- Exercises are given at the end of each chapter. The reader is encouraged to go through these, because concepts are sometimes explored
  further in them. Exercises are computational in nature, which is in
  keeping with the philosophy of the book.
- Many data sets are included with the book, so the reader can apply
  the methods to real problems and verify the results shown in the
  book. The data can also be downloaded separately from the toolbox
  at `http://www.infinityassociates.com`. The data are pro-

vided in MATLAB binary files (`.mat`) as well as text, for those who want to use them with other software.

- Typing in all of the commands in the examples can be frustrating. So, MATLAB scripts containing the commands used in the examples are also available for download at

    `http://www.infinityassociates.com`.

- A brief introduction to MATLAB is provided in Appendix A. Most of the constructs and syntax that are needed to understand the programming contained in the book are explained.

- An index of notation is given in Appendix B. Definitions and page numbers are provided, so the user can find the corresponding explanation in the text.

- Where appropriate, we provide references to internet resources for computer code implementing the algorithms described in the chapter. These include code for MATLAB, S-plus, Fortran, etc.

We would like to acknowledge the invaluable help of the reviewers: Noel Cressie, James Gentle, Thomas Holland, Tom Lane, David Marchette, Christian Posse, Carey Priebe, Adrian Raftery, David Scott, Jeffrey Solka, and Clifton Sutton. Their many helpful comments made this book a much better product. Any shortcomings are the sole responsibility of the authors. We owe a special thanks to Jeffrey Solka for some programming assistance with finite mixtures. We greatly appreciate the help and patience of those at CRC Press: Bob Stern, Joanne Blake, and Evelyn Meany. We also thank Harris Quesnell and James Yanchak for their help with resolving font problems. Finally, we are indebted to Naomi Fernandes and Tom Lane at The MathWorks, Inc. for their special assistance with MATLAB.

### Disclaimers

1. Any MATLAB programs and data sets that are included with the book are provided in good faith. The authors, publishers or distributors do not guarantee their accuracy and are not responsible for the consequences of their use.

2. The views expressed in this book are those of the authors and do not necessarily represent the views of DoD or its components.

Wendy L. and Angel R. Martinez
August 2001

# Chapter 1

## Introduction

## 1.1 What Is Computational Statistics?

Obviously, computational statistics relates to the traditional discipline of statistics. So, before we define computational statistics proper, we need to get a handle on what we mean by the field of statistics. At a most basic level, statistics is concerned with the transformation of raw data into knowledge [Wegman, 1988].

When faced with an application requiring the analysis of raw data, any scientist must address questions such as:

- What data should be collected to answer the questions in the analysis?
- How much data should be collected?
- What conclusions can be drawn from the data?
- How far can those conclusions be trusted?

Statistics is concerned with the science of uncertainty and can help the scientist deal with these questions. Many classical methods (regression, hypothesis testing, parameter estimation, confidence intervals, etc.) of statistics developed over the last century are familiar to scientists and are widely used in many disciplines [Efron and Tibshirani, 1991].

Now, what do we mean by computational statistics? Here we again follow the definition given in Wegman [1988]. Wegman defines computational statistics as a collection of techniques that have a strong "focus on the exploitation of computing in the creation of new statistical methodology."

Many of these methodologies became feasible after the development of inexpensive computing hardware since the 1980's. This computing revolution has enabled scientists and engineers to store and process massive amounts of data. However, these data are typically collected without a clear idea of what they will be used for in a study. For instance, in the practice of data analysis today, we often collect the data and then we design a study to gain some useful information from them. In contrast, the traditional approach has been to first design the study based on research questions and then collect the required data.

Because the storage and collection is so cheap, the data sets that analysts must deal with today tend to be very large and high-dimensional. It is in situations like these where many of the classical methods in statistics are inadequate. As examples of computational statistics methods, Wegman [1988] includes parallel coordinates for high dimensional data representation, nonparametric functional inference, and data set mapping where the analysis techniques are considered fixed.

Efron and Tibshirani [1991] refer to what we call computational statistics as *computer-intensive statistical methods*. They give the following as examples for these types of techniques: bootstrap methods, nonparametric regression, generalized additive models and classification and regression trees. They note that these methods differ from the classical methods in statistics because they substitute computer algorithms for the more traditional mathematical method of obtaining an answer. An important aspect of computational statistics is that the methods free the analyst from choosing methods mainly because of their mathematical tractability.

Volume 9 of the *Handbook of Statistics: Computational Statistics* [Rao, 1993] covers topics that illustrate the "... trend in modern statistics of basic methodology supported by the state-of-the-art computational and graphical facilities..." It includes chapters on computing, density estimation, Gibbs sampling, the bootstrap, the jackknife, nonparametric function estimation, statistical visualization, and others.

We mention the topics that can be considered part of computational statistics to help the reader understand the difference between these and the more traditional methods of statistics. Table 1.1 [Wegman, 1988] gives an excellent comparison of the two areas.

## 1.2 An Overview of the Book

### Philosophy

The focus of this book is on methods of computational statistics and how to implement them. We leave out much of the theory, so the reader can concentrate on how the techniques may be applied. In many texts and journal articles, the theory obscures implementation issues, contributing to a loss of interest on the part of those needing to apply the theory. The reader should not misunderstand, though; the methods presented in this book are built on solid mathematical foundations. Therefore, at the end of each chapter, we

**TABLE 1.1**

Comparison Between Traditional Statistics and Computational Statistics [Wegman, 1988]. Reprinted with permission from the *Journal of the Washington Academy of Sciences.*

| Traditional Statistics | Computational Statistics |
|---|---|
| Small to moderate sample size | Large to very large sample size |
| Independent, identically distributed data sets | Nonhomogeneous data sets |
| One or low dimensional | High dimensional |
| Manually computational | Computationally intensive |
| Mathematically tractable | Numerically tractable |
| Well focused questions | Imprecise questions |
| Strong unverifiable assumptions: <br> Relationships (linearity, additivity) <br> Error structures (normality) | Weak or no assumptions: <br> Relationships (nonlinearity) <br> Error structures (distribution free) |
| Statistical inference | Structural inference |
| Predominantly closed form algorithms | Iterative algorithms possible |
| Statistical optimality | Statistical robustness |

include a section containing references that explain the theoretical concepts associated with the methods covered in that chapter.

## What Is Covered

In this book, we cover some of the most commonly used techniques in computational statistics. While we cannot include all methods that might be a part of computational statistics, we try to present those that have been in use for several years.

Since the focus of this book is on the implementation of the methods, we include algorithmic descriptions of the procedures. We also provide examples that illustrate the use of the algorithms in data analysis. It is our hope that seeing how the techniques are implemented will help the reader understand the concepts and facilitate their use in data analysis.

Some background information is given in Chapters 2, 3, and 4 for those who might need a refresher in probability and statistics. In Chapter 2, we discuss some of the general concepts of probability theory, focusing on how they will be used in later chapters of the book. Chapter 3 covers some of the basic ideas of statistics and sampling distributions. Since many of the methods in computational statistics are concerned with estimating distributions via simulation, this chapter is fundamental to the rest of the book. For the same reason, we present some techniques for generating random variables in Chapter 4.

Some of the methods in computational statistics enable the researcher to explore the data before other analyses are performed. These techniques are especially important with high dimensional data sets or when the questions to be answered using the data are not well focused. In Chapter 5, we present some graphical exploratory data analysis techniques that could fall into the category of traditional statistics (e.g., box plots, scatterplots). We include them in this text so statisticians can see how to implement them in MATLAB and to educate scientists and engineers as to their usage in exploratory data analysis. Other graphical methods in this chapter do fall into the category of computational statistics. Among these are isosurfaces, parallel coordinates, the grand tour and projection pursuit.

In Chapters 6 and 7, we present methods that come under the general heading of resampling. We first cover some of the general concepts in hypothesis testing and confidence intervals to help the reader better understand what follows. We then provide procedures for hypothesis testing using simulation, including a discussion on evaluating the performance of hypothesis tests. This is followed by the bootstrap method, where the data set is used as an estimate of the population and subsequent sampling is done from the sample. We show how to get bootstrap estimates of standard error, bias and confidence intervals. Chapter 7 continues with two closely related methods called jackknife and cross-validation.

One of the important applications of computational statistics is the estimation of probability density functions. Chapter 8 covers this topic, with an emphasis on the nonparametric approach. We show how to obtain estimates using probability density histograms, frequency polygons, averaged shifted histograms, kernel density estimates, finite mixtures and adaptive mixtures.

Chapter 9 uses some of the concepts from probability density estimation and cross-validation. In this chapter, we present some techniques for statistical pattern recognition. As before, we start with an introduction of the classical methods and then illustrate some of the techniques that can be considered part of computational statistics, such as classification trees and clustering.

In Chapter 10 we describe some of the algorithms for nonparametric regression and smoothing. One nonparametric technique is a tree-based method called regression trees. Another uses the kernel densities of Chapter 8. Finally, we discuss smoothing using loess and its variants.

An approach for simulating a distribution that has become widely used over the last several years is called Markov chain Monte Carlo. Chapter 11 covers this important topic and shows how it can be used to simulate a posterior distribution. Once we have the posterior distribution, we can use it to estimate statistics of interest (means, variances, etc.).

We conclude the book with a chapter on spatial statistics as a way of show-ing how some of the methods can be employed in the analysis of spatial data. We provide some background on the different types of spatial data analysis, but we concentrate on spatial point patterns only. We apply kernel density estimation, exploratory data analysis, and simulation-based hypothesis test-ing to the investigation of spatial point processes.

We also include several appendices to aid the reader. Appendix A contains a brief introduction to MATLAB, which should help readers understand the code in the examples and exercises. Appendix B is an index to notation, with definitions and references to where it is used in the text. Appendices C and D include some further information about projection pursuit and MATLAB source code that is too lengthy for the body of the text. In Appendices E and F, we provide a list of the functions that are contained in the MATLAB Statis-tics Toolbox and the Computational Statistics Toolbox, respectively. Finally, in Appendix G, we include a brief description of the data sets that are men-tioned in the book.

### A Word About Notation

The explanation of the algorithms in computational statistics (and the under-standing of them!) depends a lot on notation. In most instances, we follow the notation that is used in the literature for the corresponding method. Rather than try to have unique symbols throughout the book, we think it is more important to be faithful to the convention to facilitate understanding of the theory and to make it easier for readers to make the connection between the theory and the text. Because of this, the same symbols might be used in sev-eral places.

In general, we try to stay with the convention that random variables are capital letters, whereas small letters refer to realizations of random variables. For example, $X$ is a random variable, and $x$ is an observed value of that ran-dom variable. When we use the term *log*, we are referring to the natural log-arithm.

A symbol that is in bold refers to an array. Arrays can be row vectors, col-umn vectors or matrices. Typically, a matrix is represented by a bold capital letter such as $\mathbf{B}$, while a vector is denoted by a bold lowercase letter such as $\mathbf{b}$. When we are using explicit matrix notation, then we specify the dimen-sions of the arrays. Otherwise, we do not hold to the convention that a vector always has to be in a column format. For example, we might represent a vec-tor of observed random variables as $(x_1, x_2, x_3)$ or a vector of parameters as $(\mu, \sigma)$.

## 1.3 MATLAB Code

Along with the algorithmic explanation of the procedures, we include MATLAB commands to show how they are implemented. Any MATLAB commands, functions or data sets are in courier bold font. For example, **plot** denotes the MATLAB plotting function. The commands that are in the examples can be typed in at the command line to execute the examples. However, we note that due to typesetting considerations, we often have to continue a MATLAB command using the continuation punctuation (**...**). However, users do not have to include that with their implementations of the algorithms. See Appendix A for more information on how this punctuation is used in MATLAB.

Since this is a book about computational statistics, we assume the reader has the MATLAB Statistics Toolbox. In Appendix E, we include a list of functions that are in the toolbox and try to note in the text what functions are part of the main MATLAB software package and what functions are available only in the Statistics Toolbox.

The choice of MATLAB for implementation of the methods is due to the following reasons:

- The commands, functions and arguments in MATLAB are not cryptic. It is important to have a programming language that is easy to understand and intuitive, since we include the programs to help teach the concepts.
- It is used extensively by scientists and engineers.
- Student versions are available.
- It is easy to write programs in MATLAB.
- The source code or M-files can be viewed, so users can learn about the algorithms and their implementation.
- User-written MATLAB programs are freely available.
- The graphics capabilities are excellent.

It is important to note that the MATLAB code given in the body of the book is for *learning purposes*. In many cases, it is not the most efficient way to program the algorithm. One of the purposes of including the MATLAB code is to help the reader understand the algorithms, especially how to implement them. So, we try to have the code match the procedures and to stay away from cryptic programming constructs. For example, we use **for** loops at times (when unnecessary!) to match the procedure. We make no claims that our code is the best way or the only way to program the algorithms.

In some cases, the MATLAB code is contained in an appendix, rather than in the corresponding chapter. These are applications where the MATLAB program does not provide insights about the algorithms. For example, with classification and regression trees, the code can be quite complicated in places, so the functions are relegated to an appendix (Appendix D). Including these in the body of the text would distract the reader from the important concepts being presented.

## Computational Statistics Toolbox

The majority of the algorithms covered in this book are not available in MATLAB. So, we provide functions that implement most of the procedures that are given in the text. Note that these functions are a little different from the MATLAB code provided in the examples. In most cases, the functions allow the user to implement the algorithms for the general case. A list of the functions and their purpose is given in Appendix F. We also give a summary of the appropriate functions at the end of each chapter.

The MATLAB functions for the book are part of what we are calling the Computational Statistics Toolbox. To make it easier to recognize these functions, we put the letters '**cs**' in front. The toolbox can be downloaded from

- `http://lib.stat.cmu.edu`
- `http://www.infinityassociates.com`

Information on installing the toolbox is given in the **readme** file and on the website.

## Internet Resources

One of the many strong points about MATLAB is the availability of functions written by users, most of which are freely available on the internet. With each chapter, we provide information about internet resources for MATLAB programs (and other languages) that pertain to the techniques covered in the chapter.

The following are some internet sources for MATLAB code. Note that these are not necessarily specific to statistics, but are for all areas of science and engineering.

- The main website at The MathWorks, Inc. has code written by users and technicians of the company. The website for user contributed M-files is:

  `http://www.mathworks.com/support/ftp/`

  The website for M-files contributed by The MathWorks, Inc. is:

  `ftp://ftp.mathworks.com/pub/mathworks/`
- Another excellent resource for MATLAB programs is

`http://www.mathtools.net`.

At this site, you can sign up to be notified of new submissions.

- The main website for user contributed statistics programs is StatLib at Carnegie Mellon University. They have a new section containing MATLAB code. The home page for StatLib is

  `http://lib.stat.cmu.edu`

- We also provide the following internet sites that contain a list of MATLAB code available for purchase or download.

  `http://dmoz.org/Science/Math/Software/MATLAB/`

  `http://directory.google.com/Top/`
  `Science/Math/Software/MATLAB/`

---

## 1.4 Further Reading

To gain more insight on what is computational statistics, we refer the reader to the seminal paper by Wegman [1988]. Wegman discusses many of the differences between traditional and computational statistics. He also includes a discussion on what a graduate curriculum in computational statistics should consist of and contrasts this with the more traditional course work. A later paper by Efron and Tibshirani [1991] presents a summary of the new focus in statistical data analysis that came about with the advent of the computer age. Other papers in this area include Hoaglin and Andrews [1975] and Efron [1979]. Hoaglin and Andrews discuss the connection between computing and statistical theory and the importance of properly reporting the results from simulation experiments. Efron's article presents a survey of computational statistics techniques (the jackknife, the bootstrap, error estimation in discriminant analysis, nonparametric methods, and more) for an audience with a mathematics background, but little knowledge of statistics. Chambers [1999] looks at the concepts underlying computing with data, including the challenges this presents and new directions for the future.

There are very few general books in the area of computational statistics. One is a compendium of articles edited by C. R. Rao [1993]. This is a fairly comprehensive overview of many topics pertaining to computational statistics. The new text by Gentle [2001] is an excellent resource in computational statistics for the student or researcher. A good reference for statistical computing is Thisted [1988].

For those who need a resource for learning MATLAB, we recommend a wonderful book by Hanselman and Littlefield [1998]. This gives a comprehensive overview of MATLAB Version 5 and has been updated for Version 6 [Hanselman and Littlefield, 2001]. These books have information about the many capabilities of MATLAB, how to write programs, graphics and GUIs, and much more. For the beginning user of MATLAB, these are a good place to start.

# Chapter 2

## Probability Concepts

## 2.1 Introduction

A review of probability is covered here at the outset because it provides the foundation for what is to follow: computational statistics. Readers who understand probability concepts may safely skip over this chapter.

Probability is the mechanism by which we can manage the uncertainty that underlies all real world data and phenomena. It enables us to gauge our degree of belief and to quantify the lack of certitude that is inherent in the process that generates the data we are analyzing. For example:

- To understand and use statistical hypothesis testing, one needs knowledge of the sampling distribution of the test statistic.
- To evaluate the performance (e.g., standard error, bias, etc.) of an estimate, we must know its sampling distribution.
- To adequately simulate a real system, one needs to understand the probability distributions that correctly model the underlying processes.
- To build classifiers to predict what group an object belongs to based on a set of features, one can estimate the probability density function that describes the individual classes.

In this chapter, we provide a brief overview of probability concepts and distributions as they pertain to computational statistics. In Section 2.2, we define probability and discuss some of its properties. In Section 2.3, we cover conditional probability, independence and Bayes' Theorem. Expectations are defined in Section 2.4, and common distributions and their uses in modeling physical phenomena are discussed in Section 2.5. In Section 2.6, we summarize some MATLAB functions that implement the ideas from Chapter 2. Finally, in Section 2.7 we provide additional resources for the reader who requires a more theoretical treatment of probability.

## 2.2 Probability

### Background

A *random experiment* is defined as a process or action whose outcome cannot be predicted with certainty and would likely change when the experiment is repeated. The variability in the outcomes might arise from many sources: slight errors in measurements, choosing different objects for testing, etc. The ability to model and analyze the outcomes from experiments is at the heart of statistics. Some examples of random experiments that arise in different disciplines are given below.

- Engineering: Data are collected on the number of failures of piston rings in the legs of steam-driven compressors. Engineers would be interested in determining the probability of piston failure in each leg and whether the failure varies among the compressors [Hand, et al., 1994].
- Medicine: The oral glucose tolerance test is a diagnostic tool for early diabetes mellitus. The results of the test are subject to variation because of different rates at which people absorb the glucose, and the variation is particularly noticeable in pregnant women. Scientists would be interested in analyzing and modeling the variation of glucose before and after pregnancy [Andrews and Herzberg, 1985].
- Manufacturing: Manufacturers of cement are interested in the tensile strength of their product. The strength depends on many factors, one of which is the length of time the cement is dried. An experiment is conducted where different batches of cement are tested for tensile strength after different drying times. Engineers would like to determine the relationship between drying time and tensile strength of the cement [Hand, et al., 1994].
- Software Engineering: Engineers measure the failure times in CPU seconds of a command and control software system. These data are used to obtain models to predict the reliability of the software system [Hand, et al., 1994].

The *sample space* is the set of all outcomes from an experiment. It is possible sometimes to list all outcomes in the sample space. This is especially true in the case of some discrete random variables. Examples of these sample spaces are:

- When observing piston ring failures, the sample space is $\{1, 0\}$, where 1 represents a failure and 0 represents a non-failure.
- If we roll a six-sided die and count the number of dots on the face, then the sample space is $\{1, 2, 3, 4, 5, 6\}$.

The outcomes from random experiments are often represented by an uppercase variable such as $X$. This is called a *random variable*, and its value is subject to the uncertainty intrinsic to the experiment. Formally, a random variable is a real-valued function defined on the sample space. As we see in the remainder of the text, a random variable can take on different values according to a probability distribution. Using our examples of experiments from above, a random variable $X$ might represent the failure time of a software system or the glucose level of a patient. The observed value of a random variable $X$ is denoted by a lowercase $x$. For instance, a random variable $X$ might represent the number of failures of piston rings in a compressor, and $x = 5$ would indicate that we observed 5 piston ring failures.

Random variables can be discrete or continuous. A *discrete random variable* can take on values from a finite or countably infinite set of numbers. Examples of discrete random variables are the number of defective parts or the number of typographical errors on a page. A *continuous random variable* is one that can take on values from an interval of real numbers. Examples of continuous random variables are the inter-arrival times of planes at a runway, the average weight of tablets in a pharmaceutical production line or the average voltage of a power plant at different times.

We cannot list all outcomes from an experiment when we observe a continuous random variable, because there are an infinite number of possibilities. However, we could specify the interval of values that $X$ can take on. For example, if the random variable $X$ represents the tensile strength of cement, then the sample space might be $(0, \infty)$ kg/cm$^2$.

An *event* is a subset of outcomes in the sample space. An event might be that a piston ring is defective or that the tensile strength of cement is in the range 40 to 50 kg/cm$^2$. The probability of an event is usually expressed using the random variable notation illustrated below.

- Discrete Random Variables: Letting 1 represent a defective piston ring and letting 0 represent a good piston ring, then the probability of the event that a piston ring is defective would be written as

$$P(X = 1).$$

- Continuous Random Variables: Let $X$ denote the tensile strength of cement. The probability that an observed tensile strength is in the range 40 to 50 kg/cm$^2$ is expressed as

$$P(40 \text{ kg/cm}^2 \le X \le 50 \text{ kg/cm}^2).$$

Some events have a special property when they are considered together. Two events that cannot occur simultaneously or jointly are called *mutually exclusive events*. This means that the intersection of the two events is the empty set and the probability of the events occurring together is zero. For example, a piston ring cannot be both defective and good at the same time. So, the event of getting a defective part and the event of getting a good part are mutually exclusive events. The definition of mutually exclusive events can be extended to any number of events by considering all pairs of events. Every pair of events must be mutually exclusive for all of them to be mutually exclusive.

### Probability

*Probability* is a measure of the likelihood that some event will occur. It is also a way to quantify or to gauge the likelihood that an observed measurement or random variable will take on values within some set or range of values. Probabilities always range between 0 and 1. A *probability distribution* of a random variable describes the probabilities associated with each possible value for the random variable.

We first briefly describe two somewhat classical methods for assigning probabilities: the *equal likelihood model* and the *relative frequency method*. When we have an experiment where each of $n$ outcomes is equally likely, then we assign a probability mass of $1/n$ to each outcome. This is the equal likelihood model. Some experiments where this model can be used are flipping a fair coin, tossing an unloaded die or randomly selecting a card from a deck of cards.

When the equal likelihood assumption is not valid, then the relative frequency method can be used. With this technique, we conduct the experiment $n$ times and record the outcome. The probability of event $E$ is assigned by $P(E) = f/n$, where $f$ denotes the number of experimental outcomes that satisfy event $E$.

Another way to find the desired probability that an event occurs is to use a *probability density function* when we have continuous random variables or a *probability mass function* in the case of discrete random variables. Section 2.5 contains several examples of probability density (mass) functions. In this text, $f(x)$ is used to represent the probability mass or density function for either discrete or continuous random variables, respectively. We now discuss how to find probabilities using these functions, first for the continuous case and then for discrete random variables.

To find the probability that a continuous random variable falls in a particular interval of real numbers, we have to calculate the appropriate area under the curve of $f(x)$. Thus, we have to evaluate the integral of $f(x)$ over the interval of random variables corresponding to the event of interest. This is represented by

$$P(a \le X \le b) = \int_a^b f(x)dx. \tag{2.1}$$

The area under the curve of $f(x)$ between $a$ and $b$ represents the probability that an observed value of the random variable $X$ will assume a value between $a$ and $b$. This concept is illustrated in Figure 2.1 where the shaded area represents the desired probability.

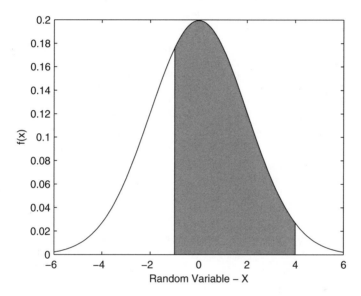

**FIGURE 2.1**
The area under the curve of $f(x)$ between -1 and 4 is the same as the probability that an observed value of the random variable will assume a value in the same interval.

It should be noted that a valid probability density function should be non-negative, and the total area under the curve must equal 1. If this is not the case, then the probabilities will not be properly restricted to the interval $[0, 1]$. This will be an important consideration in Chapter 8 where we discuss probability density estimation techniques.

The *cumulative distribution function* $F(x)$ is defined as the probability that the random variable $X$ assumes a value less than or equal to a given $x$. This is calculated from the probability density function, as follows

$$F(x) = P(X \le x) = \int_{-\infty}^x f(t)dt. \tag{2.2}$$

It is obvious that the cumulative distribution function takes on values between 0 and 1, so $0 \le F(x) \le 1$. A probability density function, along with its associated cumulative distribution function are illustrated in Figure 2.2.

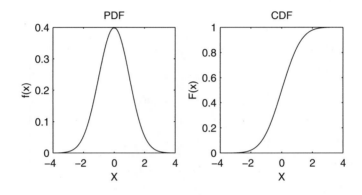

**FIGURE 2.2**
This shows the probability density function on the left with the associated cumulative distribution function on the right. Notice that the cumulative distribution function takes on values between 0 and 1.

For a discrete random variable $X$, that can take on values $x_1, x_2, \ldots$, the probability mass function is given by

$$f(x_i) = P(X = x_i); \qquad i = 1, 2, \ldots, \qquad (2.3)$$

and the cumulative distribution function is

$$F(a) = \sum_{x_i \le a} f(x_i); \qquad i = 1, 2, \ldots. \qquad (2.4)$$

## Axioms of Probability

Probabilities follow certain axioms that can be useful in computational statistics. We let $S$ represent the sample space of an experiment and $E$ represent some event that is a subset of $S$.

*AXIOM 1*
*The probability of event E must be between 0 and 1:*

$$0 \leq P(E) \leq 1.$$

*AXIOM 2*

$$P(S) = 1.$$

*AXIOM 3*
*For mutually exclusive events, $E_1, E_2, ..., E_k$,*

$$P(E_1 \cup E_2 \cup ... \cup E_k) = \sum_{i=1}^{k} P(E_i).$$

Axiom 1 has been discussed before and simply states that a probability must be between 0 and 1. Axiom 2 says that an outcome from our experiment must occur, and the probability that the outcome is in the sample space is 1. Axiom 3 enables us to calculate the probability that at least one of the mutually exclusive events $E_1, E_2, ..., E_k$ occurs by summing the individual probabilities.

## 2.3 Conditional Probability and Independence

### Conditional Probability

Conditional probability is an important concept. It is used to define independent events and enables us to revise our degree of belief given that another event has occurred. Conditional probability arises in situations where we need to calculate a probability based on some partial information concerning the experiment.

The *conditional probability* of event $E$ given event $F$ is defined as follows:

*CONDITIONAL PROBABILITY*

$$P(E|F) = \frac{P(E \cap F)}{P(F)}; \quad P(F) > 0. \tag{2.5}$$

Here $P(E \cap F)$ represents the **joint probability** that both $E$ and $F$ occur together and $P(F)$ is the probability that event $F$ occurs. We can rearrange Equation 2.5 to get the following rule:

*MULTIPLICATION RULE*

$$P(E \cap F) = P(F)P(E|F). \tag{2.6}$$

## Independence

Often we can assume that the occurrence of one event does not affect whether or not some other event happens. For example, say a couple would like to have two children, and their first child is a boy. The gender of their second child does not depend on the gender of the first child. Thus, the fact that we know they have a boy already does not change the probability that the second child is a boy. Similarly, we can sometimes assume that the value we observe for a random variable is not affected by the observed value of other random variables.

These types of events and random variables are called **independent**. If *events* are independent, then knowing that one event has occurred does not change our degree of belief or the likelihood that the other event occurs. If *random variables* are independent, then the observed value of one random variable does not affect the observed value of another.

In general, the conditional probability $P(E|F)$ is not equal to $P(E)$. In these cases, the events are called **dependent**. Sometimes we can assume independence based on the situation or the experiment, which was the case with our example above. However, to show independence mathematically, we must use the following definition.

*INDEPENDENT EVENTS*
*Two events E and F are said to be independent if and only if any of the following is true:*

$$\begin{aligned} P(E \cap F) &= P(E)P(F), \\ P(E) &= P(E|F). \end{aligned} \tag{2.7}$$

Note that if events $E$ and $F$ are independent, then the Multiplication Rule in Equation 2.6 becomes

$$P(E \cap F) = P(F)P(E),$$

which means that we simply multiply the individual probabilities for each event together. This can be extended to $k$ events to give

$$P(E_1 \cap E_2 \cap \ldots \cap E_k) = \prod_{i=1}^{k} P(E_i), \tag{2.8}$$

where events $E_i$ and $E_j$ (for all $i$ and $j$, $i \neq j$) are independent.

## Bayes Theorem

Sometimes we start an analysis with an initial degree of belief that an event will occur. Later on, we might obtain some additional information about the event that would change our belief about the probability that the event will occur. The initial probability is called a *prior probability*. Using the new information, we can update the prior probability using Bayes' Theorem to obtain the *posterior probability*.

The experiment of recording piston ring failure in compressors is an example of where Bayes' Theorem might be used, and we derive Bayes' Theorem using this example. Suppose our piston rings are purchased from two manufacturers: 60% from manufacturer A and 40% from manufacturer B.

Let $M_A$ denote the event that a part comes from manufacturer A, and $M_B$ represent the event that a piston ring comes from manufacturer B. If we select a part at random from our supply of piston rings, we would assign probabilities to these events as follows:

$$P(M_A) = 0.6,$$
$$P(M_B) = 0.4.$$

These are our prior probabilities that the piston rings are from the individual manufacturers.

Say we are interested in knowing the probability that a piston ring that subsequently failed came from manufacturer A. This would be the posterior probability that it came from manufacturer A, given that the piston ring failed. The additional information we have about the piston ring is that it failed, and we use this to update our degree of belief that it came from manufacturer A.

Bayes' Theorem can be derived from the definition of conditional probability (Equation 2.5). Writing this in terms of our events, we are interested in the following probability:

$$P(M_A|F) = \frac{P(M_A \cap F)}{P(F)}, \tag{2.9}$$

where $P(M_A|F)$ represents the posterior probability that the part came from manufacturer A, and $F$ is the event that the piston ring failed. Using the Multiplication Rule (Equation 2.6), we can write the numerator of Equation 2.9 in terms of event $F$ and our prior probability that the part came from manufacturer A, as follows

$$P(M_A|F) = \frac{P(M_A \cap F)}{P(F)} = \frac{P(M_A)P(F|M_A)}{P(F)}. \tag{2.10}$$

The next step is to find $P(F)$. The only way that a piston ring will fail is if: 1) it failed and it came from manufacturer A or 2) it failed and it came from manufacturer B. Thus, using the third axiom of probability, we can write

$$P(F) = P(M_A \cap F) + P(M_B \cap F).$$

Applying the Multiplication Rule as before, we have

$$P(F) = P(M_A)P(F|M_A) + P(M_B)P(F|M_B). \tag{2.11}$$

Substituting this for $P(F)$ in Equation 2.10, we write the posterior probability as

$$P(M_A|F) = \frac{P(M_A)P(F|M_A)}{P(M_A)P(F|M_A) + P(M_B)P(F|M_B)}. \tag{2.12}$$

Note that we need to find the probabilities $P(F|M_A)$ and $P(F|M_B)$. These are the probabilities that a piston ring will fail given it came from the corresponding manufacturer. These must be estimated in some way using available information (e.g., past failures). When we revisit Bayes' Theorem in the context of statistical pattern recognition (Chapter 9), these are the probabilities that are estimated to construct a certain type of classifier.

Equation 2.12 is Bayes' Theorem for a situation where only two outcomes are possible. In general, Bayes' Theorem can be written for any number of mutually exclusive events, $E_1, \ldots, E_k$, whose union makes up the entire sample space. This is given below.

*BAYES' THEOREM*

$$P(E_i|F) = \frac{P(E_i)P(F|E_i)}{P(E_1)P(F|E_1) + \ldots + P(E_k)P(F|E_k)}. \qquad (2.13)$$

## 2.4 Expectation

Expected values and variances are important concepts in statistics. They are used to describe distributions, to evaluate the performance of estimators, to obtain test statistics in hypothesis testing, and many other applications.

### Mean and Variance

The *mean* or *expected value* of a random variable is defined using the probability density (mass) function. It provides a measure of central tendency of the distribution. If we observe many values of the random variable and take the average of them, we would expect that value to be close to the mean. The expected value is defined below for the discrete case.

*EXPECTED VALUE - DISCRETE RANDOM VARIABLES*

$$\mu = E[X] = \sum_{i=1}^{\infty} x_i f(x_i). \qquad (2.14)$$

We see from the definition that the expected value is a sum of all possible values of the random variable where each one is weighted by the probability that X will take on that value.

The *variance* of a discrete random variable is given by the following definition.

*VARIANCE - DISCRETE RANDOM VARIABLES*
*For $\mu < \infty$,*

$$\sigma^2 = V(X) = E[(X-\mu)^2] = \sum_{i=1}^{\infty} (x_i - \mu)^2 f(x_i). \qquad (2.15)$$

From Equation 2.15, we see that the variance is the sum of the squared distances, each one weighted by the probability that $X = x_i$. Variance is a measure of dispersion in the distribution. If a random variable has a large variance, then an observed value of the random variable is more likely to be far from the mean $\mu$. The standard deviation $\sigma$ is the square root of the variance.

The mean and variance for continuous random variables are defined similarly, with the summation replaced by an integral. The mean and variance of a continuous random variable are given below.

*EXPECTED VALUE - CONTINUOUS RANDOM VARIABLES*

$$\mu = E[X] = \int_{-\infty}^{\infty} xf(x)dx.$$

(2.16)

*VARIANCE - CONTINUOUS RANDOM VARIABLES*
*For* $\mu < \infty$,

$$\sigma^2 = V(X) = E[(X-\mu)^2] = \int_{-\infty}^{\infty} (x-\mu)^2 f(x)dx.$$

(2.17)

We note that Equation 2.17 can also be written as

$$V(X) = E[X^2] - \mu^2 = E[X^2] - (E[X])^2.$$

Other expected values that are of interest in statistics are the *moments* of a random variable. These are the expectation of powers of the random variable. In general, we define the *r-th moment* as

$$\mu'_r = E[X^r],$$

(2.18)

and the *r-th central moment* as

$$\mu_r = E[(X-\mu)^r].$$

(2.19)

The mean corresponds to $\mu'_1$ and the variance is given by $\mu_2$.

## Skewness

The third central moment $\mu_3$ is often called a measure of asymmetry or skewness in the distribution. The uniform and the normal distribution are examples of symmetric distributions. The gamma and the exponential are examples of skewed or asymmetric distributions. The following ratio is called the *coefficient of skewness*, which is often used to measure this characteristic:

$$\gamma_1 = \frac{\mu_3}{\mu_2^{3/2}}.$$
(2.20)

Distributions that are skewed to the left will have a negative coefficient of skewness, and distributions that are skewed to the right will have a positive value [Hogg and Craig, 1978]. The coefficient of skewness is zero for symmetric distributions. However, a coefficient of skewness equal to zero does not mean that the distribution must be symmetric.

## Kurtosis

Skewness is one way to measure a type of departure from normality. *Kurtosis* measures a different type of departure from normality by indicating the extent of the peak (or the degree of flatness near its center) in a distribution. The *coefficient of kurtosis* is given by the following ratio:

$$\gamma_2 = \frac{\mu_4}{\mu_2^2}.$$
(2.21)

We see that this is the ratio of the fourth central moment divided by the square of the variance. If the distribution is normal, then this ratio is equal to 3. A ratio greater than 3 indicates more values in the neighborhood of the mean (is more peaked than the normal distribution). If the ratio is less than 3, then it is an indication that the curve is flatter than the normal.

Sometimes the *coefficient of excess kurtosis* is used as a measure of kurtosis. This is given by

$$\gamma_2' = \frac{\mu_4}{\mu_2^2} - 3.$$
(2.22)

In this case, distributions that are more peaked than the normal correspond to a positive value of $\gamma_2'$, and those with a flatter top have a negative coefficient of excess kurtosis.

---

## 2.5 Common Distributions

In this section, we provide a review of some useful probability distributions and briefly describe some applications to modeling data. Most of these distributions are used in later chapters, so we take this opportunity to define them and to fix our notation. We first cover two important discrete distributions: the binomial and the Poisson. These are followed by several continuous distributions: the uniform, the normal, the exponential, the gamma, the chi-square, the Weibull, the beta and the multivariate normal.

### Binomial

Let's say that we have an experiment, whose outcome can be labeled as a 'success' or a 'failure'. If we let $X = 1$ denote a successful outcome and $X = 0$ represent a failure, then we can write the probability mass function as

$$
\begin{aligned}
f(0) &= P(X = 0) = 1 - p, \\
f(1) &= P(X = 1) = p,
\end{aligned}
\tag{2.23}
$$

where $p$ represents the probability of a successful outcome. A random variable that follows the probability mass function in Equation 2.23 for $0 < p < 1$ is called a Bernoulli random variable.

Now suppose we repeat this experiment for $n$ trials, where each trial is independent (the outcome from one trial does not influence the outcome of another) and results in a success with probability $p$. If $X$ denotes the number of successes in these $n$ trials, then $X$ follows the binomial distribution with parameters $(n, p)$. Examples of binomial distributions with different parameters are shown in Figure 2.3.

To calculate a binomial probability, we use the following formula:

$$
f(x;n, p) = P(X = x) = \binom{n}{x}p^x(1 - p)^{n - x}; \qquad x = 0, 1, ..., n. \tag{2.24}
$$

The mean and variance of a binomial distribution are given by

$$
E[X] = np,
$$

and

$$
V(X) = np(1 - p).
$$

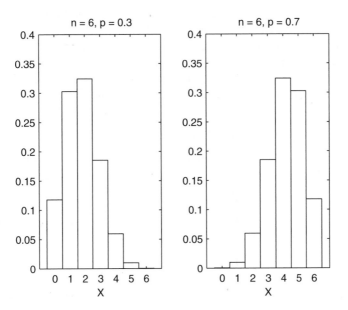

**FIGURE 2.3**
Examples of the binomial distribution for different success probabilities.

Some examples where the results of an experiment can be modeled by a binomial random variable are:

- A drug has probability 0.90 of curing a disease. It is administered to 100 patients, where the outcome for each patient is either cured or not cured. If $X$ is the number of patients cured, then $X$ is a binomial random variable with parameters (100, 0.90).

- The National Institute of Mental Health estimates that there is a 20% chance that an adult American suffers from a psychiatric disorder. Fifty adult Americans are randomly selected. If we let $X$ represent the number who have a psychiatric disorder, then $X$ takes on values according to the binomial distribution with parameters (50, 0.20).

- A manufacturer of computer chips finds that on the average 5% are defective. To monitor the manufacturing process, they take a random sample of size 75. If the sample contains more than five defective chips, then the process is stopped. The binomial distribution with parameters (75, 0.05) can be used to model the random variable $X$, where $X$ represents the number of defective chips.

## Example 2.1

Suppose there is a 20% chance that an adult American suffers from a psychiatric disorder. We randomly sample 25 adult Americans. If we let $X$ represent the number of people who have a psychiatric disorder, then $X$ is a binomial random variable with parameters $(25, 0.20)$. We are interested in the probability that at most 3 of the selected people have such a disorder. We can use the MATLAB Statistics Toolbox function **binocdf** to determine $P(X \leq 3)$, as follows:

```
prob = binocdf(3,25,0.2);
```

We could also sum up the individual values of the probability mass function from $X = 0$ to $X = 3$:

```
prob2 = sum(binopdf(0:3,25,0.2));
```

Both of these commands return a probability of 0.234. We now show how to generate the binomial distributions shown in Figure 2.3.

```
% Get the values for the domain, x.
x = 0:6;
% Get the values of the probability mass function.
% First for n = 6, p = 0.3:
pdf1 = binopdf(x,6,0.3);
% Now for n = 6, p = 0.7:
pdf2 = binopdf(x,6,0.7);
```

Now we have the values for the probability mass function (or the heights of the bars). The plots are obtained using the following code.

```
% Do the plots.
subplot(1,2,1),bar(x,pdf1,1,'w')
title(' n = 6, p = 0.3')
xlabel('X'),ylabel('f(X)')
axis square
subplot(1,2,2),bar(x,pdf2,1,'w')
title(' n = 6, p = 0.7')
xlabel('X'),ylabel('f(X)')
axis square
```

□

## Poisson

A random variable $X$ is a Poisson random variable with parameter $\lambda$, $\lambda > 0$, if it follows the probability mass function given by

$$f(x;\lambda) = P(X = x) = e^{-\lambda}\frac{\lambda^x}{x!}; \quad x = 0, 1, \dots \tag{2.25}$$

The expected value and variance of a Poisson random variable are both $\lambda$, thus,

$$E[X] = \lambda,$$

and

$$V(X) = \lambda.$$

The Poisson distribution can be used in many applications. Examples of situations where a discrete random variable might follow a Poisson distribution are:

- the number of typographical errors on a page,
- the number of vacancies in a company during a month, or
- the number of defects in a length of wire.

The Poisson distribution is often used to approximate the binomial. When $n$ is large and $p$ is small (so $np$ is moderate), then the number of successes occurring can be approximated by the Poisson random variable with parameter $\lambda = np$.

The Poisson distribution is also appropriate for some applications where events occur at points in time or space. We see it used in this context in Chapter 12, where we look at modeling spatial point patterns. Some other examples include the arrival of jobs at a business, the arrival of aircraft on a runway, and the breakdown of machines at a manufacturing plant. The number of events in these applications can be described by a *Poisson process*.

Let $N(t)$, $t \geq 0$, represent the number of events that occur in the time interval $[0, t]$. For each interval $[0, t]$, $N(t)$ is a random variable that can take on values 0, 1, 2, ... . If the following conditions are satisfied, then the counting process $\{N(t), t \geq 0\}$ is said to be a Poisson process with mean rate $\lambda$ [Ross, 2000]:

1. $N(0) = 0$.

2. The process has independent increments.

3. The number $N(t)$ of events in an interval of length $t$ follows a Poisson distribution with mean $\lambda t$. Thus, for $s \geq 0$, $t \geq 0$,

$$P(N(t + s) - N(s) = k) = e^{-\lambda t}\frac{(\lambda t)^k}{k!}; \qquad k = 0, 1, \ldots . \tag{2.26}$$

From the third condition, we know that the process has stationary increments. This means that the distribution of the number of events in an interval depends only on the length of the interval and not on the starting point. The second condition specifies that the number of events in one interval does not affect the number of events in other intervals. The first condition states that the counting starts at time $t = 0$. The expected value of $N(t)$ is given by

$$E[N(t)] = \lambda t.$$

## Example 2.2

In preparing this text, we executed the spell check command, and the editor reviewed the manuscript for typographical errors. In spite of this, some mistakes might be present. Assume that the number of typographical errors per page follows the Poisson distribution with parameter $\lambda = 0.25$. We calculate the probability that a page will have at least two errors as follows:

$$P(X \geq 2) = 1 - \{P(X = 0) + P(X = 1)\} = 1 - e^{-0.25} - e^{-0.25}0.25 \approx 0.0265.$$

We can get this probability using the MATLAB Statistics Toolbox function **poisscdf**. Note that $P(X = 0) + P(X = 1)$ is the Poisson cumulative distribution function for $a = 1$ (see Equation 2.4), which is why we use **1** as the argument to **poisscdf**.

```
prob = 1-poisscdf(1,0.25);
```

❑

## Example 2.3

Suppose that accidents at a certain intersection occur in a manner that satisfies the conditions for a Poisson process with a rate of 2 per week ($\lambda = 2$). What is the probability that at most 3 accidents will occur during the next 2 weeks? Using Equation 2.26, we have

$$P(N(2) \leq 3) = \sum_{k=0}^{3} P(N(2) = k).$$

Expanding this out yields

$$P(N(2) \leq 3) = e^{-4} + 4e^{-4} + \frac{4^2}{2!}e^{-4} + \frac{4^3}{3!}e^{-4} \approx 0.4335.$$

As before, we can use the **poisscdf** function with parameter given by $\lambda t = 2 \cdot 2$.

```
prob = poisscdf(3,2*2);
```

❑

## Uniform

Perhaps one of the most important distributions is the uniform distribution for continuous random variables. One reason is that the uniform (0, 1) distribution is used as the basis for simulating most random variables as we discuss in Chapter 4.

A random variable that is uniformly distributed over the interval (*a*, *b*) follows the probability density function given by

$$f(x;a, b) = \frac{1}{b-a}; \quad a < x < b. \tag{2.27}$$

The parameters for the uniform are the interval endpoints, *a* and *b*. The mean and variance of a uniform random variable are given by

$$E[X] = \frac{a+b}{2},$$

and

$$V(X) = \frac{(b-a)^2}{12}.$$

The cumulative distribution function for a uniform random variable is

$$F(x) = \begin{cases} 0; & x \le a \\ \dfrac{x-a}{b-a}; & a < x < b \\ 1; & x \ge b. \end{cases} \tag{2.28}$$

## Example 2.4

In this example, we illustrate the uniform probability density function over the interval (0, 10), along with the corresponding cumulative distribution function. The MATLAB Statistics Toolbox functions **unifpdf** and **unifcdf** are used to get the desired functions over the interval.

```
% First get the domain over which we will
% evaluate the functions.
x = -1:.1:11;
% Now get the probability density function
% values at x.
pdf = unifpdf(x,0,10);
% Now get the cdf.
cdf = unifcdf(x,0,10);
```

Plots of the functions are provided in Figure 2.4, where the probability density function is shown in the left plot and the cumulative distribution on the right. These plots are constructed using the following MATLAB commands.

```
% Do the plots.
subplot(1,2,1),plot(x,pdf)
title('PDF')
xlabel('X'),ylabel('f(X)')
axis([-1 11 0 0.2])
axis square
subplot(1,2,2),plot(x,cdf)
title('CDF')
xlabel('X'),ylabel('F(X)')
axis([-1 11 0 1.1])
axis square
```

❑

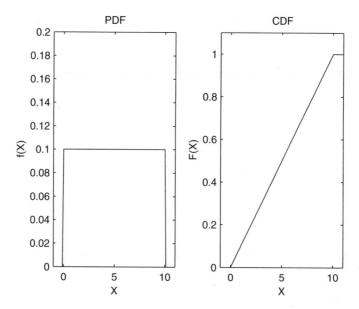

**FIGURE 2.4**
On the left is a plot of the probability density function for the uniform (0, 10). Note that the height of the curve is given by $1/(b-a) = 1/10 = 0.10$. The corresponding cumulative distribution function is shown on the right.

## Normal

A well known distribution in statistics and engineering is the normal distribution. Also called the Gaussian distribution, it has a continuous probability density function given by

$$f(x;\mu, \sigma^2) = \frac{1}{\sigma\sqrt{2\pi}} \exp\left\{-\frac{(x-\mu)^2}{2\sigma^2}\right\}, \tag{2.29}$$

where $-\infty < x < \infty$; $-\infty < \mu < \infty$; $\sigma^2 > 0$. The normal distribution is completely determined by its parameters ($\mu$ and $\sigma^2$), which are also the expected value and variance for a normal random variable. The notation $X \sim N(\mu, \sigma^2)$ is used to indicate that a random variable $X$ is normally distributed with mean $\mu$ and variance $\sigma^2$. Several normal distributions with different parameters are shown in Figure 2.5.

Some special properties of the normal distribution are given here.

- The value of the probability density function approaches zero as $x$ approaches positive and negative infinity.
- The probability density function is centered at the mean $\mu$, and the maximum value of the function occurs at $x = \mu$.
- The probability density function for the normal distribution is symmetric about the mean $\mu$.

The special case of a standard normal random variable is one whose mean is zero ($\mu = 0$), and whose standard deviation is one ($\sigma = 1$). If $X$ is normally distributed, then

$$Z = \frac{X-\mu}{\sigma} \tag{2.30}$$

is a standard normal random variable.

Traditionally, the cumulative distribution function of a standard normal random variable is denoted by

$$\Phi(z) = \frac{1}{\sqrt{2\pi}} \int_{-\infty}^{z} \exp\left\{-\frac{y^2}{2}\right\} dy. \tag{2.31}$$

The cumulative distribution function for a standard normal random variable can be calculated using the error function, denoted by *erf*. The relationship between these functions is given by

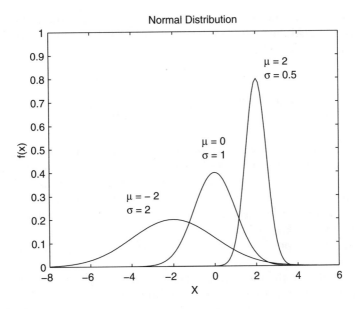

**FIGURE 2.5**
Examples of probability density functions for normally distributed random variables. Note that as the variance increases, the height of the probability density function at the mean decreases.

$$\Phi(z) = \frac{1}{2}erf\left(\frac{z}{\sqrt{2}}\right) + \frac{1}{2}. \tag{2.32}$$

The error function can be calculated in MATLAB using **erf(x)**. The MATLAB Statistics Toolbox has a function called **normcdf(x,mu,sigma)** that will calculate the cumulative distribution function for values in $x$. Its use is illustrated in the example given below.

### Example 2.5

Similar to the uniform distribution, the functions **normpdf** and **normcdf** are available in the MATLAB Statistics Toolbox for calculating the probability density function and cumulative distribution function for the normal. There is another special function called **normspec** that determines the probability that a random variable $X$ assumes a value between two limits, where $X$ is normally distributed with mean $\mu$ and standard deviation $\sigma$. This function also plots the normal density, where the area between the specified limits is shaded. The syntax is shown below.

```
% Set up the parameters for the normal distribution.
mu = 5;
sigma = 2;
% Set up the upper and lower limits. These are in
% the two element vector 'specs'.
specs = [2, 8];
prob = normspec(specs, mu, sigma);
```

The resulting plot is shown in Figure 2.6. Note that the default title and axes labels are shown, but these can be changed easily using the **title, xlabel,** and **ylabel** functions. You can also obtain tail probabilities by using -**Inf** as the first element of **specs** to designate no lower limit or **Inf** as the second element to indicate no upper limit.
❑

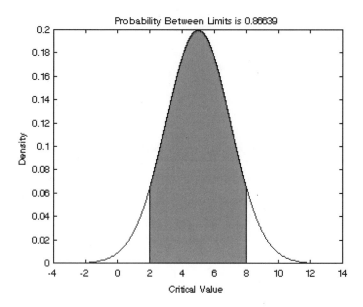

**FIGURE 2.6**
This shows the output from the function **normspec**. Note that it shades the area between the lower and upper limits that are specified as input arguments.

## Exponential

The exponential distribution can be used to model the amount of time until a specific event occurs or to model the time between independent events. Some examples where an exponential distribution could be used as the model are:

- • the time until the computer locks up,
- • the time between arrivals of telephone calls, or
- • the time until a part fails.

The exponential probability density function with parameter $\lambda$ is

$$f(x;\lambda) = \lambda e^{-\lambda x}; \quad x \geq 0; \quad \lambda > 0. \tag{2.33}$$

The mean and variance of an exponential random variable are given by the following:

$$E[X] = \frac{1}{\lambda},$$

and

$$V(X) = \frac{1}{\lambda^2}.$$

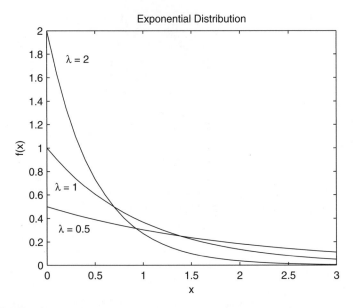

**FIGURE 2.7**
Exponential probability density functions for various values of $\lambda$.

The cumulative distribution function of an exponential random variable is given by

$$F(x) = \begin{cases} 0; & x < 0 \\ 1 - e^{-\lambda x}; & x \geq 0. \end{cases} \tag{2.34}$$

The exponential distribution is the only continuous distribution that has the memoryless property. This property describes the fact that the remaining lifetime of an object (whose lifetime follows an exponential distribution) does not depend on the amount of time it has already lived. This property is represented by the following equality, where $s \geq 0$ and $t \geq 0$:

$$P(X > s + t \,|\, X > s) = P(X > t).$$

In words, this means that the probability that the object will operate for time $s + t$, given it has already operated for time $s$, is simply the probability that it operates for time $t$.

When the exponential is used to represent interarrival times, then the parameter $\lambda$ is a rate with units of arrivals per time period. When the exponential is used to model the time until a failure occurs, then $\lambda$ is the failure rate. Several examples of the exponential distribution are shown in Figure 2.7.

## Example 2.6

The time between arrivals of vehicles at an intersection follows an exponential distribution with a mean of 12 seconds. What is the probability that the time between arrivals is 10 seconds or less? We are given the average interarrival time, so $\lambda = 1/12$. The required probability is obtained from Equation 2.34 as follows

$$P(X \leq 10) = 1 - e^{-(1/12)10} \approx 0.57.$$

You can calculate this using the MATLAB Statistics Toolbox function **expocdf(x, 1/$\lambda$)**. Note that this MATLAB function is based on a different definition of the exponential probability density function, which is given by

$$f(x;\mu) = \frac{1}{\mu} e^{-\frac{x}{\mu}}; \quad x \geq 0; \quad \mu > 0. \tag{2.35}$$

In the Computational Statistics Toolbox, we include a function called **csex-poc(x,$\lambda$)** that calculates the exponential cumulative distribution function using Equation 2.34.
❑

## Gamma

The gamma probability density function with parameters $\lambda > 0$ and $t > 0$ is

$$f(x;\lambda, t) = \frac{\lambda e^{-\lambda x}(\lambda x)^{t-1}}{\Gamma(t)}; \quad x \geq 0, \tag{2.36}$$

where $t$ is a shape parameter, and $\lambda$ is the scale parameter. The gamma function $\Gamma(t)$ is defined as

$$\Gamma(t) = \int_0^\infty e^{-y} y^{t-1} dy. \tag{2.37}$$

For integer values of $t$, Equation 2.37 becomes

$$\Gamma(t) = (t-1)!. \tag{2.38}$$

Note that for $t = 1$, the gamma density is the same as the exponential. When $t$ is a positive integer, the gamma distribution can be used to model the amount of time one has to wait until $t$ events have occurred, if the inter-arrival times are exponentially distributed.

The mean and variance of a gamma random variable are

$$E[X] = \frac{t}{\lambda},$$

and

$$V(X) = \frac{t}{\lambda^2}.$$

The cumulative distribution function for a gamma random variable is calculated using [Meeker and Escobar, 1998; Banks, et al., 2001]

$$F(x;\lambda, t) = \begin{cases} 0; & x \le 0 \\ \dfrac{1}{\Gamma(t)} \displaystyle\int_0^{\lambda x} y^{t-1} e^{-y} dy; & x > 0 . \end{cases} \tag{2.39}$$

Equation 2.39 can be evaluated easily in MATLAB using the **gammainc** $(\lambda * x, t)$ function, where the above notation is used for the arguments.

## Example 2.7

We plot the gamma probability density function for $\lambda = t = 1$ (this should look like the exponential), $\lambda = t = 2$, and $\lambda = t = 3$. You can use the MATLAB Statistics Toolbox function **gampdf** $(x, t, 1/\lambda)$ or the function **csgammp** $(x, t, \lambda)$.

```
% First get the domain over which to
% evaluate the functions.
x = 0:.1:3;
% Now get the functions values for
% different values of lambda.
y1 = gampdf(x,1,1/1);
y2 = gampdf(x,2,1/2);
y3 = gampdf(x,3,1/3);
% Plot the functions.
plot(x,y1,'r',x,y2,'g',x,y3,'b')
title('Gamma Distribution')
xlabel('X')
ylabel('f(x)')
```

The resulting curves are shown in Figure 2.8.
❑

### Chi-Square

A gamma distribution where $\lambda = 0.5$ and $t = v/2$, with $v$ a positive integer, is called a chi-square distribution (denoted as $\chi_v^2$) with $v$ degrees of freedom. The chi-square distribution is used to derive the distribution of the sample variance and is important for goodness-of-fit tests in statistical analysis [Mood, Graybill, and Boes, 1974].

The probability density function for a chi-square random variable with $v$ degrees of freedom is

$$f(x;v) = \frac{1}{\Gamma(v/2)} \left(\frac{1}{2}\right)^{v/2} x^{v/2-1} e^{-\frac{1}{2}x}; \qquad x \ge 0 . \tag{2.40}$$

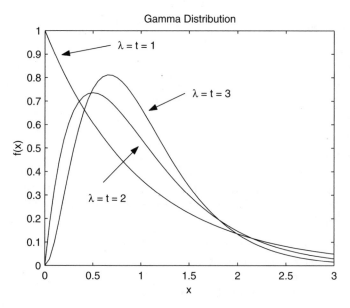

**FIGURE 2.8**
We show three examples of the gamma probability density function. We see that when $\lambda = t = 1$, we have the same probability density function as the exponential with parameter $\lambda = 1$.

The mean and variance of a chi-square random variable can be obtained from the gamma distribution. These are given by

$$E[X] = \nu,$$

and

$$V(X) = 2\nu.$$

## Weibull

The Weibull distribution has many applications in engineering. In particular, it is used in reliability analysis. It can be used to model the distribution of the amount of time it takes for objects to fail. For the special case where $\nu = 0$ and $\beta = 1$, the Weibull reduces to the exponential with $\lambda = 1/\alpha$.

The Weibull density for $\alpha > 0$ and $\beta > 0$ is given by

$$f(x;\nu, \alpha, \beta) = \left(\frac{\beta}{\alpha}\right)\left(\frac{x-\nu}{\alpha}\right)^{\beta-1} e^{-\left(\frac{x-\nu}{\alpha}\right)^{\beta}}; \qquad x > \nu, \qquad (2.41)$$

and the cumulative distribution is

$$F(x;\nu, \alpha, \beta) = \begin{cases} 0; & x \le \nu \\ 1 - e^{-\left(\frac{x-\nu}{\alpha}\right)^{\beta}}; & x > \nu. \end{cases} \qquad (2.42)$$

The location parameter is denoted by $\nu$, and the scale parameter is given by $\alpha$. The shape of the Weibull distribution is governed by the parameter $\beta$.

The mean and variance [Banks, et al., 2001] of a random variable from a Weibull distribution are given by

$$E[X] = \nu + \alpha\Gamma(1/\beta + 1),$$

and

$$V(X) = \alpha^2\left\{\Gamma(2/\beta + 1) - [\Gamma(1/\beta + 1)]^2\right\}.$$

## Example 2.8

Suppose the time to failure of piston rings for stream-driven compressors can be modeled by the Weibull distribution with a location parameter of zero, $\beta = 1/3$, and $\alpha = 500$. We can find the mean time to failure using the expected value of a Weibull random variable, as follows

$$E[X] = \nu + \alpha\Gamma(1/\beta + 1) = 500 \times \Gamma(3 + 1) = 3000 \text{ hours.}$$

Let's say we want to know the probability that a piston ring will fail before 2000 hours. We can calculate this probability using

$$F(2000;0, 500, 1/3) = 1 - \exp\left\{-\left(\frac{2000}{500}\right)^{1/3}\right\} \approx 0.796.$$

❑

You can use the MATLAB Statistics Toolbox function for applications where the location parameter is zero ($\nu = 0$). This function is called

**weibcdf** (for the cumulative distribution function), and the input arguments are: $(x, \alpha^{-\beta}, \beta)$. The reason for the different parameters is that MATLAB uses an alternate definition for the Weibull probability density function given by

$$f(x; a, b) = abx^{b-1}e^{-ax^b}; \qquad x > 0. \tag{2.43}$$

Comparing this with Equation 2.41, we can see that $v = 0$, $a = \alpha^{-\beta}$ and $b = \beta$. You can also use the function **csweibc** $(x, v, \alpha, \beta)$ to evaluate the cumulative distribution function for a Weibull.

## Beta

The beta distribution is very flexible because it covers a range of different shapes depending on the values of the parameters. It can be used to model a random variable that takes on values over a bounded interval and assumes one of the shapes governed by the parameters. A random variable has a beta distribution with parameters $\alpha > 0$ and $\beta > 0$ if its probability density function is given by

$$f(x; \alpha, \beta) = \frac{1}{B(\alpha, \beta)} x^{\alpha - 1}(1 - x)^{\beta - 1}; \qquad 0 < x < 1, \tag{2.44}$$

where

$$B(\alpha, \beta) = \int_0^1 x^{\alpha - 1}(1 - x)^{\beta - 1}dx = \frac{\Gamma(\alpha)\Gamma(\beta)}{\Gamma(\alpha + \beta)}. \tag{2.45}$$

The function $B(\alpha, \beta)$ can be calculated in MATLAB using the **beta** $(\alpha, \beta)$ function. The mean and variance of a beta random variable are

$$E[X] = \frac{\alpha}{\alpha + \beta},$$

and

$$V(X) = \frac{\alpha\beta}{(\alpha + \beta)^2(\alpha + \beta + 1)}.$$

The cumulative distribution function for a beta random variable is given by integrating the beta probability density function as follows

$$F(x;\alpha, \beta) = \int\limits_{0}^{x} \frac{1}{B(\alpha, \beta)} y^{\alpha - 1} (1 - y)^{\beta - 1} dy . \qquad (2.46)$$

The integral in Equation 2.46 is called the incomplete beta function. This can be calculated in MATLAB using the function `betainc(x,alpha,beta)`.

## EXAMPLE 2.9

We use the following MATLAB code to plot the beta density over the interval (0,1). We let $\alpha = \beta = 0.5$ and $\alpha = \beta = 3$.

```
% First get the domain over which to evaluate
% the density function.
x = 0.01:.01:.99;
% Now get the values for the density function.
y1 = betapdf(x,0.5,0.5);
y2 = betapdf(x,3,3);
% Plot the results.
plot(x,y1,'r',x,y2,'g')
title('Beta Distribution')
xlabel('x')
ylabel('f(x)')
```

The resulting curves are shown in Figure 2.9. You can use the MATLAB Statistics Toolbox function `betapdf(x,`$\alpha$`,`$\beta$`)`, as we did in the example, or the function `csbetap(x,`$\alpha$`,`$\beta$`)`.
□

## Multivariate Normal

So far, we have discussed several univariate distributions for discrete and continuous random variables. In this section, we describe one of the important and most commonly used multivariate densities: the multivariate normal distribution. This distribution is used throughout the rest of the text. Some examples of where we use it are in exploratory data analysis, in probability density estimation, and in statistical pattern recognition.

The probability density function for a general multivariate normal density for $d$ dimensions is given by

$$f(\mathbf{x};\boldsymbol{\mu}, \boldsymbol{\Sigma}) = \frac{1}{(2\pi)^{d/2} |\boldsymbol{\Sigma}|^{1/2}} \exp\left\{ -\frac{1}{2} (\mathbf{x} - \boldsymbol{\mu})^T \boldsymbol{\Sigma}^{-1} (\mathbf{x} - \boldsymbol{\mu}) \right\}, \qquad (2.47)$$

where $\mathbf{x}$ is a $d$-component column vector, $\boldsymbol{\mu}$ is the $d \times 1$ column vector of means, and $\boldsymbol{\Sigma}$ is the $d \times d$ covariance matrix. The superscript $T$ represents the

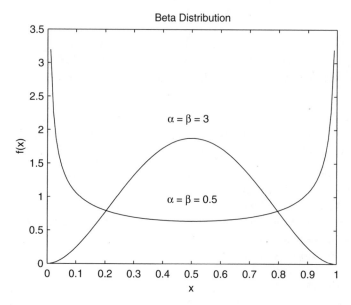

**FIGURE 2.9.**
Beta probability density functions for various parameters.

transpose of an array, and the notation || denotes the determinant of a matrix.

The mean and covariance are calculated using the following formulas:

$$\mu = E[\mathbf{x}], \tag{2.48}$$

and

$$\Sigma = E[(\mathbf{x} - \mu)(\mathbf{x} - \mu)^T], \tag{2.49}$$

where the expected value of an array is given by the expected values of its components. Thus, if we let $X_i$ represent the $i$-th component of $\mathbf{x}$ and $\mu_i$ the $i$-th component of $\mu$, then the elements of Equation 2.48 can be written as

$$\mu_i = E[X_i].$$

If $\sigma_{ij}$ represents the $ij$-th element of $\Sigma$, then the elements of the covariance matrix (Equation 2.49) are given by

$$\sigma_{ij} = E[(X_i - \mu_i)(X_j - \mu_j)].$$

The covariance matrix is symmetric ($\Sigma^T = \Sigma$) positive definite (all eigenvalues of $\Sigma$ are greater than zero) for most applications of interest to statisticians and engineers.

We illustrate some properties of the multivariate normal by looking at the bivariate ($d = 2$) case. The probability density function for a bivariate normal is represented by a bell-shaped surface. The center of the surface is determined by the mean $\mu$ and the shape of the surface is determined by the covariance $\Sigma$. If the covariance matrix is diagonal (all of the off-diagonal elements are zero), and the diagonal elements are equal, then the shape is circular. If the diagonal elements are not equal, then we get an ellipse with the major axis vertical or horizontal. If the covariance matrix is not diagonal, then the shape is elliptical with the axes at an angle. Some of these possibilities are illustrated in the next example.

## Example 2.10

We first provide the following MATLAB function to calculate the multivariate normal probability density function and illustrate its use in the bivariate case. The function is called **csevalnorm**, and it takes input arguments **x,mu,cov_mat**. The input argument **x** is a matrix containing the points in the domain where the function is to be evaluated, **mu** is a $d$-dimensional row vector, and **cov_mat** is the $d \times d$ covariance matrix.

```
function prob = csevalnorm(x,mu,cov_mat);
[n,d] = size(x);
% center the data points
x = x-ones(n,1)*mu;
a = (2*pi)^(d/2)*sqrt(det(cov_mat));
arg = diag(x*inv(cov_mat)*x');
prob = exp((-.5)*arg);
prob = prob/a;
```

We now call this function for a bivariate normal centered at zero and covariance matrix equal to the identity matrix. The density surface for this case is shown in Figure 2.10.

```
% Get the mean and covariance.
mu = zeros(1,2);
cov_mat = eye(2);% Identity matrix
% Get the domain.
% Should range (-4,4) in both directions.
[x,y] = meshgrid(-4:.2:4,-4:.2:4);
% Reshape into the proper format for the function.
X = [x(:),y(:)];
Z = csevalnorm(X,mu,cov_mat);
% Now reshape the matrix for plotting.
z = reshape(Z,size(x));
subplot(1,2,1) % plot the surface
```

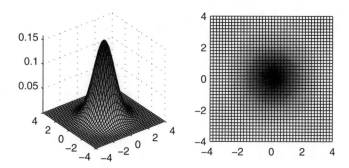

**FIGURE 2.10**
This figure shows a standard bivariate normal probability density function that is centered at the origin. The covariance matrix is given by the identity matrix. Notice that the shape of the surface looks circular. The plot on the right is for a viewpoint looking down on the surface.

```
surf(x,y,z),axis square, axis tight
title('BIVARIATE STANDARD NORMAL')
```

Next, we plot the surface for a bivariate normal centered at the origin with non-zero off-diagonal elements in the covariance matrix. Note the elliptical shape of the surface shown in Figure 2.11.

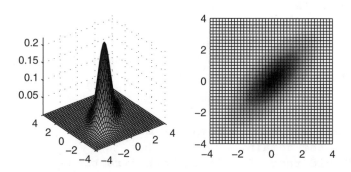

**FIGURE 2.11**
This shows a bivariate normal density where the covariance matrix has non-zero off-diagonal elements. Note that the surface has an elliptical shape. The plot on the right is for a viewpoint looking down on the surface.

```
subplot(1,2,2) % look down on the surface
pcolor(x,y,z),axis square
title('BIVARIATE STANDARD NORMAL')
% Now do the same thing for a covariance matrix
% with non-zero off-diagonal elements.
cov_mat = [1 0.7 ; 0.7 1];
Z = csevalnorm(X,mu,cov_mat);
z = reshape(Z,size(x));
subplot(1,2,1)
surf(x,y,z),axis square, axis tight
title('BIVARIATE NORMAL')
subplot(1,2,2)
pcolor(x,y,z),axis square
title('BIVARIATE NORMAL')
```

❑

The probability that a point $\mathbf{x} = (x_1, x_2)^T$ will assume a value in a region $R$ can be found by integrating the bivariate probability density function over the region. Any plane that cuts the surface parallel to the $x_1$-$x_2$ plane intersects in an elliptic (or circular) curve, yielding a curve of constant density. Any plane perpendicular to the $x_1$-$x_2$ plane cuts the surface in a normal curve. This property indicates that in each dimension, the multivariate normal is a univariate normal distribution. This is discussed further in Chapter 5.

---

## 2.6 MATLAB Code

The MATLAB Statistics Toolbox has many functions for the more common distributions. It has functions for finding the value of the probability density (mass) function and the value of the cumulative distribution function. The reader is cautioned to remember that the definitions of the distributions (exponential, gamma, and Weibull) differ from what we describe in the text. For example, the exponential and the gamma distributions are parameterized differently in the MATLAB Statistics Toolbox. For a complete list of what is available in the toolbox for calculating probability density (mass) functions or cumulative distribution functions, see Appendix E.

The Computational Statistics Toolbox contains functions for several of the distributions, as defined in this chapter. In general, those functions that end in **p** correspond to the probability density (mass) function, and those ending with a **c** calculate the cumulative distribution function. Table 2.1 provides a summary of the functions.

We note that a different function for evaluating the multivariate normal probability density function is available for download at

**TABLE 2.1**

List of Functions from Chapter 2 Included in the
Computational Statistics Toolbox

| Distribution | MATLAB Function |
|---|---|
| Beta | `csbetap, csbetac` |
| Binomial | `csbinop, csbinoc` |
| Chi-square | `cschip, cschic` |
| Exponential | `csexpop, csexpoc` |
| Gamma | `csgammp, csgammc` |
| Normal - univariate | `csnormp, csnormc` |
| Normal - multivariate | `csevalnorm` |
| Poisson | `cspoisp, cspoisc` |
| Continuous Uniform | `csunifp, csunifc` |
| Weibull | `csweibp, csweibc` |

`ftp://ftp.mathworks.com/pub/mathworks/`

under the **stats** directory. This function can be substituted for
`csevalnorm`.

## 2.7 Further Reading

There are many excellent books on probability theory at the undergraduate
and graduate levels. Ross [1994; 1997; 2000] is the author of several books on
probability theory and simulation. These texts contain many examples and
are appropriate for advanced undergraduate students in statistics, engineering and science. Rohatgi [1976] provides a solid theoretical introduction to
probability theory. This text can be used by advanced undergraduate and
beginning graduate students. It has recently been updated with many new
examples and special topics [Rohatgi and Saleh, 2000]. For those who want to
learn about probability, but do not want to be overwhelmed with the theory,
then we recommend Durrett [1994].

At the graduate level, there is a book by Billingsley [1995] on probability and measure theory. He uses probability to motivate measure theory and then uses measure theory to generate more probability concepts. Another good reference is a text on probability and real analysis by Ash [1972]. This is suitable for graduate students in mathematics and statistics. For a book that can be used by graduate students in mathematics, statistics and engineering, see Port [1994]. This text provides a comprehensive treatment of the subject and can also be used as a reference by professional data analysts. Finally, Breiman [1992] provides an overview of probability theory that is accessible to statisticians and engineers.

---

**Exercises**

2.1. Write a function using MATLAB's functions for numerical integration such as **quad** or **quadl** (MATLAB 6) that will find $P(X \leq x)$ when the random variable is exponentially distributed with parameter $\lambda$. See **help** for information on how to use these functions.

2.2. Verify that the exponential probability density function with parameter $\lambda$ integrates to 1. Use the MATLAB functions **quad** or **quadl** (MATLAB 6). See **help** for information on how to use these functions.

2.3. Radar and missile detection systems warn of enemy attacks. Suppose that a radar detection system has a probability 0.95 of detecting a missile attack.

    a. What is the probability that one detection system will detect an attack? What distribution did you use?

    b. Suppose three detection systems are located together in the same area and the operation of each system is independent of the others. What is the probability that at least one of the systems will detect the attack? What distribution did you use in this case?

2.4. When a random variable is equally likely to be either positive or negative, then the Laplacian or the double exponential distribution can be used to model it. The Laplacian probability density function for $\lambda > 0$ is given by

$$f(x) = \frac{1}{2}\lambda e^{-\lambda|x|}; \quad -\infty < x < \infty.$$

    a. Derive the cumulative distribution function for the Laplacian.

    b. Write a MATLAB function that will evaluate the Laplacian probability density function for given values in the domain.

    c. Write a MATLAB function that will evaluate the Laplacian cumulative distribution function.

    d. Plot the probability density function when $\lambda = 1$.

2.5. Suppose $X$ follows the exponential distribution with parameter $\lambda$. Show that for $s \geq 0$ and $t \geq 0$,

$$P(X > s + t | X > s) = P(X > t).$$

2.6. The lifetime in years of a flat panel display is a random variable with the exponential probability density function given by

$$f(x;0.1) = 0.1e^{-0.1x}.$$

    a. What is the mean lifetime of the flat panel display?

    b. What is the probability that the display fails within the first two years?

    c. Given that the display has been operating for one year, what is the probability that it will fail within the next year?

2.7. The time to failure for a widget follows a Weibull distribution, with $v = 0$, $\beta = 1/2$, and $\alpha = 750$ hours.

    a. What is the mean time to failure of the widget?

    b. What percentage of the widgets will fail by 2500 hours of operation? That is, what is the probability that a widget will fail within 2500 hours?

2.8. Let's say the probability of having a boy is 0.52. Using the Multiplication Rule, find the probability that a family's first and second children are boys. What is the probability that the first child is a boy and the second child is a girl?

2.9. Repeat Example 2.1 for $n = 6$ and $p = 0.5$. What is the shape of the distribution?

2.10. Recall that in our piston ring example, $P(M_A) = 0.6$ and $P(M_B) = 0.4$. From prior experience with the two manufacturers, we know that 2% of the parts supplied by manufacturer A are likely to fail and 6% of the parts supplied by manufacturer B are likely to fail. Thus, $P(F|M_A) = 0.02$ and $P(F|M_B) = 0.06$. If we observe a piston ring failure, what is the probability that it came from manufacturer A?

2.11. Using the functions **fminbnd** or **fmin** (available in the standard MATLAB package), find the value for $x$ where the maximum of the $N(3, 1)$ probability density occurs. Note that you have to find the minimum of $-f(x)$ to find the maximum of $f(x)$ using these functions. Refer to the **help** files on these functions for more information on how to use them.

2.12. Using **normpdf** or **csnormp**, find the value of the probability density for $N(0, 1)$ at $\pm\infty$. Use a small (large) value of $x$ for $-\infty$ ($\infty$).

2.13. Verify Equation 2.38 using the MATLAB functions **factorial** and **gamma**.

2.14. Find the height of the curve for a normal probability density function at $x = \mu$, where $\sigma = 0.5, 1, 2$. What happens to the height of the curve as $\sigma$ gets larger? Does the height change for different values of $\mu$?

2.15. Write a function that calculates the Bayes' posterior probability given a vector of conditional probabilities and a vector of prior probabilities.

2.16. Compare the Poisson approximation to the actual binomial probability $P(X = 4)$, using $n = 9$ and $p = 0.1, 0.2, ..., 0.9$.

2.17. Using the function **normspec**, find the probability that the random variable defined in Example 2.5 assumes a value that is less than 3. What is the probability that the same random variable assumes a value that is greater than 5? Find these probabilities again using the function **normcdf**.

2.18. Find the probability for the Weibull random variable of Example 2.8 using the MATLAB Statistics Toolbox function **weibcdf** or the Computational Statistics Toolbox function **csweibc**.

2.19. The MATLAB Statistics Toolbox has a GUI demo called **disttool**. First view the **help** file on **disttool**. Then run the demo. Examine the probability density (mass) and cumulative distribution functions for the distributions discussed in the chapter.

# Chapter 3

## Sampling Concepts

### 3.1 Introduction

In this chapter, we cover the concepts associated with random sampling and the sampling distribution of statistics. These notions are fundamental to computational statistics and are needed to understand the topics covered in the rest of the book. As with Chapter 2, those readers who have a basic understanding of these ideas may safely move on to more advanced topics.

In Section 3.2, we discuss the terminology and concepts associated with random sampling and sampling distributions. Section 3.3 contains a brief discussion of the Central Limit Theorem. In Section 3.4, we describe some methods for deriving estimators (maximum likelihood and the method of moments) and introduce criteria for evaluating their performance. Section 3.5 covers the empirical distribution function and how it is used to estimate quantiles. Finally, we conclude with a section on the MATLAB functions that are available for calculating the statistics described in this chapter and a section on further readings.

### 3.2 Sampling Terminology and Concepts

In Chapter 2, we introduced the idea of a random experiment. We typically perform an experiment where we collect data that will provide information on the phenomena of interest. Using these data, we draw conclusions that are usually beyond the scope of our particular experiment. The researcher generalizes from that experiment to the class of all similar experiments. This is the heart of inferential statistics. The problem with this sort of generalization is that we cannot be absolutely certain about our conclusions. However, by using statistical techniques, we can measure and manage the degree of uncertainty in our results.

*Inferential statistics* is a collection of techniques and methods that enable researchers to observe a subset of the objects of interest and using the information obtained from these observations make statements or inferences about the entire population of objects. Some of these methods include the estimation of population parameters, statistical hypothesis testing, and probability density estimation.

The *target population* is defined as the entire collection of objects or individuals about which we need some information. The target population must be well defined in terms of what constitutes membership in the population (e.g., income level, geographic area, etc.) and what characteristics of the population we are measuring (e.g., height, IQ, number of failures, etc.).

The following are some examples of populations, where we refer back to those described at the beginning of Chapter 2.

- For the piston ring example, our population is all piston rings contained in the legs of steam-driven compressors. We would be observing the time to failure for each piston ring.

- In the glucose example, our population might be all pregnant women, and we would be measuring the glucose levels.

- For cement manufacturing, our population would be batches of cement, where we measure the tensile strength and the number of days the cement is cured.

- In the software engineering example, our population consists of all executions of a particular command and control software system, and we observe the failure time of the system in seconds.

In most cases, it is impossible or unrealistic to observe the entire population. For example, some populations have members that do not exist yet (e.g., future batches of cement) or the population is too large (e.g., all pregnant women). So researchers measure only a part of the target population, called a *sample.* If we are going to make inferences about the population using the information obtained from a sample, then it is important that the sample be representative of the population. This can usually be accomplished by selecting a *simple random sample*, where all possible samples are equally likely to be selected.

A random sample of size $n$ is said to be *independent and identically distributed* (iid) when the random variables $X_1, X_2, ..., X_n$ each have a common probability density (mass) function given by $f(x)$. Additionally, when they are both independent and identically distributed (iid), the joint probability density (mass) function is given by

$$f(x_1, ..., x_n) = f(x_1) \times ... \times f(x_n),$$

which is simply the product of the individual densities (or mass functions) evaluated at each sample point.

There are two types of simple random sampling: sampling with replacement and sampling without replacement. When we sample with replacement, we select an object, observe the characteristic we are interested in, and return the object to the population. In this case, an object can be selected for the sample more than once. When the sampling is done without replacement, objects can be selected at most one time. These concepts will be used in Chapters 6 and 7 where the bootstrap and other resampling methods are discussed.

Alternative sampling methods exist. In some situations, these methods are more practical and offer better random samples than simple random sampling. One such method, called *stratified random sampling*, divides the population into levels, and then a simple random sample is taken from each level. Usually, the sampling is done in such a way that the number sampled from each level is proportional to the number of objects of that level that are in the population. Other sampling methods include cluster sampling and systematic random sampling. For more information on these and others, see the book by Levy and Lemeshow [1999].

Sometimes the goal of inferential statistics is to use the sample to estimate or make some statements about a population parameter. Recall from Chapter 2 that a *parameter* is a descriptive measure for a population or a distribution of random variables. For example, population parameters that might be of interest include the mean ($\mu$), the standard deviation ($\sigma$), quantiles, proportions, correlation coefficients, etc.

A *statistic* is a function of the observed random variables obtained in a random sample and does not contain any unknown population parameters. Often the statistic is used for the following purposes:

- as a point estimate for a population parameter,
- to obtain a confidence interval estimate for a parameter, or
- as a test statistic in hypothesis testing.

Before we discuss some of the common methods for deriving statistics, we present some of the statistics that will be encountered in the remainder of the text. In most cases, we assume that we have a random sample, $X_1, \ldots, X_n$, of independent, identically (iid) distributed random variables.

## Sample Mean and Sample Variance

A familiar statistic is the *sample mean* given by

$$\bar{X} = \frac{1}{n}\sum_{i=1}^{n} X_i. \tag{3.1}$$

To calculate this in MATLAB, one can use the function called **mean**. If the argument to this function is a matrix, then it provides a vector of means, each one corresponding to the mean of a column. One can find the mean along any dimension (**dim**) of multi-dimensional arrays using the syntax: **mean(x,dim)**.

Another statistic that we will see again is the *sample variance*, calculated from

$$S^2 = \frac{1}{n-1}\sum_{i=1}^{n}(X_i - \bar{X})^2 = \frac{1}{n(n-1)}\left(n\sum_{i=1}^{n}X_i^2 - \left(\sum_{i=1}^{n}X_i\right)^2\right). \tag{3.2}$$

The sample standard deviation is given by the square root of the variance (Equation 3.2) and is denoted by $S$. These statistics can be calculated in MATLAB using the functions **std(x)** and **var(x)**, where **x** is an array containing the sample values. As with the function **mean**, these can have matrices or multi-dimensional arrays as input arguments.

### Sample Moments

The sample moments can be used to estimate the population moments described in Chapter 2. The *r-th sample moment* about zero is given by

$$M'_r = \frac{1}{n}\sum_{i=1}^{n} X_i^r. \tag{3.3}$$

Note that the sample mean is obtained when $r = 1$. The $r$-th sample moments about the sample mean are statistics that estimate the population central moments and can be found using the following

$$M_r = \frac{1}{n}\sum_{i=1}^{n}(X_i - \bar{X})^r. \tag{3.4}$$

We can use Equation 3.4 to obtain estimates for the coefficient of skewness $\gamma_1$ and the coefficient of kurtosis $\gamma_2$. Recall that these are given by

$$\gamma_1 = \frac{\mu_3}{\mu_2^{3/2}}, \tag{3.5}$$

and

$$\gamma_2 = \frac{\mu_4}{\mu_2^2}. \tag{3.6}$$

Substituting the sample moments for the population moments in Equations 3.5 and 3.6, we have

$$\hat{\gamma}_1 = \frac{\frac{1}{n}\sum_{i=1}^{n}(X_i - \bar{X})^3}{\left(\frac{1}{n}\sum_{i=1}^{n}(X_i - \bar{X})^2\right)^{3/2}}, \tag{3.7}$$

and

$$\hat{\gamma}_2 = \frac{\frac{1}{n}\sum_{i=1}^{n}(X_i - \bar{X})^4}{\left(\frac{1}{n}\sum_{i=1}^{n}(X_i - \bar{X})^2\right)^2}. \tag{3.8}$$

We are using the 'hat' notation to denote an estimate. Thus, $\hat{\gamma}_1$ is an estimate for $\gamma_1$. The following example shows how to use MATLAB to obtain the sample coefficient of skewness and sample coefficient of kurtosis.

## Example 3.1
In this example, we will generate a random sample that is uniformly distributed over the interval (0, 1). We would expect this sample to have a coefficient of skewness close to zero because it is a symmetric distribution. We would expect the kurtosis to be different from 3, because the random sample is not generated from a normal distribution.

```
% Generate a random sample from the uniform
% distribution.
n = 200;
x = rand(1,200);
% Find the mean of the sample.
```

```
mu = mean(x);
% Find the numerator and denominator for gamma_1.
num = (1/n)*sum((x-mu).^3);
den = (1/n)*sum((x-mu).^2);
gam1 = num/den^(3/2);
```

This results in a coefficient of skewness of **gam1** = -0.0542, which is not too far from zero. Now we find the kurtosis using the following MATLAB commands:

```
% Find the kurtosis.
num = (1/n)*sum((x-mu).^4);
den = (1/n)*sum((x-mu).^2);
gam2 = num/den^2;
```

This gives a kurtosis of **gam2** = 1.8766, which is not close to 3, as expected.
❑

We note that these statistics might not be the best to use in terms of bias (see Section 3.4). However, they will prove to be useful as examples in Chapters 6 and 7, where we look at bootstrap methods for estimating the bias in a statistic. The MATLAB Statistics Toolbox function called **skewness** returns the coefficient of skewness for a random sample. The function **kurtosis** calculates the sample coefficient of kurtosis (*not* the coefficient of excess kurtosis).

## Covariance

In the definitions given below (Equations 3.9 and 3.10), we assume that all expectations exist. The *covariance* of two random variables $X$ and $Y$, with joint probability density function $f(x, y)$, is defined as

$$Cov(X, Y) = \sigma_{X, Y} = E[(X - \mu_X)(Y - \mu_Y)]. \tag{3.9}$$

The *correlation coefficient* of $X$ and $Y$ is given by

$$Corr(X, Y) = \rho_{X, Y} = \frac{Cov(X, Y)}{\sigma_X \sigma_Y} = \frac{\sigma_{X, Y}}{\sigma_X \sigma_Y}, \tag{3.10}$$

where $\sigma_X > 0$ and $\sigma_Y > 0$.

The correlation is a measure of the linear relationship between two random variables. If the joint distribution of two variables has a correlation coefficient, then $-1 \le \rho_{X, Y} \le 1$. When $\rho_{X, Y} = 1$, then $X$ and $Y$ are perfectly positively correlated. This means that the possible values for $X$ and $Y$ lie on a line with positive slope. On the other hand, when $\rho_{X, Y} = -1$, then the situation is the opposite: $X$ and $Y$ are perfectly negatively correlated. If $X$ and $Y$ are independent, then $\rho_{X, Y} = 0$. Note that the converse of this statement does not necessarily hold.

There are statistics that can be used to estimate these quantities. Let's say we have a random sample of size $n$ denoted as $(X_1, Y_1), ..., (X_n, Y_n)$. The sample covariance is typically calculated using the following statistic

$$\hat{\sigma}_{X, Y} = \frac{1}{n-1} \sum_{i=1}^{n} (X_i - \bar{X})(Y_i - \bar{Y}) . \tag{3.11}$$

This is the definition used in the MATLAB function **cov**. In some instances, the empirical covariance is used [Efron and Tibshirani, 1993]. This is similar to Equation 3.11, except that we divide by $n$ instead of $n - 1$. The sample correlation coefficient for two variables is given by

$$\hat{\rho}_{X, Y} = \frac{\displaystyle\sum_{i=1}^{n} (X_i - \bar{X})(Y_i - \bar{Y})}{\left(\displaystyle\sum_{i=1}^{n} (X_i - \bar{X})^2\right)^{1/2} \left(\displaystyle\sum_{i=1}^{n} (Y_i - \bar{Y})^2\right)^{1/2}} . \tag{3.12}$$

In the next example, we investigate the commands available in MATLAB that return the statistics given in Equations 3.11 and 3.12. It should be noted that the quantity in Equation 3.12 is also bounded below by $-1$ and above by 1.

### Example 3.2

In this example, we show how to use the MATLAB **cov** function to find the covariance between two variables and the **corrcoef** function to find the correlation coefficient. Both of these functions are available in the standard MATLAB language. We use the **cement** data [Hand, et al., 1994], which were analyzed by Hald [1952], to illustrate the basic syntax of these functions. The relationship between the two variables is nonlinear, so Hald looked at the log of the tensile strength as a function of the reciprocal of the drying time. When the **cement** data are loaded, we get a vector **x** representing the drying times and a vector **y** that contains the tensile strength. A scatterplot of the transformed data is shown in Figure 3.1.

```
% First load the data.
load cement
% Now get the transformations.
xr = 1./x;
logy = log(y);
% Now get a scatterplot of the data to see if
% the relationship is linear.
```

```
plot(xr,logy,'x')
axis([0 1.1 2.4 4])
xlabel('Reciprocal of Drying Time')
ylabel('Log of Tensile Strength')
```

We now show how to get the covariance matrix and the correlation coefficient for these two variables.

```
% Now get the covariance and
% the correlation coefficient.
cmat = cov(xr,logy);
cormat = corrcoef(xr,logy);
```

The results are:

```
cmat =
      0.1020    -0.1169
     -0.1169     0.1393
cormat =
      1.0000    -0.9803
     -0.9803     1.0000
```

Note that the sample correlation coefficient (Equation 3.12) is given by the off-diagonal element of **cormat**, $\hat{\rho} = -0.9803$. We see that the variables are negatively correlated, which is what we expect from Figure 3.1 (the log of the tensile strength decreases with increasing reciprocal of drying time).
❏

## 3.3 Sampling Distributions

It was stated in the previous section that we sometimes use a statistic calculated from a random sample as a point estimate of a population parameter. For example, we might use $\overline{X}$ to estimate $\mu$ or use $S$ to estimate $\sigma$. Since we are using a sample and not observing the entire population, there will be some error in our estimate. In other words, it is unlikely that the statistic will equal the parameter. To manage the uncertainty and error in our estimate, we must know the sampling distribution for the statistic. The *sampling distribution* is the underlying probability distribution for a statistic. To understand the remainder of the text, it is important to remember that *a statistic is a random variable*.

The sampling distributions for many common statistics are known. For example, if our random variable is from the normal distribution, then we know how the sample mean is distributed. Once we know the sampling distribution of our statistic, we can perform statistical hypothesis tests and calculate confidence intervals. If we do not know the distribution of our statistic,

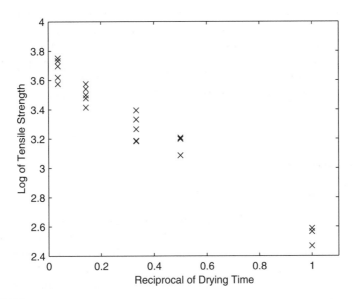

**FIGURE 3.1**
This scatterplot shows the observed drying times and corresponding tensile strength of the cement. Since the relationship is nonlinear, the variables are transformed as shown here. A linear relationship seems to be a reasonable model for these data.

then we must use Monte Carlo simulation techniques or bootstrap methods to estimate the sampling distribution (see Chapter 6).

To illustrate the concept of a sampling distribution, we discuss the sampling distribution for $\overline{X}$, where the random variable $X$ follows a distribution given by the probability density function $f(x)$. It turns out that the distribution for the sample mean can be found using the Central Limit Theorem.

### CENTRAL LIMIT THEOREM
*Let $f(x)$ represent a probability density with finite variance $\sigma^2$ and mean $\mu$. Also, let $\overline{X}$ be the sample mean for a random sample of size n drawn from this distribution. For large n, the distribution of $\overline{X}$ is approximately normally distributed with mean $\mu$ and variance given by $\sigma^2/n$.*

❏

The Central Limit Theorem states that as the sample size gets large, the distribution of the sample mean approaches the normal distribution regardless of how the random variable $X$ is distributed. However, if we are sampling from a normal population, then the distribution of the sample mean is exactly normally distributed with mean $\mu$ and variance $\sigma^2/n$.

This information is important, because we can use it to determine how much error there is in using $\overline{X}$ as an estimate of the population mean $\mu$. We can also perform statistical hypothesis tests using $\overline{X}$ as a test statistic and can calculate confidence intervals for $\mu$. In this book, we are mainly concerned with computational (rather than theoretical) methods for finding sampling distributions of statistics (e.g., Monte Carlo simulation or resampling). The sampling distribution of $\overline{X}$ is used to illustrate the concepts covered in remaining chapters.

## 3.4 Parameter Estimation

One of the first tasks a statistician or an engineer undertakes when faced with data is to try to summarize or describe the data in some manner. Some of the statistics (sample mean, sample variance, coefficient of skewness, etc.) we covered in Section 3.2 can be used as descriptive measures for our sample. In this section, we look at methods to derive and to evaluate estimates of population parameters.

There are several methods available for obtaining parameter estimates. These include the method of moments, maximum likelihood estimation, Bayes estimators, minimax estimation, Pitman estimators, interval estimates, robust estimation, and many others. In this book, we discuss the maximum likelihood method and the method of moments for deriving estimates for population parameters. These somewhat classical techniques are included as illustrative examples only and are not meant to reflect the state of the art in this area. Many useful (and computationally intensive!) methods are not covered here, but references are provided in Section 3.7. However, we do present some alternative methods for calculating interval estimates using Monte Carlo simulation and resampling methods (see Chapters 6 and 7).

Recall that a sample is drawn from a population that is distributed according to some function whose characteristics are governed by certain parameters. For example, our sample might come from a population that is normally distributed with parameters $\mu$ and $\sigma^2$. Or, it might be from a population that is exponentially distributed with parameter $\lambda$. The goal is to use the sample to estimate the corresponding population parameters. If the sample is representative of the population, then a function of the sample should provide a useful estimate of the parameters.

Before we undertake our discussion of maximum likelihood, we need to define what an estimator is. Typically, population parameters can take on values from a subset of the real line. For example, the population mean can be any real number, $-\infty < \mu < \infty$, and the population standard deviation can be any positive real number, $\sigma > 0$. The set of all possible values for a parameter $\theta$ is called the *parameter space*. The *data space* is defined as the set of all possible values of the random sample of size $n$. The estimate is calculated from the sample data as a function of the random sample. An *estimator* is a function or mapping from the data space to the parameter space and is denoted as

$$T = t(X_1, ..., X_n). \tag{3.13}$$

Since an estimator is calculated using the sample alone, it is a statistic. Furthermore, if we have a random sample, then an estimator is also a random variable. This means that the value of the estimator varies from one sample to another based on its sampling distribution. In order to assess the usefulness of our estimator, we need to have some criteria to measure the performance. We discuss four criteria used to assess estimators: bias, mean squared error, efficiency, and standard error. In this discussion, we only present the definitional aspects of these criteria.

## Bias

The bias in an estimator gives a measure of how much error we have, on average, in our estimate when we use $T$ to estimate our parameter $\theta$. The *bias* is defined as

$$\text{bias}(T) = E[T] - \theta. \tag{3.14}$$

If the estimator is unbiased, then the expected value of our estimator equals the true parameter value, so $E[T] = \theta$.

To determine the expected value in Equation 3.14, we must know the distribution of the statistic $T$. In these situations, the bias can be determined analytically. When the distribution of the statistic is not known, then we can use methods such as the jackknife and the bootstrap (see Chapters 6 and 7) to estimate the bias of $T$.

## Mean Squared Error

Let $\theta$ denote the parameter we are estimating and $T$ denote our estimate, then the *mean squared error* (MSE) of the estimator is defined as

$$\text{MSE}(T) = E[(T - \theta)^2]. \tag{3.15}$$

Thus, the MSE is the expected value of the squared error. We can write this in more useful quantities such as the bias and variance of $T$. (The reader will see this again in Chapter 8 in the context of probability density estimation.) If we expand the expected value on the right hand side of Equation 3.15, then we have

$$\text{MSE}(T) = E[(T^2 - 2T\theta + \theta^2)] = E[T^2] - 2\theta E[T] + \theta^2. \qquad (3.16)$$

By adding and subtracting $(E[T])^2$ to the right hand side of Equation 3.16, we have the following

$$\text{MSE}(T) = E[T^2] - (E[T])^2 + (E[T])^2 - 2\theta E[T] + \theta^2. \qquad (3.17)$$

The first two terms of Equation 3.17 are the variance of $T$, and the last three terms equal the squared bias of our estimator. Thus, we can write the mean squared error as

$$\begin{aligned} \text{MSE}(T) &= E[T^2] - (E[T])^2 + (E[T] - \theta)^2 \\ &= V(T) + [\text{bias}(T)]^2. \end{aligned} \qquad (3.18)$$

Since the mean squared error is based on the variance and the squared bias, the error will be small when the variance and the bias are both small. When $T$ is unbiased, then the mean squared error is equal to the variance only. The concepts of bias and variance are important for assessing the performance of any estimator.

## Relative Efficiency

Another measure we can use to compare estimators is called efficiency, which is defined using the MSE. For example, suppose we have two estimators $T_1 = t_1(X_1, \ldots, X_n)$ and $T_2 = t_2(X_1, \ldots, X_n)$ for the same parameter. If the MSE of one estimator is less than the other (e.g., $\text{MSE}(T_1) < \text{MSE}(T_2)$), then $T_1$ is said to be more efficient than $T_2$.

The *relative efficiency* of $T_1$ to $T_2$ is given by

$$\textit{eff}(T_1, T_2) = \frac{\text{MSE}(T_2)}{\text{MSE}(T_1)}. \qquad (3.19)$$

If this ratio is greater than one, then $T_1$ is a more efficient estimator of the parameter.

## Standard Error

We can get a measure of the precision of our estimator by calculating the standard error. The *standard error* of an estimator (or a statistic) is defined as the standard deviation of its sampling distribution:

$$SE(T) = \sqrt{V(T)} = \sigma_T.$$

To illustrate this concept, let's use the sample mean as an example. We know that the variance of the estimator is

$$V(\overline{X}) = \frac{1}{n}\sigma^2 \, ,$$

for large $n$. So, the standard error is given by

$$SE(\overline{X}) = \sigma_{\overline{X}} = \frac{\sigma}{\sqrt{n}} \, . \tag{3.20}$$

If the standard deviation $\sigma$ for the underlying population is unknown, then we can substitute an estimate for the parameter. In this case, we call it the estimated standard error:

$$\hat{SE}(\overline{X}) = \hat{\sigma}_{\overline{X}} = \frac{S}{\sqrt{n}} \, . \tag{3.21}$$

Note that the estimate in Equation 3.21 is also a random variable and has a probability distribution associated with it.

If the bias in an estimator is small, then the variance of the estimator is approximately equal to the MSE, $V(T) \approx MSE(T)$. Thus, we can also use the square root of the MSE as an estimate of the standard error.

## Maximum Likelihood Estimation

A *maximum likelihood estimator* is that value of the parameter (or parameters) that maximizes the likelihood function of the sample. The *likelihood function* of a random sample of size $n$ from density (mass) function $f(x;\theta)$ is the joint probability density (mass) function, denoted by

$$L(\theta;x_1, \ldots, x_n) = f(x_1, \ldots, x_n;\theta) \, . \tag{3.22}$$

Equation 3.22 provides the likelihood that the random variables take on a particular value $x_1, \ldots, x_n$. Note that the likelihood function $L$ is a function of the unknown parameter $\theta$, and that we allow $\theta$ to represent a vector of parameters.

If we have a random sample (independent, identically distributed random variables), then we can write the likelihood function as

$$L(\theta) = L(\theta;x_1, \ldots, x_n) = f(x_1;\theta) \times \ldots \times f(x_n;\theta) \, , \tag{3.23}$$

which is the product of the individual density functions evaluated at each $x_i$ or sample point.

In most cases, to find the value $\hat{\theta}$ that maximizes the likelihood function, we take the derivative of $L$, set it equal to 0 and solve for $\theta$. Thus, we solve the following likelihood equation

$$\frac{d}{d\theta}L(\theta) = 0. \tag{3.24}$$

It can be shown that the likelihood function, $L(\theta)$, and logarithm of the likelihood function, $\ln L(\theta)$, have their maxima at the same value of $\theta$. It is sometimes easier to find the maximum of $\ln L(\theta)$, especially when working with an exponential function. However, keep in mind that a solution to the above equation does not imply that it is a maximum; it could be a minimum. It is important to ensure this is the case before using the result as a maximum likelihood estimator.

When a distribution has more than one parameter, then the likelihood function is a function of all parameters that pertain to the distribution. In these situations, the maximum likelihood estimates are obtained by taking the partial derivatives of the likelihood function (or $\ln L(\theta)$), setting them all equal to zero, and solving the system of equations. The resulting estimators are called the joint maximum likelihood estimators. We see an example of this below, where we derive the maximum likelihood estimators for $\mu$ and $\sigma^2$ for the normal distribution.

## Example 3.3

In this example, we derive the maximum likelihood estimators for the parameters of the normal distribution. We start off with the likelihood function for a random sample of size $n$ given by

$$L(\theta) = \prod_{i=1}^{n} \frac{1}{\sigma\sqrt{2\pi}} \exp\left\{-\frac{(x_i - \mu)^2}{2\sigma^2}\right\} = \left(\frac{1}{2\pi\sigma^2}\right)^{n/2} \exp\left(-\frac{1}{2\sigma^2}\sum_{i=1}^{n}(x_i - \mu)^2\right).$$

Since this has the exponential function in it, we will take the logarithm to obtain

$$\ln[L(\theta)] = \ln\left[\left(\frac{1}{2\pi\sigma^2}\right)^{\frac{n}{2}}\right] + \ln\left[\exp\left(-\frac{1}{2\sigma^2}\sum_{i=1}^{n}(x_i - \mu)^2\right)\right].$$

This simplifies to

$$\ln[L(\theta)] = -\frac{n}{2}\ln[2\pi] - \frac{n}{2}\ln[\sigma^2] - \frac{1}{2\sigma^2}\sum_{i=1}^{n}(x_i - \mu)^2, \qquad (3.25)$$

with $\sigma > 0$ and $-\infty < \mu < \infty$. The next step is to take the partial derivative of Equation 3.25 with respect to $\mu$ and $\sigma^2$. These derivatives are

$$\frac{\partial}{\partial\mu}\ln L = \frac{1}{\sigma^2}\sum_{i=1}^{n}(x_i - \mu), \qquad (3.26)$$

and

$$\frac{\partial}{\partial\sigma^2}\ln L = -\frac{n}{2\sigma^2} + \frac{1}{2\sigma^4}\sum_{i=1}^{n}(x_i - \mu)^2. \qquad (3.27)$$

We then set Equations 3.26 and 3.27 equal to zero and solve for $\mu$ and $\sigma^2$. Solving the first equation for $\mu$, we get the familiar sample mean for the estimator.

$$\frac{1}{\sigma^2}\sum_{i=1}^{n}(x_i - \mu) = 0,$$

$$\sum_{i=1}^{n}x_i = n\mu,$$

$$\hat{\mu} = \bar{x} = \frac{1}{n}\sum_{i=1}^{n}x_i.$$

Substituting $\hat{\mu} = \bar{x}$ into Equation 3.27, setting it equal to zero, and solving for the variance, we get

$$-\frac{n}{2\sigma^2} + \frac{1}{2\sigma^4}\sum_{i=1}^{n}(x_i - \bar{x})^2 = 0$$

$$\hat{\sigma}^2 = \frac{1}{n}\sum_{i=1}^{n}(x_i - \bar{x})^2. \qquad (3.28)$$

These are the sample moments about the sample mean, and it can be verified that these solutions jointly maximize the likelihood function [Lindgren, 1993].
❑

We know that the $E[\overline{X}] = \mu$ [Mood, Graybill and Boes, 1974], so the sample mean is an unbiased estimator for the population mean. However, that is not the case for the maximum likelihood estimate for the variance. It can be shown [Hogg and Craig, 1978] that

$$E[\hat{\sigma}^2] = \frac{(n-1)\sigma^2}{n},$$

so we know (from Equation 3.14) that the maximum likelihood estimate, $\hat{\sigma}^2$, for the variance is biased. If we want to obtain an unbiased estimator for the variance, we simply multiply our maximum likelihood estimator by $n/(n-1)$. This yields the familiar statistic for the sample variance given by

$$s^2 = \frac{1}{n-1}\sum_{i=1}^{n}(x_i - \bar{x})^2.$$

**Method of Moments**

In some cases, it is difficult finding the maximum of the likelihood function. For example, the gamma distribution has the unknown parameter $t$ that is used in the gamma function, $\Gamma(t)$. This makes it hard to take derivatives and solve the equations for the unknown parameters. The method of moments is one way to approach this problem.

In general, we write the unknown population parameters in terms of the population moments. We then replace the population moments with the corresponding sample moments. We illustrate these concepts in the next example, where we find estimates for the parameters of the gamma distribution.

**Example 3.4**
The gamma distribution has two parameters, $t$ and $\lambda$. Recall that the mean and variance are given by $t/\lambda$ and $t/\lambda^2$, respectively. Writing these in terms of the population moments, we have

$$E[X] = \frac{t}{\lambda}, \tag{3.29}$$

and

$$V(X) = E[X^2] - (E[X])^2 = \frac{t}{\lambda^2}. \qquad (3.30)$$

The next step is to solve Equations 3.29 and 3.30 for $t$ and $\lambda$. From Equation 3.29, we have $t = \lambda E[X]$, and substituting this in the second equation yields

$$E[X^2] - (E[X])^2 = \frac{\lambda E[X]}{\lambda^2}. \qquad (3.31)$$

Rearranging Equation 3.31 gives the following expression for $\lambda$

$$\lambda = \frac{E[X]}{E[X^2] - (E[X])^2}. \qquad (3.32)$$

We can now obtain the parameter $t$ in terms of the population moments (substitute Equation 3.32 for $\lambda$ in Equation 3.29) as

$$t = \frac{(E[X])^2}{E[X^2] - (E[X])^2}. \qquad (3.33)$$

To get our estimates, we substitute the sample moments for $E[X]$ and $E[X^2]$ in Equations 3.32 and 3.33. This yields

$$\hat{t} = \frac{\bar{X}^2}{\frac{1}{n}\sum_{i=1}^{n} X_i^2 - \bar{X}^2}, \qquad (3.34)$$

and

$$\hat{\lambda} = \frac{\bar{X}}{\frac{1}{n}\sum_{i=1}^{n} X_i^2 - \bar{X}^2}. \qquad (3.35)$$

❑

In Table 3.1, we provide some suggested point estimates for several of the distributions covered in Chapter 2. This table also contains the names of functions to calculate the estimators. In Section 3.6, we discuss the MATLAB code available in the Statistics Toolbox for calculating maximum likelihood estimates of distribution parameters. The reader is cautioned that the estimators discussed in this chapter are not necessarily the best in terms of bias, variance, etc.

TABLE 3.1

Suggested Point Estimators for Parameters

| Distribution | Suggested Estimator | MATLAB Function |
|---|---|---|
| Binomial<br>Note: $X$ is the number of<br>   successes in $n$ trials | $\hat{p} = \dfrac{X}{n}$ | `csbinpar` |
| Exponential | $\hat{\lambda} = 1/\overline{X}$ | `csexpar` |
| Gamma | $\hat{t} = \overline{X}^2 \Big/ \left( \dfrac{1}{n}\sum X_i^2 - \overline{X}^2 \right)$<br><br>$\hat{\lambda} = \overline{X} \Big/ \left( \dfrac{1}{n}\sum X_i^2 - \overline{X}^2 \right)$ | `csgampar` |
| Normal | $\hat{\mu} = \overline{X}$<br>$\hat{\sigma}^2 = S^2$ | `mean`<br>`var` |
| Multivariate Normal | $\hat{\mu}_j = \dfrac{1}{n}\sum_{i=1}^{n} X_{ij}$<br><br>$\hat{\Sigma}_{ij} = \dfrac{n\sum_{k=1}^{n} X_{ik}X_{jk} - \sum_{k=1}^{n} X_{ik}\sum_{k=1}^{n} X_{jk}}{n(n-1)}$ | `mean`<br>`cov` |
| Poisson | $\hat{\lambda} = \overline{X}$ | `cspoipar` |

## 3.5 Empirical Distribution Function

Recall from Chapter 2 that the cumulative distribution function is given by

$$F(x) = P(X \le x) = \int_{-\infty}^{x} f(t)\,dt \qquad (3.36)$$

for a continuous random variable and by

$$F(a) = \sum_{x_i \leq a} f(x_i) \tag{3.37}$$

for a discrete random variable. In this section, we examine the sample analog of the cumulative distribution function called the *empirical distribution function*. When it is not suitable to assume a distribution for the random variable, then we can use the empirical distribution function as an estimate of the underlying distribution. One can call this a *nonparametric* estimate of the distribution function, because we are not assuming a specific parametric form for the distribution that generates the random phenomena. In a *parametric* setting, we would assume a particular distribution generated the sample and estimate the cumulative distribution function by estimating the appropriate parameters.

The empirical distribution function is based on the *order statistics*. The order statistics for a sample are obtained by putting the data in ascending order. Thus, for a random sample of size $n$, the order statistics are defined as

$$X_{(1)} \leq X_{(2)} \leq \ldots \leq X_{(n)},$$

with $X_{(i)}$ denoting the $i$-th order statistic. The order statistics for a random sample can be calculated easily in MATLAB using the **sort** function.

The empirical distribution function $\hat{F}_n(x)$ is defined as the number of data points less than or equal to $x$ ($\#(X_i \leq x)$) divided by the sample size $n$. It can be expressed in terms of the order statistics as follows

$$\hat{F}_n(x) = \begin{cases} 0; & x < X_{(1)} \\ j/n; & X_{(j)} \leq x < X_{(j+1)} \\ 1; & x \geq X_{(n)}. \end{cases} \tag{3.38}$$

Figure 3.2 illustrates these concepts. We show the empirical cumulative distribution function for a standard normal and include the theoretical distribution function to verify the results. In the following section, we describe a descriptive measure for a population called a quantile, along with its corresponding estimate. Quantiles are introduced here, because they are based on the cumulative distribution function.

## Quantiles

Quantiles have a fundamental role in statistics. For example, they can be used as a measure of central tendency and dispersion, they provide the critical val-

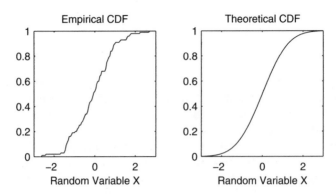

**FIGURE 3.2**
This shows the theoretical and empirical distribution functions for a standard normal distribution.

ues in hypothesis testing (see Chapter 6), and they are used in exploratory data analysis for assessing distributions (see Chapter 5).

The *quantile* $q_p$ of a random variable (or equivalently of its distribution) is defined as the smallest number $q$ such that the cumulative distribution function is greater than or equal to some $p$, where $0 < p < 1$. This can be calculated for a continuous random variable with density function $f(x)$ by solving

$$p = \int_{-\infty}^{q_p} f(x)dx \tag{3.39}$$

for $q_p$, or by using the inverse of the cumulative distribution function,

$$q_p = F^{-1}(p). \tag{3.40}$$

Stating this another way, the $p$-th quantile of a random variable $X$ is the value $q_p$ such that

$$F(q_p) = P(X \le q_p) = p \tag{3.41}$$

for $0 < p < 1$.

Some well known examples of quantiles are the *quartiles*. These are denoted by $q_{0.25}$, $q_{0.5}$, and $q_{0.75}$. In essence, these divide the distribution into four equal (in terms of probability or area under the curve) segments. The second quartile is also called the *median* and satisfies

$$0.5 = \int_{-\infty}^{q_{0.5}} f(x)dx. \tag{3.42}$$

We can get a measure of the dispersion of the random variable by looking at the *interquartile range* (IQR) given by

$$\text{IQR} = q_{0.75} - q_{0.25}. \tag{3.43}$$

One way to obtain an estimate of the quantiles is based on the empirical distribution function. If we let $X_{(1)}, X_{(2)}, \ldots, X_{(n)}$ denote the order statistics for a random sample of size $n$, then $X_{(j)}$ is an estimate of the $(j - 0.5)/n$ quantile [Banks, 2001; Cleveland, 1993]:

$$X_{(j)} \approx F^{-1}\left(\frac{j-0.5}{n}\right). \tag{3.44}$$

We are not limited to a value of 0.5 in Equation 3.44. In general, we can estimate the $p$-th quantile using the following

$$\hat{q}_p = X_{(j)}; \quad \frac{j-1}{n} < p \le \frac{j}{n}; \quad j = 1, \ldots, n. \tag{3.45}$$

As already stated, Equation 3.45 is not the only way to estimate quantiles. For more information on other methods, see Kotz and Johnson [Vol. 7, 1986]. The analyst should exercise caution when calculating quartiles (or other quantiles) using computer packages. Statistical software packages define them differently [Frigge, Hoaglin, and Iglewicz, 1989], so these statistics might vary depending on the formulas that are used.

### EXAMPLE 3.5
In this example, we will show one way to determine the sample quartiles. The second sample quartile $\hat{q}_{0.5}$ is the sample median of the data set. We can calculate this using the function **median**. We could calculate the first quartile $\hat{q}_{0.25}$ as the median of the ordered data that are at the median or below. The third quartile $\hat{q}_{0.75}$ would be calculated as the median of the data that are at $\hat{q}_{0.5}$ or above. The following MATLAB code illustrates these concepts.

```
% Generate the random sample and sort.
x = sort(rand(1,100));
% Find the median of the lower half - first quartile.
q1 = median(x(1:50));
% Find the median.
q2 = median(x);
```

```
% Find the median of the upper half - third quartile.
q3 = median(x(51:100));
```

The quartiles obtained from this random sample are:

```
q1 = 0.29, q2 = 0.53, q3 = 0.79
```

The theoretical quartiles for the uniform distribution are $q_{0.25} = 0.25$, $q_{0.5} = 0.5$, and $q_{0.75} = 0.75$. So we see that the estimates seem reasonable.
❑

Equation 3.44 provides one way to estimate the quantiles from a random sample. In some situations, we might need to determine an estimate of a quantile that does not correspond to $(j - 0.5)/n$. For instance, this is the case when we are constructing q-q plots (see Chapter 5), and the sample sizes differ. We can use interpolation to find estimates of quantiles that are not represented by Equation 3.44.

## Example 3.5

The MATLAB function **interp1** (in the standard package) returns the interpolated value $Y_I$ at a given $X_I$, based on some observed values $X_{obs}$ and $Y_{obs}$. The general syntax is

```
yint = interp1(xobs, yobs, xint);
```

In our case, the argument of $F^{-1}$ in Equation 3.44 represents the observed values $X_{obs}$, and the order statistics $X_{(j)}$ correspond to the $Y_{obs}$. The MATLAB code for this procedure is shown below.

```
% First generate some standard normal data.
x = randn(500,1);
% Now get the order statistics. These will serve
% as the observed values for the ordinate (Y_obs).
xs = sort(x);
% Now get the observed values for the abscissa (X_obs).
n=length(x);
phat = ((1:n)-0.5)/n;
% We want to get the quartiles.
p = [0.25, 0.5, 0.75];
% The following provides the estimates of the quartiles
% using linear interpolation.
qhat = interp1(phat,xs,p);
```

The resulting estimates are

```
qhat = -0.6928    0.0574    0.6453.
```

The reader is asked to explore this further in the exercises.
❑

## 3.6 MATLAB Code

The MATLAB Statistics Toolbox has functions for calculating the maximum likelihood estimates for most of the common distributions, including the gamma and the Weibull distributions. It is important to remember that the parameters estimated for some of the distributions (e.g., exponential and gamma) are different from those defined in Chapters 2 and 3. We refer the reader to Appendix E for a complete list of the functions appropriate to this chapter. Table 3.2 provides a partial list of MATLAB functions for calculating statistics. We also provide some functions for statistics with the Computational Statistics Toolbox. These are summarized in Table 3.3.

TABLE 3.2

List of MATLAB functions for calculating statistics

| Purpose | MATLAB Function |
|---|---|
| These functions are available in the standard MATLAB package. | mean |
| | var |
| | std |
| | cov |
| | median |
| | corrcoef |
| | max, min |
| | sort |
| These functions for calculating descriptive statistics are available in the MATLAB Statistics Toolbox. | harmmean |
| | iqr |
| | kurtosis |
| | mad |
| | moment |
| | prctile |
| | range |
| | skewness |
| | trimmean |
| These MATLAB Statistics Toolbox functions provide the maximum likelihood estimates for distributions. | betafit |
| | binofit |
| | expfit |
| | gamfit |
| | normfit |
| | poissfit |
| | weibfit |
| | unifit |
| | mle |

TABLE 3.3

List of Functions from Chapter 3 Included in the Computational Statistics Toolbox

| Purpose | MATLAB Function |
|---|---|
| These functions are used to obtain parameter estimates for a distribution. | `csbinpar` |
| | `csexpar` |
| | `csgampar` |
| | `cspoipar` |
| | `csunipar` |
| These functions return the quantiles. | `csbinoq` |
| | `csexpoq` |
| | `csunifq` |
| | `csweibq` |
| | `csnormq` |
| | `csquantiles` |
| Other descriptive statistics | `csmomentc` |
| | `cskewness` |
| | `cskurtosis` |
| | `csmoment` |
| | `csecdf` |

## 3.7 Further Reading

Many books discuss sampling distributions and parameter estimation. These topics are covered at an undergraduate level in most introductory statistics books for engineers or non-statisticians. For the advanced undergraduate and beginning graduate student, we recommend the text on mathematical statistics by Hogg and Craig [1978]. Another excellent introductory book on mathematical statistics that contains many applications and examples is written by Mood, Graybill and Boes [1974]. Other texts at this same level include Bain and Engelhardt [1992], Bickel and Doksum [2001], and Lindgren [1993]. For the reader interested in the theory of point estimation on a more advanced graduate level, the book by Lehmann and Casella [1998] and Lehmann [1994] are classics.

Most of the texts already mentioned include descriptions of other methods (Bayes methods, minimax methods, Pitman estimators, etc.) for estimating parameters. For an introduction to robust estimation methods, see the books by Wilcox [1997], Launer and Wilkinson [1979], Huber [1981], or Rousseeuw and Leroy [1987] or see the survey paper by Hogg [1974]. Finally, the text by

Keating, Mason and Sen [1993] provides an introduction to Pitman's measure of closeness as a way to assess the performance of competing estimators.

**Exercises**

3.1. Generate 500 random samples from the standard normal distribution for sample sizes of $n = 2$, 15, and 45. At each sample size, calculate the sample mean for all 500 samples. How are the means distributed as $n$ gets large? Look at a histogram of the sample means to help answer this question. What is the mean and variance of the sample means for each $n$? Is this what you would expect from the Central Limit Theorem? Here is some MATLAB code to get you started.

For each $n$:

```
% Generate 500 random samples of size n:
x = randn(n, 500);
% Get the mean of each sample:
xbar = mean(x);
% Do a histogram with superimposed normal density.
% This function is in the MATLAB Statistics Toolbox.
% If you do not have this, then just use the
% function hist instead of histfit.
histfit(xbar);
```

3.2. Repeat problem 3.1 for random samples drawn from a uniform distribution. Use the MATLAB function **rand** to get the samples.

3.3. We have two unbiased estimators $T_1$ and $T_2$ of the parameter $\theta$. The variances of the estimators are given by $V(T_2) = 8$ and $V(T_1) = 4$. What is the MSE of the estimators? Which estimator is better and why? What is the relative efficiency of the two estimators?

3.4. Repeat Example 3.1 using different sample sizes. What happens to the coefficient of skewness and kurtosis as the sample size gets large?

3.5. Repeat Example 3.1 using samples generated from a standard normal distribution. You can use the MATLAB function **randn** to generate your samples. What happens to the coefficient of skewness and kurtosis as the sample size gets large?

3.6. Generate a random sample that is uniformly distributed over the interval $(0, 1)$. Plot the empirical distribution function over the interval $(-0.5, 1.5)$. There is also a function in the Statistics Toolbox called **cdfplot** that will do this.

3.7. Generate a random sample of size 100 from a normal distribution with mean 10 and variance of 2 (use **randn(1,100)*sqrt(2)+10**). Plot the empirical cumulative distribution function. What is the value of the empirical distribution function evaluated at a point less than the smallest observation in your random sample? What is the value of the empirical cumulative distribution function evaluated at a point that is greater than the largest observation in your random sample?

3.8. Generate a random sample of size 100 from a normal distribution. What are the estimated quartiles?

3.9. Generate a random sample of size 100 from a uniform distribution (use the MATLAB function **rand** to generate the samples). What are the sample quantiles for $p = 0.33, 0.40, 0.63, 0.90$? Is this what you would expect from theory?

3.10. Write a MATLAB function that will return the sample quartiles based on the general definition given for sample quantiles (Equation 3.44).

3.11. Repeat Examples 3.5 and 3.6 for larger sample sizes. Do your estimates for the quartiles get closer to the theoretical values?

3.12. Derive the median for an exponential random variable.

3.13. Calculate the quartiles for the exponential distribution.

3.14. Compare the values obtained for the estimated quartiles in Example 3.6 with the theoretical quantities. You can find the theoretical quantities using **norminv**. Increase the sample size to $n = 1000$. Does your estimate get better?

3.15. Another measure of skewness, called the *quartile coefficient of skewness*, for a sample is given by

$$\hat{\gamma}_{1_q} = \frac{\hat{q}_{0.75} - 2\hat{q}_{0.5} + \hat{q}_{0.25}}{\hat{q}_{0.75} - \hat{q}_{0.25}}.$$

Write a MATLAB function that returns this statistic.

3.16. Investigate the bias in the maximum likelihood estimate of the variance that is given in Equation 3.28. Generate a random sample from the standard normal distribution. You can use the **randn** function that is available in the standard MATLAB package. Calculate $\hat{\sigma}^2$ using Equation 3.28 and record the value in a vector. Repeat this process (generate a random sample from the standard normal distribution, estimate the variance, save the value) many times. Once you are done with this procedure, you should have many estimates for the variance. Take the mean of these estimates to get an estimate of the expected value of $\hat{\sigma}^2$. How does this compare with the known value of $\sigma^2 = 1$? Does this indicate that the maximum likelihood estimate for the variance is biased? What is the estimated bias from this procedure?

# Chapter 4

## Generating Random Variables

### 4.1 Introduction

Many of the methods in computational statistics require the ability to generate random variables from known probability distributions. This is at the heart of Monte Carlo simulation for statistical inference (Chapter 6), bootstrap and resampling methods (Chapters 6 and 7), Markov chain Monte Carlo techniques (Chapter 11), and the analysis of spatial point processes (Chapter 12). In addition, we use simulated random variables to explain many other topics in this book, such as exploratory data analysis (Chapter 5), density estimation (Chapter 8), and statistical pattern recognition (Chapter 9).

There are many excellent books available that discuss techniques for generating random variables and the underlying theory; references will be provided in the last section. Our purpose in covering this topic is to give the reader the tools they need to generate the types of random variables that often arise in practice and to provide examples illustrating the methods. We first discuss general techniques for generating random variables, such as the inverse transformation and acceptance-rejection methods. We then provide algorithms and MATLAB code for generating random variables for some useful distributions.

### 4.2 General Techniques for Generating Random Variables

#### Uniform Random Numbers

Most methods for generating random variables start with random numbers that are uniformly distributed on the interval $(0, 1)$. We will denote these random variables by the letter $U$. With the advent of computers, we now have the ability to generate uniform random variables very easily. However, we have to caution the reader that the numbers generated by computers are really pseudorandom because they are generated using a deterministic algorithm. The techniques used to generate uniform random variables have been widely studied in the literature, and it has been shown that some generators have serious flaws [Gentle, 1998].

The basic MATLAB program has a function **rand** for generating uniform random variables. There are several optional arguments, and we take a moment to discuss them because they will be useful in simulation. The function **rand** with no arguments returns a single instance of the random variable $U$. To get an $m \times n$ array of uniform variates, you can use the syntax **rand(m,n)**. A note of caution: if you use **rand(n)**, then you get an $n \times n$ matrix.

The sequence of random numbers that is generated in MATLAB depends on the seed or the state of the generator. The state is reset to the default when it starts up, so the same sequences of random variables are generated whenever you start MATLAB. This can sometimes be an advantage in situations where we would like to obtain a specific random sample, as we illustrate in the next example. If you call the function using **rand('state',0)**, then MATLAB resets the generator to the initial state. If you want to specify another state, then use the syntax **rand('state',j)** to set the generator to the $j$-th state. You can obtain the current state using **S = rand('state')**, where **S** is a 35 element vector. To reset the state to this one, use **rand('state',S)**.

It should be noted that random numbers that are uniformly distributed over an interval $a$ to $b$ may be generated by a simple transformation, as follows

$$X = (b-a) \cdot U + a. \qquad (4.1)$$

### Example 4.1

In this example, we illustrate the use of MATLAB's function **rand**.

```
% Obtain a vector of uniform random variables in (0,1).
x = rand(1,1000);
% Do a histogram to plot.
% First get the height of the bars.
[N,X] = hist(x,15);
% Use the bar function to plot.
bar(X,N,1,'w')
title('Histogram of Uniform Random Variables')
xlabel('X')
ylabel('Frequency')
```

The resulting histogram is shown in Figure 4.1. In some situations, the analyst might need to reproduce results from a simulation, say to verify a con-

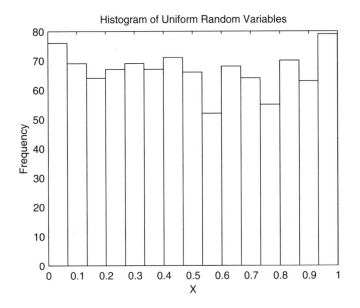

**FIGURE 4.1**
This figure shows a histogram of a random sample from the uniform distribution on the interval (0, 1).

clusion or to illustrate an interesting sample. To accomplish this, the state of the uniform random number generator should be specified at each iteration of the loop. This is accomplished in MATLAB as shown below.

```
% Generate 3 random samples of size 5.
x = zeros(3,5);    % Allocate the memory.
for i = 1:3
    rand('state',i) % set the state
    x(i,:) = rand(1,5);
end
```

The three sets of random variables are

| | | | | |
|---|---|---|---|---|
| 0.9528 | 0.7041 | 0.9539 | 0.5982 | 0.8407 |
| 0.8752 | 0.3179 | 0.2732 | 0.6765 | 0.0712 |
| 0.5162 | 0.2252 | 0.1837 | 0.2163 | 0.4272 |

We can easily recover the five random variables generated in the second sample by setting the state of the random number generator, as follows

```
rand('state',2)
xt = rand(1,5);
```

From this, we get

    `xt = 0.8752      0.3179      0.2732      0.6765      0.0712`

which is the same as before.
❏

## Inverse Transform Method

The inverse transform method can be used to generate random variables from a continuous distribution. It uses the fact that the cumulative distribution function $F$ is uniform $(0, 1)$ [Ross, 1997]:

$$U = F(X). \tag{4.2}$$

If $U$ is a uniform $(0, 1)$ random variable, then we can obtain the desired random variable $X$ from the following relationship

$$X = F^{-1}(U). \tag{4.3}$$

We see an example of how to use the inverse transform method when we discuss generating random variables from the exponential distribution (see Example 4.6). The general procedure for the inverse transformation method is outlined here.

*PROCEDURE - INVERSE TRANSFORM METHOD (CONTINUOUS)*

  1. Derive the expression for the inverse distribution function $F^{-1}(U)$.
  2. Generate a uniform random number $U$.
  3. Obtain the desired $X$ from $X = F^{-1}(U)$.

This same technique can be adapted to the discrete case [Banks, 2001]. Say we would like to generate a discrete random variable $X$ that has a probability mass function given by

$$P(X = x_i) = p_i; \qquad x_0 < x_1 < x_2 < \ldots; \qquad \sum_i p_i = 1. \tag{4.4}$$

We get the random variables by generating a random number $U$ and then deliver the random number $X$ according to the following

$$X = x_i, \qquad \text{if} \quad F(x_{i-1}) < U \le F(x_i). \tag{4.5}$$

We illustrate this procedure using a simple example.

## Example 4.2

We would like to simulate a discrete random variable $X$ that has probability mass function given by

$$P(X = 0) = 0.3,$$
$$P(X = 1) = 0.2,$$
$$P(X = 2) = 0.5.$$

The cumulative distribution function is

$$F(x) = \begin{cases} 0; & x < 0 \\ 0.3; & 0 \le x < 1 \\ 0.5; & 1 \le x < 2 \\ 1.0; & 2 \le x. \end{cases}$$

We generate random variables for $X$ according to the following scheme

$$X = \begin{cases} 0; & U \le 0.3 \\ 1; & 0.3 < U \le 0.5 \\ 2; & 0.5 < U \le 1. \end{cases}$$

This is easily implemented in MATLAB and is left as an exercise. The procedure is illustrated in Figure 4.2, for the situation where a uniform random variable 0.73 was generated. Note that this would return the variate $x = 2$. ❑

We now outline the algorithmic technique for this procedure. This will be useful when we describe a method for generating Poisson random variables.

*PROCEDURE - INVERSE TRANSFORM (DISCRETE)*

1. Define a probability mass function for $x_i$, $i = 1, ..., k$. Note that $k$ could grow infinitely.
2. Generate a uniform random number $U$.
3. If $U \le p_0$ deliver $X = x_0$
4. else if $U \le p_0 + p_1$ deliver $X = x_1$
5. else if $U \le p_0 + p_1 + p_2$ deliver $X = x_2$

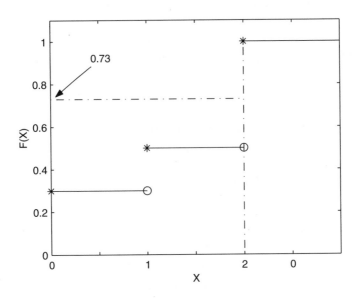

**FIGURE 4.2**
This figure illustrates the inverse transform procedure for generating discrete random variables. If we generate a uniform random number of $u = 0.73$, then this yields a random variable of $x = 2$.

6. ... else if $U \leq p_0 + ... + p_k$ deliver $X = x_k$.

## Example 4.3

We repeat the previous example using this new procedure and implement it in MATLAB. We first generate 100 variates from the desired probability mass function.

```
% Set up storage space for the variables.
X = zeros(1,100);
% These are the x's in the domain.
x = 0:2;
% These are the probability masses.
pr = [0.3 0.2 0.5];
% Generate 100 rv's from the desired distribution.
for i = 1:100
    u = rand;   % Generate the U.
    if u <= pr(1)
        X(i) = x(1);
    elseif u <= sum(pr(1:2))
        % It has to be between 0.3 and 0.5.
        X(i) = x(2);
```

```
    else
        X(i) = x(3); % It has to be between 0.5 and 1.
    end
end
```

One way to verify that our random variables are from the desired distribution is to look at the relative frequency of each $x$.

```
% Find the proportion of each number.
x0 = length(find(X==0))/100;
x1 = length(find(X==1))/100;
x2 = length(find(X==2))/100;
```

The resulting estimated probabilities are

$$\hat{P}(x = x_0) = 0.26$$
$$\hat{P}(x = x_1) = 0.21$$
$$\hat{P}(x = x_2) = 0.53.$$

These values are reasonable when compared with the desired probability mass values.
□

## Acceptance-Rejection Method

In some cases, we might have a simple method for generating a random variable from one density, say $g(y)$, instead of the density we are seeking. We can use this density to generate from the desired continuous density $f(x)$. We first generate a random number $Y$ from $g(y)$ and accept the value with a probability proportional to the ratio $f(Y)/(g(Y))$.

If we define $c$ as a constant that satisfies

$$\frac{f(y)}{g(y)} \leq c; \qquad \text{for all } y, \tag{4.6}$$

then we can generate the desired variates using the procedure outlined below. The constant $c$ is needed because we might have to adjust the height of $g(y)$ to ensure that it is above $f(y)$. We generate points from $cg(y)$, and those points that are inside the curve $f(y)$ are accepted as belonging to the desired density. Those that are outside are rejected. It is best to keep the number of rejected variates small for maximum efficiency.

*PROCEDURE - ACCEPTANCE-REJECTION METHOD (CONTINUOUS)*

1. Choose a density $g(y)$ that is easy to sample from.
2. Find a constant $c$ such that Equation 4.6 is satisfied.
3. Generate a random number $Y$ from the density $g(y)$.
4. Generate a uniform random number $U$.
5. If

$$U \le \frac{f(Y)}{cg(Y)},$$

then accept $X = Y$, else go to step 3.

## Example 4.4

We shall illustrate the acceptance-rejection method by generating random variables from the beta distribution with parameters $\alpha = 2$ and $\beta = 1$ [Ross, 1997]. This yields the following probability density function

$$f(x) = 2x; \qquad 0 < x < 1. \tag{4.7}$$

Since the domain of this density is 0 to 1, we use the uniform distribution for our $g(y)$. We must find a constant that we can use to inflate the uniform so it is above the desired beta density. This constant is given by the maximum value of the density function, and from Equation 4.7, we see that $c = 2$. For more complicated functions, techniques from calculus or the MATLAB function **fminsearch** may be used. The following MATLAB code generates 100 random variates from the desired distribution. We save both the accepted and the rejected variates for display purposes only.

```
c = 2;    % constant
n = 100;  % Generate 100 random variables.
% Set up the arrays to store variates.
x = zeros(1,n);   % random variates
xy = zeros(1,n);% corresponding y values
rej = zeros(1,n);% rejected variates
rejy = zeros(1,n); % corresponding y values
irv = 1;
irej = 1;
while irv <= n
    y = rand(1);   % random number from g(y)
    u = rand(1);   % random number for comparison
    if u <= 2*y/c;
        x(irv) = y;
        xy(irv) = u*c;
```

```
        irv = irv+1
    else
        rej(irej) = y;
        rejy(irej) = u*c; % really comparing u*c<=2*y
        irej = irej + 1
    end
end
```

In Figure 4.3, we show the accepted and rejected random variates that were generated in this process. Note that the accepted variates are those that are less than $f(x)$.
❑

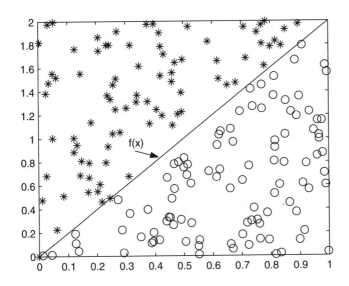

**FIGURE 4.3**
This shows the points that were accepted ('o') as being generated by $f(x) = 2x$ and those points that were rejected ('*'). The curve represents $f(x)$, so we see that the accepted variates are the ones below the curve.

We can easily adapt this method to generate random variables from a discrete distribution. Here we have a method for simulating a random variable with a probability mass function $q_i = P(Y = i)$, and we would like to obtain a random variable $X$ having a probability mass function $p_i = P(X = i)$. As in the continuous case, we generate a random variable $Y$ from $q_i$ and accept this value with probability $p_Y/(cq_Y)$.

PROCEDURE - REJECTION METHOD (DISCRETE)

    1. Choose a probability mass function $q_i$ that is easy to sample from.

    2. Find a constant $c$ such that $p_Y < cq_Y$.

    3. Generate a random number $Y$ from the density $q_i$.

    4. Generate a uniform random number $U$.

    5. If

$$U \le \frac{p_Y}{cq_Y},$$

    then deliver $X = Y$, else go to step 3.

## Example 4.5

In this example, we use the discrete form of the acceptance-rejection method to generate random variables according to the probability mass function defined as follows

$$P(X = 1) = 0.15,$$
$$P(X = 2) = 0.22,$$
$$P(X = 3) = 0.33,$$
$$P(X = 4) = 0.10,$$
$$P(X = 5) = 0.20.$$

We let $q_Y$ be the discrete uniform distribution on $1, \ldots, 5$, where the probability mass function is given by

$$q_y = \frac{1}{5}; \qquad y = 1, \ldots, 5.$$

We describe a method for generating random variables from the discrete uniform distribution in a later section. The value for $c$ is obtained as the maximum value of $p_y/q_y$, which is 1.65. This quantity is obtained by taking the maximum $p_y$, which is $P(X = 3) = 0.33$, and dividing by $1/5$:

$$\frac{max(p_y)}{1/5} = 0.33 \times 5 = 1.65.$$

The steps for generating the variates are:

1. Generate a variate $Y$ from the discrete uniform density on $1, \ldots, 5$. (One could use the MATLAB Statistics Toolbox function **unidrnd** or **csdunrnd**.)
2. Generate a uniform random number $U$.
3. If

$$U \leq \frac{p_Y}{cq_Y} = \frac{p_Y}{1.65 \cdot 1/5} = \frac{p_Y}{0.33},$$

then deliver $X = Y$, else return to step 1.

The implementation of this example in MATLAB is left as an exercise.
□

## 4.3 Generating Continuous Random Variables

### Normal Distribution

The main MATLAB program has a function that will generate numbers from the standard normal distribution, so we do not discuss any techniques for generating random variables from the normal distribution. For the reader who is interested in how normal random variates can be generated, most of the references provided in Section 4.6 contain this information.

The MATLAB function for generating standard normal random variables is called **randn**, and its functionality is similar to the function **rand** that was discussed in the previous section. As with the uniform random variable $U$, we can obtain a normal random variable $X$ with mean $\mu$ and variance $\sigma^2$ by means of a transformation. Letting $Z$ represent a standard normal random variable (possibly generated from **randn**), we get the desired $X$ from the relationship

$$X = Z \cdot \sigma + \mu. \tag{4.8}$$

### Exponential Distribution

The inverse transform method can be used to generate random variables from the exponential distribution and serves as an example of this procedure. The distribution function for an exponential random variable with parameter $\lambda$ is given by

$$F(x) = 1 - e^{-\lambda x}; \qquad 0 < x < \infty. \tag{4.9}$$

Letting

$$u = F(x) = 1 - e^{-\lambda x}, \tag{4.10}$$

we can solve for $x$, as follows

$$u = 1 - e^{-\lambda x}$$
$$e^{-\lambda x} = 1 - u$$
$$-\lambda x = \log(1 - u)$$
$$x = -\frac{1}{\lambda}\log(1 - u).$$

By making note of the fact that $1 - u$ is also uniformly distributed over the interval $(0,1)$, we can generate exponential random variables with parameter $\lambda$ using the transformation

$$X = -\frac{1}{\lambda}\log(U). \tag{4.11}$$

### Example 4.6

The following MATLAB code will generate exponential random variables for a given $\lambda$.

```
% Set up the parameters.
lam = 2;
n = 1000;
% Generate the random variables.
uni = rand(1,n);
X = -log(uni)/lam;
```

We can generate a set of random variables and plot them to verify that the function does yield exponentially distributed random variables. We plot a histogram of the results along with the theoretical probability density function in Figure 4.4. The MATLAB code given below shows how we did this.

```
% Get the values to draw the theoretical curve.
x = 0:.1:5;
% This is a function in the Statistics Toolbox.
y = exppdf(x,1/2);
% Get the information for the histogram.
[N,h] = hist(X,10);
% Change bar heights to make it correspond to
```

```
% the theoretical density - see Chapter 5.
N = N/(h(2)-h(1))/n;
% Do the plots.
bar(h,N,1,'w')
hold on
plot(x,y)
hold off
xlabel('X')
ylabel('f(x) - Exponential')
```

❑

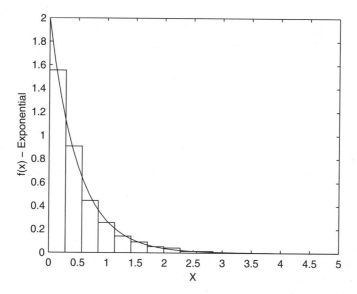

**FIGURE 4.4**
This shows a probability density histogram of the random variables generated in Example 4.6. We also superimpose the curve corresponding to the theoretical probability density function with $\lambda = 2$. The histogram and the curve match quite well.

## Gamma

In this section, we present an algorithm for generating a gamma random variable with parameters $(t, \lambda)$, where $t$ is an integer. Recall that it has the following distribution function

$$F(x) = \int_0^{\lambda x} \frac{e^{-y} y^{t-1}}{(t-1)!} dy. \qquad (4.12)$$

The inverse transform method cannot be used in this case, because a simple closed form solution for its inverse is not possible. It can be shown [Ross, 1997] that the sum of $t$ independent exponentials with the same parameter $\lambda$ is a gamma random variable with parameters $t$ and $\lambda$. This leads to the following transformation based on $t$ uniform random numbers,

$$X = -\frac{1}{\lambda} \log U_1 - \ldots - \frac{1}{\lambda} \log U_t. \qquad (4.13)$$

We can simplify this and compute only one logarithm by using a familiar relationship of logarithms. This yields the following

$$X = -\frac{1}{\lambda} \log(U_1 \times \ldots \times U_t) = -\frac{1}{\lambda} \log\left(\prod_{i=1}^{t} U_i\right). \qquad (4.14)$$

## Example 4.7

The MATLAB code given below implements the algorithm described above for generating gamma random variables, when the parameter $t$ is an integer.

```
n = 1000;
t = 3;
lam = 2;
% Generate the uniforms needed. Each column
% contains the t uniforms for a realization of a
% gamma random variable.
U = rand(t,n);
% Transform according to Equation 4.13.
% See Example 4.8 for an illustration of Equation 4.14.
logU = -log(U)/lam;
X = sum(logU);
```

To see whether the implementation of the algorithm is correct, we plot them in a probability density histogram.

```
% Now do the histogram.
[N,h] = hist(X,10);
% Change bar heights.
N = N/(h(2)-h(1))/n;
% Now get the theoretical probability density.
% This is a function in the Statistics Toolbox.
x = 0:.1:6;
```

```
y = gampdf(x,t,1/lam);
bar(h,N,1,'w')
hold on
plot(x,y,'k')
hold off
```

The histogram and the corresponding theoretical probability density function are shown in Figure 4.5.
❑

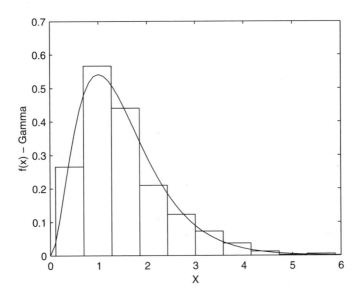

**FIGURE 4.5**
This shows the probability density histogram for a set of gamma random variables with $t = 3$ and $\lambda = 2$.

## Chi-Square

A chi-square random variable with $v$ degrees of freedom is a special case of the gamma distribution, where $\lambda = 1/2$, $t = v/2$ and $v$ is a positive integer. This can be generated using the gamma distribution method described above with one change. We have to make this change, because the method we presented for generating gamma random variables is for integer $t$, which works for even values of $v$.

When $v$ is even, say $2k$, we can obtain a chi-square random variable from

$$X = -2\log\left(\prod_{i=1}^{k} U_i\right). \tag{4.15}$$

When $\nu$ is odd, say $2k + 1$, we can use the fact that the chi-square distribution with $\nu$ degrees of freedom is the sum of $\nu$ squared independent standard normals [Ross, 1997]. We obtain the required random variable by first simulating a chi-square with $2k$ degrees of freedom and adding a squared standard normal variate $Z$, as follows

$$X = Z^2 - 2\log\left(\prod_{i=1}^{k} U_i\right). \tag{4.16}$$

## Example 4.8

In this example, we provide a function that will generate chi-square random variables.

```
% function X = cschirnd(n,nu)
% This function will return n chi-square
% random variables with degrees of freedom nu.

function X = cschirnd(n,nu)
% Generate the uniforms needed.
rm = rem(nu,2);
k = floor(nu/2);
if rm == 0    % then even degrees of freedom
    U = rand(k,n);
    if k ~= 1
        X = -2*log(prod(U));
    else
        X = -2*log(U);
    end
else          % odd degrees of freedom
    U = rand(k,n);
    Z = randn(1,n);
    if k ~= 1
        X = Z.^2-2*log(prod(U));
    else
        X = Z.^2-2*log(U);
    end
end
```

The use of this function to generate random variables is left as an exercise.
❑

The chi-square distribution is useful in situations where we need to systematically investigate the behavior of a statistic by changing the skewness of the distribution. As the degrees of freedom for a chi-square increases, the distribution changes from being right skewed to one approaching normality and symmetry.

### Beta

The beta distribution is useful in simulations because it covers a wide range of distribution shapes, depending on the values of the parameters $\alpha$ and $\beta$. These shapes include skewed, uniform, approximately normal, and a bimodal distribution with an interior dip.

First, we describe a simple approach for generating beta random variables with parameters $\alpha$ and $\beta$, when both are integers [Rubinstein, 1981; Gentle, 1998]. It is known [David, 1981] that the $k$-th order statistic of $n$ uniform $(0,1)$ variates is distributed according to a beta distribution with parameters $k$ and $n - k + 1$. This means that we can generate random variables from the beta distribution using the following procedure.

*PROCEDURE - BETA RANDOM VARIABLES (INTEGER PARAMETERS)*

    1. Generate $\alpha + \beta - 1$ uniform random numbers: $U_1, ..., U_{\alpha + \beta - 1}$

    2. Deliver $X = U_{(\alpha)}$ which is the $\alpha$-th order statistic.

One simple way to generate random variates from the beta distribution is to use the following result from Rubinstein [1981]. If $Y_1$ and $Y_2$ are independent random variables, where $Y_1$ has a gamma distribution with parameters $\alpha$ and 1, and $Y_2$ follows a gamma distribution with parameters $\beta$ and 1, then

$$X = \frac{Y_1}{Y_1 + Y_2} \qquad (4.17)$$

is from a beta distribution with parameters $\alpha$ and $\beta$. This is the method that is used in the MATLAB Statistics Toolbox function **betarnd** that generates random variates from the beta distribution. We illustrate the use of **betarnd** in the following example.

### Example 4.9

We use this example to illustrate the use of the MATLAB Statistics Toolbox function that generates beta random variables. In general, most of these toolbox functions for generating random variables use the following general syntax:

```
rvs = pdfrnd(par1,par2,nrow,ncol);
```

Here, **pdf** refers to the type of distribution (see Table 4.1, on page 106). The first several arguments represent the appropriate parameters of the distribution, so the number of them might change. The last two arguments denote the number of rows and the number of columns in the array of random variables that are returned by the function. We use the function **betarnd** to generate random variables from two beta distributions with different parameters $\alpha$ and $\beta$. First we look at the case where $\alpha = 3$ and $\beta = 3$. So, to generate $n = 500$ beta random variables (that are returned in a row vector), we use the following commands:

```
% Let a = 3, b = 3
n = 500;
a = 3;
b = 3;
rvs = betarnd(a,b,1,n);
```

We can construct a histogram of the random variables and compare it to the corresponding beta probability density function. This is easily accomplished in MATLAB as shown below.

```
% Now do the histogram.
[N,h] = hist(rvs,10);
% Change bar heights.
N = N/(h(2)-h(1))/n;
% Now get the theoretical probability density.
x = 0:.05:1;
y = betapdf(x,a,b);
plot(x,y)
axis equal
bar(h,N,1,'w')
hold on
plot(x,y,'k')
hold off
```

The result is shown in the left plot of Figure 4.6. Notice that this density looks approximately bell-shaped. The beta density on the right has parameters $\alpha = 0.5$ and $\beta = 0.5$. We see that this curve has a dip in the middle with modes on either end. The reader is asked to construct this plot in the exercises.
❑

## Multivariate Normal

In the following chapters, we will have many applications where we need to generate multivariate random variables in order to study the algorithms of computational statistics as they apply to multivariate distributions. Thus, we need some methods for generating multivariate random variables. The easi-

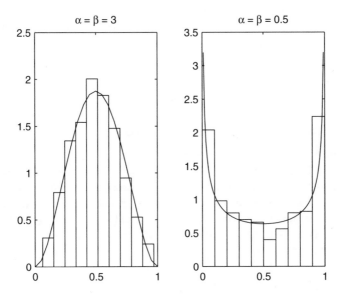

**FIGURE 4.6**
This figure shows two histograms created from random variables generated from the beta distribution. The beta distribution on the left has parameters $\alpha = 3$ and $\beta = 3$, while the one on the right has parameters $\alpha = 0.5$ and $\beta = 0.5$.

est distribution of this type to generate is the multivariate normal. We cover other methods for generating random variables from more general multivariate distributions in Chapter 11.

The method is similar to the one used to generate random variables from a univariate normal distribution. One starts with a $d$-dimensional vector of standard normal random numbers. These can be transformed to the desired distribution using

$$\mathbf{x} = \mathbf{R}^T \mathbf{z} + \mathbf{\mu}. \tag{4.18}$$

Here $\mathbf{z}$ is a $d \times 1$ vector of standard normal random numbers, $\mathbf{\mu}$ is a $d \times 1$ vector representing the mean, and $\mathbf{R}$ is a $d \times d$ matrix such that $\mathbf{R}^T \mathbf{R} = \Sigma$. The matrix $\mathbf{R}$ can be obtained in several ways, one of which is the Cholesky factorization of the covariance matrix $\Sigma$. This is the method we illustrate below. Another possibility is to factor the matrix using singular value decomposition, which will be shown in the examples provided in Chapter 5.

## Example 4.10

The function **csmvrnd** generates multivariate normal random variables using the Cholesky factorization. Note that we are transposing the transformation given in Equation 4.18, yielding the following

$$\mathbf{X} = \mathbf{ZR} + \mu^T,$$

where $\mathbf{X}$ is an $n \times d$ matrix of $d$-dimensional random variables and $\mathbf{Z}$ is an $n \times d$ matrix of standard normal random variables.

```
% function X = csmvrnd(mu,covm,n);
% This function will return n multivariate random
% normal variables with d-dimensional mean mu and
% covariance matrix covm. Note that the covariance
% matrix must be positive definite (all eigenvalues
% are greater than zero), and the mean
% vector is a column

function X = csmvrnd(mu,covm,n)
d = length(mu);
% Get Cholesky factorization of covariance.
R = chol(covm);
% Generate the standard normal random variables.
Z = randn(n,d);
X = Z*R + ones(n,1)*mu';
```

We illustrate its use by generating some multivariate normal random variables with $\mu^T = (-2, 3)$ and covariance

$$\Sigma = \begin{bmatrix} 1 & 0.7 \\ 0.7 & 1 \end{bmatrix}.$$

```
% Generate the multivariate random normal variables.
mu = [-2;3];
covm = [1 0.7 ; 0.7 1];
X = csmvrnd(mu,covm,500);
```

To check the results, we plot the random variables in a scatterplot in Figure 4.7. We can also calculate the sample mean and sample covariance matrix to compare with what we used as input arguments to **csmvrnd**. By typing **mean(X)** at the command line, we get

```
-2.0629    2.9394
```

Similarly, entering **corrcoef(X)** at the command line yields

```
1.0000     0.6957
0.6957     1.0000
```

We see that these values for the sample statistics correspond to the desired mean and covariance. We note that you could also use the **cov** function to compare the variances.
❑

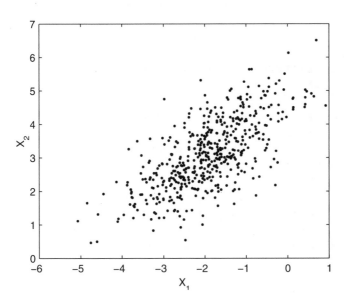

**FIGURE 4.7**
This shows the scatter plot of the random variables generated using the function **csmvrnd**.

## Generating Variates on a Sphere

In some applications, we would like to generate $d$-dimensional random variables that are distributed on the surface of the unit hypersphere $S^d$, $d = 2, \ldots$ . Note that when $d = 2$ the surface is a circle, and for $d = 3$ the surface is a sphere. We will be using this technique in Chapter 5, where we present an algorithm for exploratory data analysis using projection pursuit. The easiest method is to generate $d$ standard normal random variables and then to scale them such that the magnitude of the vector is one. This is illustrated in the following example.

## Example 4.11

The following function **cssphrnd** generates random variables on a $d$-dimensional unit sphere. We illustrate its use by generating random variables that are on the unit circle $S^2$.

```
% function X = cssphrnd(n,d);
% This function will generate n d-dimensional
% random variates that are distributed on the
% unit d-dimensional sphere. d >= 2

function X = cssphrnd(n,d)
if d < 2
   error('ERROR - d must be greater than 1.')
   break
end
% Generate standard normal random variables.
tmp = randn(d,n);
% Find the magnitude of each column.
% Square each element, add and take the square root.
mag = sqrt(sum(tmp.^2));
% Make a diagonal matrix of them - inverses.
dm = diag(1./mag);
% Multiply to scale properly.
% Transpose so X contains the observations.
X = (tmp*dm)';
```

We can use this function to generate a set of random variables for $d = 2$ and plot the result in Figure 4.8.

```
X = cssphrnd(500,2);
plot(X(:,1),X(:,2),'x')
axis equal
xlabel('X_1'),ylabel('X_2')
```

❑

---

## 4.4 Generating Discrete Random Variables

### Binomial

A binomial random variable with parameters $n$ and $p$ represents the number of successes in $n$ independent trials. We can obtain a binomial random vari-

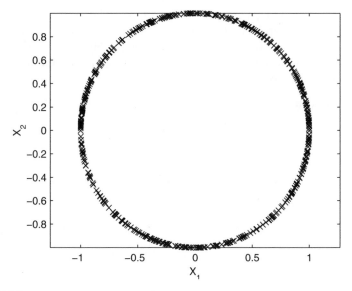

**FIGURE 4.8**
This is the scatter plot of the random variables generated in Example 4.11. These random variables are distributed on the surface of a 2-D unit sphere (i.e., a unit circle).

able by generating $n$ uniform random numbers $U_1, U_2, ..., U_n$ and letting $X$ be the number of $U_i$ that are less than or equal to $p$. This is easily implemented in MATLAB as illustrated in the following example.

## Example 4.12

We implement this algorithm for generating binomial random variables in the function **csbinrnd**.

```
% function X = csbinrnd(n,p,N)
% This function will generate N binomial
% random variables with parameters n and p.

function X = csbinrnd(n,p,N)
X = zeros(1,N);
% Generate the uniform random numbers:
% N variates of n trials.
U = rand(N,n);
% Loop over the rows, finding the number
% less than p
for i = 1:N
    ind = find(U(i,:) <= p);
    X(i) = length(ind);
```

```
    end
```

We use this function to generate a set of random variables that are distributed according to the binomial distribution with parameters $n = 6$ and $p = 0.5$. The histogram of the random variables is shown in Figure 4.9. Before moving on, we offer the following more efficient way to generate binomial random variables in MATLAB:

```
X = sum(rand(n,N) <= p);
```

❏

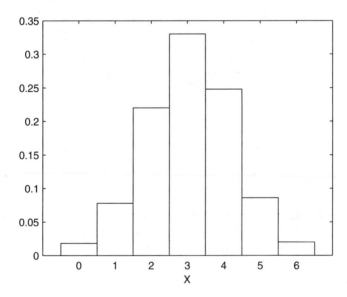

**FIGURE 4.9**
This is the histogram for the binomial random variables generated in Example 4.12. The parameters for the binomial are $n = 6$ and $p = 0.5$.

## Poisson

We use the inverse transform method for discrete random variables as described in Ross [1997] to generate variates from the Poisson distribution. We need the following recursive relationship between successive Poisson probabilities

$$p_{i+1} = P(X = i) = \frac{\lambda}{i+1} p_i; \qquad i \ge 0.$$

This leads to the following algorithm.

*PROCEDURE - GENERATING POISSON RANDOM VARIABLES*

1. Generate a uniform random number $U$.
2. Initialize the quantities: $i = 0$, $p_0 = e^{-\lambda}$, and $F_0 = p_0$.
3. If $U \le F_i$, then deliver $X = i$. Return to step 1.
4. Else increment the values: $p_{i+1} = \lambda p_i / (i+1)$, $i = i+1$, and $F_{i+1} = F_i + p_{i+1}$.
5. Return to step 3.

This algorithm could be made more efficient when $\lambda$ is large. The interested reader is referred to Ross [1997] for more details.

## Example 4.13
The following shows how to implement the procedure for generating Poisson random variables in MATLAB.

```
% function X = cspoirnd(lam,n)
% This function will generate Poisson
% random variables with parameter lambda.
% The reference for this is Ross, 1997, page 50.

function x = cspoirnd(lam,n)
x = zeros(1,n);
j = 1;
while j <= n
   flag = 1;
   % initialize quantities
   u = rand(1);
   i = 0;
   p = exp(-lam);
   F = p;
   while flag % generate the variate needed
      if u <= F % then accept
         x(j) = i;
         flag = 0;
         j = j+1;
      else % move to next probability
         p = lam*p/(i+1);
         i = i+1;
         F = F + p;
      end
   end
end
```

```
end
```

We can use this to generate a set of Poisson random variables with $\lambda = 0.5$, and show a histogram of the data in Figure 4.10.

```
% Set the parameter for the Poisson.
lam = .5;
N = 500; % Sample size
x = cspoirnd(lam,N);
edges = 0:max(x);
f = histc(x,edges);
bar(edges,f/N,1,'w')
```

As an additional check to ensure that our algorithm is working correctly, we can determine the observed relative frequency of each value of the random variable $X$ and compare that to the corresponding theoretical values.

```
% Determine the observed relative frequencies.
% These are the estimated values.
relf = zeros(1,max(x)+1);
for i = 0:max(x)
   relf(i+1) = length(find(x==i))/N;
end
% Use the Statistics Toolbox function to get the
% theoretical values.
y = poisspdf(0:4,.5);
```

When we print these to the MATLAB command window, we have the following

```
% These are the estimated values.
relf = 0.5860    0.3080    0.0840    0.0200    0.0020
% These are the theoretical values.
y = 0.6065    0.3033    0.0758    0.0126    0.0016
```

❑

### Discrete Uniform

When we implement some of the Monte Carlo methods in Chapter 6 (such as the bootstrap), we will need the ability to generate numbers that follow the discrete uniform distribution. This is a distribution where $X$ takes on values in the set $\{1, 2, ..., N\}$, and the probability that $X$ equals any of the numbers is $1/N$. This distribution can be used to randomly sample without replacement from a group of $N$ objects.

We can generate from the discrete uniform distribution using the following transform

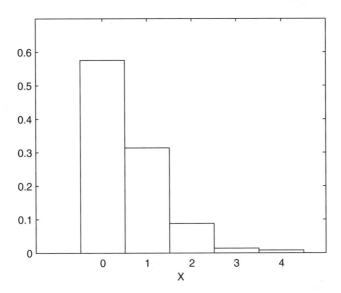

**FIGURE 4.10**
This is the histogram for random variables generated from the Poisson with $\lambda = 0.5$.

$$X = \lceil NU \rceil,$$

where the function $\lceil y \rceil$, $y \geq 0$ means to round up the argument $y$. The next example shows how to implement this in MATLAB.

### Example 4.14
The method for generating discrete uniform is implemented in the function **csdunrnd**, given below.

```
% function X = csdunrnd(N,n)
% This function will generate random variables
% from the discrete uniform distribution. It picks
% numbers uniformly between 1 and N.

function X = csdunrnd(N,n)
X = ceil(N*rand(1,n));
```

To verify that we are generating the right random variables, we can look at the observed relative frequencies. Each should have relative frequency of $1/N$. This is shown below where $N = 5$ and the sample size is 500.

```
N = 5;
n = 500;
x = csdunrnd(N,n);
```

```
% Determine the estimated relative frequencies.
relf = zeros(1,N);
for i = 1:N
   relf(i) = length(find(x==i))/n;
end
```

Printing out the observed relative frequencies, we have

> **relf** = 0.1820      0.2080      0.2040      0.1900      0.2160

which is close to the theoretical value of $1/N = 1/5 = 0.2$.
❏


## 4.5 MATLAB Code

The MATLAB Statistics Toolbox has functions that will generate random variables from all of the distributions discussed in Section 2.6. As we explained in that section, the analyst must keep in mind that probability distributions are often defined differently, so caution should be exercised when using any software package. Table 4.1 provides a partial list of the MATLAB functions that are available for random number generation. A complete list can be found in Appendix E. As before, the reader should note that the **gamrnd**, **weibrnd**, and **exprnd** functions use the alternative definition for the given distribution (see 24).

TABLE 4.1

Partial List of Functions in the MATLAB Statistics Toolbox for Generating Random Variables

| Distribution | MATLAB Function |
| --- | --- |
| Beta | betarnd |
| Binomial | binornd |
| Chi-Square | chi2rnd |
| Discrete Uniform | unidrnd |
| Exponential | exprnd |
| Gamma | gamrnd |
| Normal | normrnd |
| Poisson | poissrnd |
| Continuous Uniform | unifrnd |
| Weibull | weibrnd |

Another function that might prove useful in implementing computational statistics methods is called **randperm**. This is provided with the standard MATLAB software package, and it generates random permutations of the integers 1 to $n$. The result can be used to permute the elements of a vector. For example, to permute the elements of a vector **x** of size **n**, use the following MATLAB statements:

```
% Get the permuted indices.
ind = randperm(n);
% Now re-order based on the permuted indices.
xperm = x(ind);
```

We also provide some functions in the Computational Statistics Toolbox for generating random variables. These are outlined in Table 4.2. Note that these generate random variables using the distributions as defined in Chapter 2.

TABLE 4.2

List of Functions from Chapter 4 Included in the Computational Statistics Toolbox

| Distribution | MATLAB Function |
| --- | --- |
| Beta | csbetarnd |
| Binomial | csbinrnd |
| Chi-Square | cschirnd |
| Discrete Uniform | csdunrnd |
| Exponential | csexprnd |
| Gamma | csgamrnd |
| Multivariate Normal | csmvrnd |
| Poisson | cspoirnd |
| Points on a sphere | cssphrnd |

## 4.6 Further Reading

In this text we do not attempt to assess the computational efficiency of the methods for generating random variables. If the statistician or engineer is performing extensive Monte Carlo simulations, then the time it takes to generate random samples becomes important. In these situations, the reader is encouraged to consult Gentle [1998] or Rubinstein [1981] for efficient algorithms. Our goal is to provide methods that are easily implemented using MATLAB or other software, in case the data analyst must write his own functions for generating random variables from non-standard distributions.

There has been considerable research into methods for random number generation, and we refer the reader to the sources mentioned below for more information on the theoretical foundations. The book by Ross [1997] is an excellent resource and is suitable for advanced undergraduate students. He addresses simulation in general and includes a discussion of discrete event simulation and Markov chain Monte Carlo methods. Another text that covers the topic of random number generation and Monte Carlo simulation is Gentle [1998]. This book includes an extensive discussion of uniform random number generation and covers more advanced topics such as Gibbs sampling. Two other resources on random number generation are Rubinstein [1981] and Kalos and Whitlock [1986]. For a description of methods for generating random variables from more general multivariate distributions, see Johnson [1987]. The article by Deng and Lin [2000] offers improvements on some of the standard uniform random number generators.

A recent article in the MATLAB *News & Notes* [Spring, 2001] describes the method employed in MATLAB for obtaining normally distributed random variables. The algorithm that MATLAB uses for generating uniform random numbers is described in a similar newsletter article and is available for download at:

**www.mathworks.com/company/newsletter/pdf/Cleve.pdf.**

## Exercises

4.1. Repeat Example 4.3 using larger sample sizes. What happens to the estimated probability mass function (i.e., the relative frequencies from the random samples) as the sample size gets bigger?

4.2. Write the MATLAB code to implement Example 4.5. Generate 500 random variables from this distribution and construct a histogram (**hist** function) to verify your code.

4.3. Using the algorithm implemented in Example 4.3, write a MATLAB function that will take any probability mass function (i.e., a vector of probabilities) and return the desired number of random variables generated according to that probability function.

4.4. Write a MATLAB function that will return random numbers that are uniformly distributed over the interval $(a, b)$.

4.5. Write a MATLAB function that will return random numbers from the normal distribution with mean $\mu$ and variance $\sigma^2$. The user should be able to set values for the mean and variance as input arguments.

4.6. Write a function that will generate chi-square random variables with $\nu$ degrees of freedom by generating $\nu$ standard normals, squaring them and then adding them up. This uses the fact that

$$X = Z_1^2 + \dots + Z_\nu^2$$

is chi-square with $\nu$ degrees of freedom. Generate some random variables and plot in a histogram. The degrees of freedom should be an input argument set by the user.

4.7. An alternative method for generating beta random variables is described in Rubinstein [1981]. Generate two variates $Y_1 = U_1^{1/\alpha}$ and $Y_2 = U_2^{1/\beta}$, where the $U_i$ are from the uniform distribution. If $Y_1 + Y_2 \le 1$, then

$$X = \frac{Y_1}{Y_1 + Y_2},$$

is from a beta distribution with parameters $\alpha$ and $\beta$. Implement this algorithm.

4.8. Run Example 4.4 and generate 1000 random variables. Determine the number of variates that were rejected and the total number generated to obtain the random sample. What percentage were rejected? How efficient was it?

4.9. Run Example 4.4 and generate 500 random variables. Plot a histogram of the variates. Does it match the probability density function shown in Figure 4.3?

4.10. Implement Example 4.5 in MATLAB. Generate 100 random variables. What is the relative frequency of each value of the random variable 1, ..., 5 ? Does this match the probability mass function?

4.11. Generate four sets of random variables with $v$ = 2, 5, 15, 20, using the function **cschirnd**. Create histograms for each sample. How does the shape of the distribution depend on the degrees of freedom $v$ ?

4.12. Repeat Example 4.13 for larger sample sizes. Is the agreement better between the observed relative frequencies and the theoretical values?

4.13. Generate 1000 binomial random variables for $n$ = 5 and $p$ = 0.3, 0.5, 0.8. In each case, determine the observed relative frequencies and the corresponding theoretical probabilities. How is the agreement between them?

4.14. The MATLAB Statistics Toolbox has a GUI called **randtool**. This is an interactive demo that generates random variables from distributions that are available in the toolbox. The user can change parameter values and see the results via a histogram. There are options to change the sample size and to output the results. To start the GUI, simply type **randtool** at the command line. Run the function and experiment with the distributions that are discussed in the text (normal, exponential, gamma, beta, etc.).

4.15. The plot on the right in Figure 4.6 shows a histogram of beta random variables with parameters $\alpha = \beta = 0.5$. Construct a similar plot using the information in Example 4.9.

# Chapter 5

## Exploratory Data Analysis

### 5.1 Introduction

Exploratory data analysis (EDA) is quantitative detective work according to John Tukey [1977]. EDA is the philosophy that data should first be explored without assumptions about probabilistic models, error distributions, number of groups, relationships between the variables, etc. for the purpose of discovering what they can tell us about the phenomena we are investigating. The goal of EDA is to explore the data to reveal patterns and features that will help the analyst better understand, analyze and model the data. With the advent of powerful desktop computers and high resolution graphics capabilities, these methods and techniques are within the reach of every statistician, engineer and data analyst.

EDA is a collection of techniques for revealing information about the data and methods for visualizing them to see what they can tell us about the underlying process that generated it. In most situations, exploratory data analysis should precede confirmatory analysis (e.g., hypothesis testing, ANOVA, etc.) to ensure that the analysis is appropriate for the data set. Some examples and goals of EDA are given below to help motivate the reader.

- If we have a time series, then we would plot the values over time to look for patterns such as trends, seasonal effects or change points. In Chapter 11, we have an example of a time series that shows evidence of a change point in a Poisson process.

- We have observations that relate two characteristics or variables, and we are interested in how they are related. Is there a linear or a nonlinear relationship? Are there patterns that can provide insight into the process that relates the variables? We will see examples of this application in Chapters 7 and 10.

- We need to provide some summary statistics that describe the data set. We should look for outliers or aberrant observations that might contaminate the results. If EDA indicates extreme observations are in the data set, then robust statistical methods might be more appropriate. In Chapter 10, we illustrate an example where a graphical look at the data indicates the presence of outliers, so we use a robust method of nonparametric regression.

• We have a random sample that will be used to develop a model. This model will be included in our simulation of a process (e.g., simulating a physical process such as a queue). We can use EDA techniques to help us determine how the data might be distributed and what model might be appropriate.

In this chapter, we will be discussing graphical EDA and how these techniques can be used to gain information and insights about the data. Some experts include techniques such as smoothing, probability density estimation, clustering and principal component analysis in exploratory data analysis. We agree that these can be part of EDA, but we do not cover them in this chapter. Smoothing techniques are discussed in Chapter 10 where we present methods for nonparametric regression. Techniques for probability density estimation are presented in Chapter 8, but we do discuss simple histograms in this chapter. Methods for clustering are described in Chapter 9. Principal component analysis is not covered in this book, because the subject is discussed in many linear algebra texts [Strang, 1988; Jackson, 1991].

It is likely that some of the visualization methods in this chapter are familiar to statisticians, data analysts and engineers. As we stated in Chapter 1, one of the goals of this book is to promote the use of MATLAB for statistical analysis. Some readers might not be familiar with the extensive graphics capabilities of MATLAB, so we endeavor to describe the most useful ones for data analysis. In Section 5.2, we consider techniques for visualizing univariate data. These include such methods as stem-and-leaf plots, box plots, histograms, and quantile plots. We turn our attention to techniques for visualizing bivariate data in Section 5.3 and include a description of surface plots, scatterplots and bivariate histograms. Section 5.4 offers several methods for viewing multi-dimensional data, such as slices, isosurfaces, star plots, parallel coordinates, Andrews curves, projection pursuit, and the grand tour.

## 5.2 Exploring Univariate Data

Two important goals of EDA are: 1) to determine a reasonable model for the process that generated the data, and 2) to locate possible outliers in the sample. For example, we might be interested in finding out whether the distribution that generated the data is symmetric or skewed. We might also like to know whether it has one mode or many modes. The univariate visualization techniques presented here will help us answer questions such as these.

## Histograms

A *histogram* is a way to graphically represent the frequency distribution of a data set. Histograms are a good way to

- summarize a data set to understand general characteristics of the distribution such as shape, spread or location,
- suggest possible probabilistic models, or
- determine unusual behavior.

In this chapter, we look only at the simple, basic histogram. Variants and extensions of the histogram are discussed in Chapter 8.

A *frequency histogram* is obtained by creating a set of bins or intervals that cover the range of the data set. It is important that these bins do not overlap and that they have equal width. We then count the number of observations that fall into each bin. To visualize this, we plot the frequency as the height of a bar, with the width of the bar representing the width of the bin. The histogram is determined by two parameters, the bin width and the starting point of the first bin. We discuss these issues in greater detail in Chapter 8. *Relative frequency histograms* are obtained by representing the height of the bin by the relative frequency of the observations that fall into the bin.

The basic MATLAB package has a function for calculating and plotting a univariate histogram. This function is illustrated in the example given below.

## Example 5.1

In this example, we look at a histogram of the data in **forearm**. These data [Hand, et al., 1994; Pearson and Lee, 1903] consist of 140 measurements of the length in inches of the forearm of adult males. We can obtain a simple histogram in MATLAB using these commands:

```
load forearm
subplot(1,2,1)
% The hist function optionally returns the
% bin centers and frequencies.
[n,x] = hist(forearm);
% Plot and use the argument of width=1
% to produce bars that touch.
bar(x,n,1);
axis square
title('Frequency Histogram')
% Now create a relative frequency histogram.
% Divide each box by the total number of points.
subplot(1,2,2)
bar(x,n/140,1)
title('Relative Frequency Histogram')
axis square
```

These plots are shown in Figure 5.1. Notice that the shapes of the histograms are the same in both types of histograms, but the vertical axis is different. From the shape of the histograms, it seems reasonable to assume that the data are normally distributed.
❏

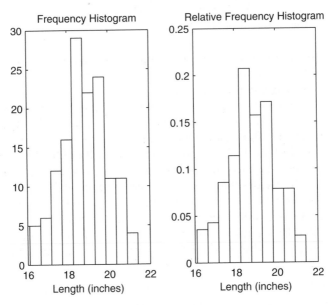

**FIGURE 5.1**
On the left is a frequency histogram of the `forearm` data, and on the right is the relative frequency histogram. These indicate that the distribution is unimodal and that the normal distribution is a reasonable model.

One problem with using a frequency or relative frequency histogram is that they do not represent meaningful probability densities, because they do not integrate to one. This can be seen by superimposing a corresponding normal distribution over the relative frequency histogram as shown in Figure 5.2.

A *density histogram* is a histogram that has been normalized so it will integrate to one. That means that if we add up the *areas* represented by the bars, then they should add up to one. A density histogram is given by the following equation

$$\hat{f}(x) = \frac{v_k}{nh} \qquad x \text{ in } B_k, \tag{5.1}$$

where $B_k$ denotes the $k$-th bin, $v_k$ represents the number of data points that fall into the $k$-th bin and $h$ represents the width of the bins. In the following

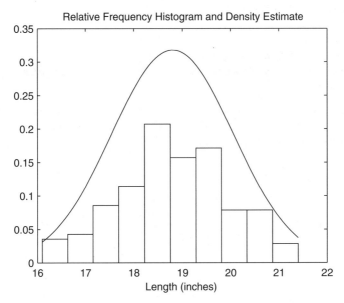

**FIGURE 5.2**
This shows a relative frequency histogram of the **forearm** data. Superimposed on the histogram is the normal probability density function using parameters estimated from the data. Note that the curve is higher than the histogram, indicating that the histogram is not a valid probability density function.

example, we reproduce the histogram of Figure 5.2 using the density histogram.

## Example 5.2

Here we explore the **forearm** data using a density histogram. Assuming a normal distribution and estimating the parameters from the data, we can superimpose a smooth curve that represents an estimated density for the normal distribution.

```
% Get parameter estimates for the normal distribution.
mu = mean(forearm);
v = var(forearm);
% Obtain normal pdf based on parameter estimates.
xp = linspace(min(forearm),max(forearm));
yp = normp(xp,mu,v);
% Get the information needed for a histogram.
[nu,x] = hist(forearm);
% Get the widths of the bins.
h = x(2)-x(1);
```

```
% Plot as density histogram - Equation 5.1.
bar(x,nu/(140*h),1)
hold on
plot(xp,yp)
xlabel('Length (inches)')
title('Density Histogram and Density Estimate')
hold off
```

The results are shown in Figure 5.3. Note that the assumption of normality for the data is not unreasonable. The estimated density function and the density histogram match up quite well.
❑

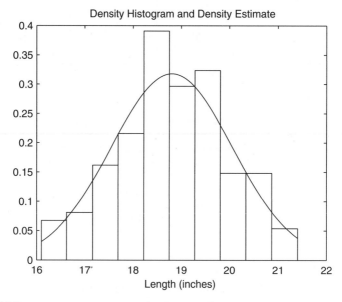

**FIGURE 5.3**
Density histogram for the **forearm** data. The curve represents a normal probability density function with parameters given by the sample mean and sample variance of the data. From this we see that the normal distribution is a reasonable probabilistic model.

## Stem-and-Leaf

Stem-and-leaf plots were introduced by Tukey [1977] as a way of displaying data in a structured list. Presenting data in a table or an ordered list does not readily convey information about how the data are distributed, as is the case with histograms.

If we have data where each observation consists of at least two digits, then we can construct a stem-and-leaf diagram. To display these, we separate each measurement into two parts: the *stem* and the *leaf*. The stems are comprised of the leading digit or digits, and the remaining digit makes up the leaf. For example, if we had the number 75, then the stem is the 7, and the leaf is the 5. If the number is 203, then the stem is 20 and the leaf is 3.

The stems are listed to the left of a vertical line with all of the leaves corresponding to that stem listed to the right. If the data contain decimal places, then they can be rounded for easier display. An alternative is to move the decimal place to specify the appropriate leaf unit. We provide a function with the text that will construct stem-and-leaf plots, and its use is illustrated in the next example.

## Example 5.3

The heights of 32 Tibetan skulls [Hand, et al. 1994; Morant, 1923] measured in millimeters is given in the file **tibetan**. These data comprise two groups of skulls collected in Tibet. One group of 17 skulls comes from graves in Sikkim and nearby areas of Tibet and the other 15 skulls come from a battlefield in Lhasa. The original data contain five measurements, but for this example, we only use the fourth measurement. This is the upper face height, and we round to the nearest millimeter. We use the function **csstemleaf** that is provided with the text.

```
load tibetan
% This loads up all 5 measurements of the skulls.
% We use the fourth characteristic to illustrate
% the stem-and-leaf plot. We first round them.
x = round(tibetan(:,4));
csstemleaf(x)
title('Height (mm) of Tibetan Skulls')
```

The resulting stem-and-leaf is shown in Figure 5.4. From this plot, we see there is not much evidence that there are two groups of skulls, if we look only at the characteristic of upper face height. We will explore these data further in Chapter 9, where we apply pattern recognition methods to the problem. ❑

It is possible that we do not see much evidence for two groups of skulls because there are too few stems. EDA is an iterative process, where the analyst should try several visualization methods in search of patterns and information in the data. An alternative approach is to plot more than one line per stem. The function **csstemleaf** has an optional argument that allows the user to specify two lines per stem. The default value is one line per stem, as we saw in Example 5.3. When we plot two lines per stem, leaves that correspond to the digits 0 through 4 are plotted on the first line and those that have digits 5 through 9 are shown on the second line. A stem-and-leaf with two lines per stem for the Tibetan skull data is shown in Figure 5.5. In practice,

Height (mm) of Tibetan Skulls

| 6 | 2 3 5 5 6 8 9 |
|---|---|
| 7 | 0 0 1 1 1 2 2 3 4 4 4 4 5 6 6 7 7 7 8 9 9 |
| 8 | 0 1 2 3 |

**FIGURE 5.4**
This shows the stem-and-leaf plot for the upper face height of 32 Tibetan skulls. The data have been rounded to the nearest millimeter.

Height (mm) of Tibetan Skulls

| 6 | 2 3 |
|---|---|
| 6 | 5 5 6 8 9 |
| 7 | 0 0 1 1 1 2 2 3 4 4 4 4 |
| 7 | 5 6 6 7 7 7 8 9 9 |
| 8 | 0 1 2 3 |
| 8 | |

**FIGURE 5.5**
This shows a stem-and-leaf plot for the upper face height of 32 Tibetan skulls where we now have two lines per stem. Note that we see approximately the same information (a unimodal distribution) as in Figure 5.4.

one could plot a stem-and-leaf with one and with two lines per stem as a way of discovering more about the data. The stem-and-leaf is useful in that it approximates the shape of the density, and it also provides a listing of the data. One can usually recover the original data set from the stem-and-leaf (if it has not been rounded), unlike the histogram. A disadvantage of the stem-and-leaf plot is that it is not useful for large data sets, while a histogram is very effective in reducing and displaying massive data sets.

## Quantile-Based Plots - Continuous Distributions

If we need to compare two distributions, then we can use the quantile plot to visually compare them. This is also applicable when we want to compare a distribution and a sample or to compare two samples. In comparing the distributions or samples, we are interested in knowing how they are shifted relative to each other. In essence, we want to know if they are distributed in the same way. This is important when we are trying to determine the distribution that generated our data, possibly with the goal of using that information to generate data for Monte Carlo simulation. Another application where this is useful is in checking model assumptions, such as normality, before we conduct our analysis.

In this part, we discuss several versions of quantile-based plots. These include *quantile-quantile plots* (q-q plots) and *quantile plots* (sometimes called a *probability plot*). Quantile plots for discrete data are discussed next. The quantile plot is used to compare a sample with a theoretical distribution. Typically, a q-q plot (sometimes called an *empirical quantile plot*) is used to determine whether two random samples are generated by the same distribution. It should be noted that the q-q plot can also be used to compare a random sample with a theoretical distribution by generating a sample from the theoretical distribution as the second sample.

### Q-Q Plot

The q-q plot was originally proposed by Wilk and Gnanadesikan [1968] to visually compare two distributions by graphing the quantiles of one versus the quantiles of the other. Say we have two data sets consisting of univariate measurements. We denote the order statistics for the first data set by

$$x_{(1)}, x_{(2)}, \ldots, x_{(n)}.$$

Let the order statistics for the second data set be

$$y_{(1)}, y_{(2)}, \ldots, y_{(m)},$$

with $m \leq n$.

We look first at the case where the sizes of the data sets are equal, so $m = n$. In this case, we plot as points the sample quantiles of one data set versus the other data set. This is illustrated in Example 5.4. If the data sets come from the same distribution, then we would expect the points to approximately follow a straight line.

A major strength of the quantile-based plots is that they do not require the two samples (or the sample and theoretical distribution) to have the same location and scale parameter. If the distributions are the same, but differ in location or scale, then we would still expect the quantile-based plot to produce a straight line.

### Example 5.4

We will generate two sets of normal random variables and construct a q-q plot. As expected, the q-q plot (Figure 5.6) follows a straight line, indicating that the samples come from the same distribution.

```
% Generate the random variables.
x = randn(1,75);
y = randn(1,75);
% Find the order statistics.
xs = sort(x);
ys = sort(y);
% Now construct the q-q plot.
plot(xs,ys,'o')
xlabel('X - Standard Normal')
ylabel('Y - Standard Normal')
axis equal
```

If we repeat the above MATLAB commands using a data set generated from an exponential distribution and one that is generated from the standard normal, then we have the plot shown in Figure 5.7. Note that the points in this q-q plot do not follow a straight line, leading us to conclude that the data are not generated from the same distribution.
❏

We now look at the case where the sample sizes are not equal. Without loss of generality, we assume that $m < n$. To obtain the q-q plot, we graph the $y_{(i)}$, $i = 1, ..., m$ against the $(i - 0.5)/m$ quantile of the other data set. Note that this definition is not unique [Cleveland, 1993]. The $(i - 0.5)/m$ quantiles of the $x$ data are usually obtained via interpolation, and we show in the next example how to use the function **csquantiles** to get the desired plot.

Users should be aware that q-q plots provide a rough idea of how similar the distribution is between two random samples. If the sample sizes are small, then a lot of variation is expected, so comparisons might be suspect. To help aid the visual comparison, some q-q plots include a reference line. These are lines that are estimated using the first and third quartiles ($q_{0.25}, q_{0.75}$) of each data set and extending the line to cover the range of the data. The

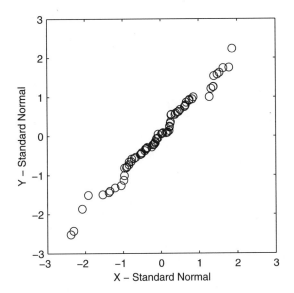

**FIGURE 5.6**
This is a q-q plot of *x* and *y* where both data sets are generated from a standard normal distribution. Note that the points follow a line, as expected.

MATLAB Statistics Toolbox provides a function called **qqplot** that displays this type of plot. We show below how to add the reference line.

**Example 5.5**

This example shows how to do a q-q plot when the samples do not have the same number of points. We use the function **csquantiles** to get the required sample quantiles from the data set that has the larger sample size. We then plot these versus the order statistics of the other sample, as we did in the previous examples. Note that we add a reference line based on the first and third quartiles of each data set, using the function **polyfit** (see Chapter 7 for more information on this function).

```
% Generate the random variables.
m = 50;
n = 75;
x = randn(1,n);
y = randn(1,m);
% Find the order statistics for y.
ys = sort(y);
% Now find the associated quantiles using the x.
% Probabilities for quantiles:
p = ((1:m) - 0.5)/m;
```

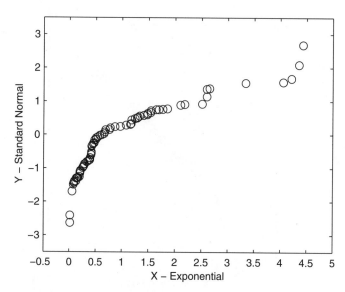

**FIGURE 5.7**
This is a q-q plot where one random sample is generated from the exponential distribution and one is generated by a standard normal distribution. Note that the points do not follow a straight line, indicating that the distributions that generated the random variables are not the same.

```
xs = csquantiles(x,p);
% Construct the plot.
plot(xs,ys,'ko')
% Get the reference line.
% Use the 1st and 3rd quartiles of each set to
% get a line.
qy = csquantiles(y,[0.25,0.75]);
qx = csquantiles(x,[0.25,0.75]);
[pol, s] = polyfit(qx,qy,1);
% Add the line to the figure.
yhat = polyval(pol,xs);
hold on
plot(xs,yhat,'k')
xlabel('Sample Quantiles - X'),
ylabel('Sorted Y Values')
hold off
```

From Figure 5.8, the assumption that each data set is generated according to the same distribution seems reasonable.
❑

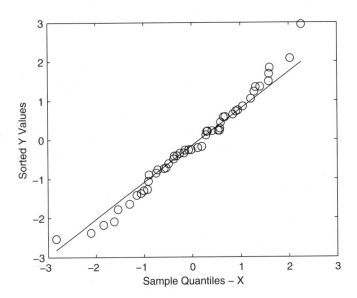

**FIGURE 5.8**
Here we show the q-q plot of Example 5.5. In this example, we also show the reference line estimated from the first and third quartiles. The q-q plot shows that the data do seem to come from the same distribution.

## Quantile Plots

A *quantile plot* or *probability plot* is one where the theoretical quantiles are plotted against the order statistics for the sample. Thus, on one axis we plot the $x_{(i)}$ and on the other axis we plot

$$F^{-1}\left(\frac{i - 0.5}{n}\right),$$

where $F^{-1}(.)$ denotes the inverse of the cumulative distribution function for the hypothesized distribution. As before, the 0.5 in the above argument can be different [Cleveland, 1993]. A well-known example of a quantile plot is the *normal probability plot*, where the ordered sample versus the quantiles of the normal distribution are plotted.

The MATLAB Statistics Toolbox has two functions for obtaining quantile plots. One is called **normplot**, and it produces a normal probability plot. So, if one would like to assess the assumption that a data set comes from a normal distribution, then this is the one to use. There is also a function for constructing a quantile plot that compares a data set to the Weibull distribution. This is called **weibplot**. For quantile plots with other theoretical distributions, one can use the MATLAB code given below, substituting the appropriate function to get the theoretical quantiles.

## Example 5.6

This example illustrates how you can display a quantile plot in MATLAB. We first generate a random sample from the standard normal distribution as our data set. The sorted sample is an estimate of the $(i - 0.5)/n$ quantile, so we next calculate these probabilities and get the corresponding theoretical quantiles. Finally, we use the function **norminv** from the Statistics Toolbox to get the theoretical quantiles for the normal distribution. The resulting quantile plot is shown in Figure 5.9.

```
% Generate a random sample from a standard normal.
x = randn(1,100);
% Get the probabilities.
prob = ((1:100)-0.5)/100;
% Now get the theoretical quantiles.
qp = norminv(prob,0,1);
% Now plot theoretical quantiles versus
% the sorted data.
plot(sort(x),qp,'ko')
xlabel('Sorted Data')
ylabel('Standard Normal Quantiles')
```

To further illustrate these concepts, let's see what happens when we generate a random sample from a uniform (0, 1) distribution and check it against the normal distribution. The MATLAB code is given below, and the quantile plot is shown in Figure 5.10. As expected, the points do not lie on a line, and we see that the data are not from a normal distribution.

```
% Generate a random sample from a
% uniform distribution.
x = rand(1,100);
% Get the probabilities.
prob = ((1:100)-0.5)/100;
% Now get the theoretical quantiles.
qp = norminv(prob,0,1);
% Now plot theoretical quantiles versus
% the sorted data.
plot(sort(x),qp,'ko')
ylabel('Standard Normal Quantiles')
xlabel('Sorted Data')
```

□

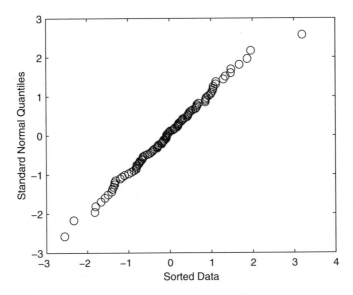

**FIGURE 5.9**
This is a quantile plot or normal probability plot of a random sample generated from a standard normal distribution. Note that the points approximately follow a straight line, indicating that the normal distribution is a reasonable model for the sample.

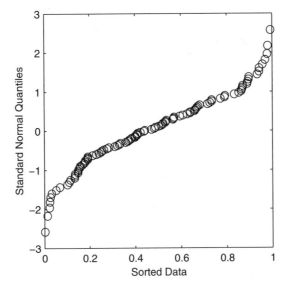

**FIGURE 5.10**
Here we have a quantile plot where the sample is generated from a uniform distribution, and the theoretical quantiles are from the normal distribution. The shape of the curve verifies that the sample is not from a normal distribution.

## Quantile Plots - Discrete Distributions

Previously, we discussed quantile plots that are primarily used for continuous data. We would like to have a similar technique for graphically comparing the shapes of discrete distributions. Hoaglin and Tukey [1985] developed several plots to accomplish this. We present two of them here: the *Poissonness plot* and the *binomialness plot*. These will enable us to search for evidence that our discrete data follow a Poisson or a binomial distribution. They also serve to highlight which points might be incompatible with the model.

### *Poissonness Plot*

Typically, discrete data are whole number values that are often obtained by counting the number of times something occurs. For example, these might be the number of traffic fatalities, the number of school-age children in a household, the number of defects on a hard drive, or the number of errors in a computer program. We sometimes have the data in the form of a frequency distribution that lists the possible count values (e.g., 0.1, 2, ... ) and the number of observations that are equal to the count values.

The counts will be denoted as $k$, with $k = 0, 1, ..., L$. We will assume that $L$ is the maximum observed value for our discrete variable or counts in the data set and that we are interested in all counts between 0 and $L$. Thus, the total number of observations in the sample is

$$N = \sum_{k=0}^{L} n_k,$$

where $n_k$ represents the number of observations that are equal to the count $k$.

A basic Poissonness plot is constructed by plotting the count values $k$ on the horizontal axis and

$$\varphi(n_k) = \ln(k! n_k / N) \tag{5.2}$$

on the vertical axis. These are plotted as symbols, similar to the quantile plot. If a Poisson distribution is a reasonable model for the data, then this should follow a straight line. Systematic curvature in the plot would indicate that these data are not consistent with a Poisson distribution. The values for $\varphi(n_k)$ tend to have more variability when $n_k$ is small, so Hoaglin and Tukey [1985] suggest plotting a special symbol or a '1' to highlight these points.

## Example 5.7

This example is taken from Hoaglin and Tukey [1985]. In the late 1700's, Alexander Hamilton, John Jay and James Madison wrote a series of 77 essays under the title of *The Federalist*. These appeared in the newspapers under a

TABLE 5.1

Frequency distribution of the word *may* in essays known to be written by James Madison. The $n_k$ represent the number of blocks of text that contained $k$ occurrences of the word *may* [Hoaglin and Tukey, 1985].

| Number of Occurrences of the Word *may* ($k$) | Number of Blocks ($n_k$) |
|:---:|:---:|
| 0 | 156 |
| 1 | 63 |
| 2 | 29 |
| 3 | 8 |
| 4 | 4 |
| 5 | 1 |
| 6 | 1 |

pseudonym. Most analysts accept that John Jay wrote 5 essays, Alexander Hamilton wrote 43, Madison wrote 14, and 3 were jointly written by Hamilton and Madison. Later, Hamilton and Madison claimed that they each solely wrote the remaining 12 papers. To verify this claim, Mosteller and Wallace [1964] used statistical methods, some of which were based on the frequency of words in blocks of text. Table 5.1 gives the frequency distribution for the word *may* in papers that were known to be written by Madison. We are not going to repeat the analysis of Mosteller and Wallace, we are simply using the data to illustrate a Poissonness plot. The following MATLAB code produces the Poissonness plot shown in Figure 5.11.

```
k = 0:6;  % vector of counts
n_k = [156 63 29 8 4 1 1];
N=sum(n_k);
% Get vector of factorials.
fact = zeros(size(k));
for i = k
    fact(i+1) = factorial(i);
end
% Get phi(n_k) for plotting.
phik = log(fact.*n_k/N);
% Find the counts that are equal to 1.
% Plot these with the symbol 1.
% Plot rest with a symbol.
ind = find(n_k~=1);
plot(k(ind),phik(ind),'o')
ind = find(n_k==1);
if ~isempty(ind)
    text(k(ind),phik(ind),'1')
```

```
end
% Add some whitespace to see better.
axis([-0.5 max(k)+1 min(phik)-1 max(phik)+1])
xlabel('Number of Occurrences - k')
ylabel('\phi (n_k)')
```

The Poissonness plot has significant curvature indicating that the Poisson distribution is not a good model for these data. There are also a couple of points with a frequency of 1 that seem incompatible with the rest of the data. Thus, if a statistical analysis of these data relies on the Poisson model, then any results are suspect.
☐

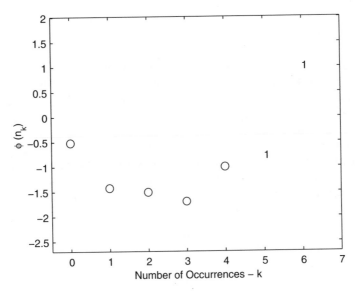

**FIGURE 5.11**
This is a basic Poissonness plot using the data in Table 5.1. The symbol 1 indicates that $n_k = 1$.

Hoaglin and Tukey [1985] suggest a modified Poissonness plot that is obtained by changing the $n_k$, which helps account for the variability of the individual values. They propose the following change:

$$n_k^* = \begin{cases} n_k - 0.67 - 0.8 n_k/N; & n_k \geq 2 \\ 1/e; & n_k = 1 \\ \text{undefined}; & n_k = 0. \end{cases} \qquad (5.3)$$

As we will see in the following example where we apply the modified Poissonness plot to the word frequency data, the main effect of the modified plot is to highlight those data points with small counts that do not behave contrary to the other observations. Thus, if a point that is plotted as a *1* in a modified Poissonness plot seems different from the rest of the data, then it should be investigated.

### Example 5.8

We return to the word frequency data in Table 5.1 and show how to get a modified Poissonness plot. In this modified version shown in Figure 5.12, we see that the points where $n_k = 1$ do not seem so different from the rest of the data.

```
% Poissonness plot - modified
k = 0:6;  % vector of counts
% Find n*_k.
n_k = [156 63 29 8 4 1 1];
N = sum(n_k);
phat = n_k/N;
nkstar = n_k-0.67-0.8*phat;
% Get vector of factorials.
fact = zeros(size(k));
for i = k
    fact(i+1) = factorial(i);
end
% Find the frequencies that are 1; nkstar=1/e.
ind1 = find(n_k==1);
nkstar(ind1)= 1/2.718;
% Get phi(n_k) for plotting.
phik = log(fact.*nkstar/N);
ind = find(n_k~=1);
plot(k(ind),phik(ind),'o')
if ~isempty(ind1)
    text(k(ind1),phik(ind1),'1')
end
% Add some whitespace to see better.
axis([-0.5 max(k)+1 min(phik)-1 max(phik)+1])
xlabel('Number of Occurrences - k')
ylabel('\phi (n^*_k)')
```

❑

### *Binomialness Plot*

A binomialness plot is obtained by plotting *k* along the horizontal axis and plotting

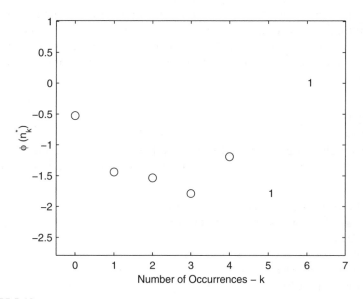

**FIGURE 5.12**
This is a modified Poissonness plot for the word frequency data in Table 5.1. Here the counts where $n_k = 1$ do not seem radically different from the rest of the observations.

$$\varphi(n_k^*) = \ln\left\{\frac{n_k^*}{N \times \binom{n}{k}}\right\}, \tag{5.4}$$

along the vertical axis. Recall that $n$ represents the number of trials, and $n_k^*$ is given by Equation 5.3. As with the Poissonness plot, we are looking for an approximate linear relationship between $k$ and $\varphi(n_k^*)$. An example of the binomialness plot is given in Example 5.9.

**Example 5.9**
Hoaglin and Tukey [1985] provide a frequency distribution representing the number of females in 100 queues of length 10. These data are given in Table 5.2. The MATLAB code to display a binomialness plot for $n = 10$ is given below. Note that we cannot display $\varphi(n_k^*)$ for $k = 10$ (in this example), because it is not defined for $n_k = 0$. The resulting binomialness plot is shown in Figure 5.13, and it indicates a linear relationship. Thus, the binomial model for these data seems adequate.

```
% Binomialness plot.
```

TABLE 5.2

Frequency Distribution for the Number of Females in a
Queue of Size 10 [Hoaglin and Tukey, 1985]

| Number of Females $(k)$ | Number of Blocks $(n_k)$ |
|:---:|:---:|
| 0 | 1 |
| 1 | 3 |
| 2 | 4 |
| 3 | 23 |
| 4 | 25 |
| 5 | 19 |
| 6 | 18 |
| 7 | 5 |
| 8 | 1 |
| 9 | 1 |
| 10 | 0 |

```
k = 0:9;
n = 10;
n_k = [1 3 4 23 25 19 18 5 1 1];
N = sum(n_k);
nCk = zeros(size(k));
for i = k
   nCk(i+1) = cscomb(n,i);
end
phat = n_k/N;
nkstar = n_k-0.67-0.8*phat;
% Find the frequencies that are 1; nkstar=1/e.
ind1 = find(n_k==1);
nkstar(ind1) = 1/2.718;
% Get phi(n_k) for plotting.
phik = log(nkstar./(N*nCk));
% Find the counts that are equal to 1.
ind = find(n_k~=1);
plot(k(ind),phik(ind),'o')
if ~isempty(ind1)
   text(k(ind1),phik(ind1),'1')
end
% Add some whitespace to see better.
axis([-0.5 max(k)+1 min(phik)-1 max(phik)+1])
xlabel('Number of Females - k')
ylabel('\phi (n^*_k)')
```

❑

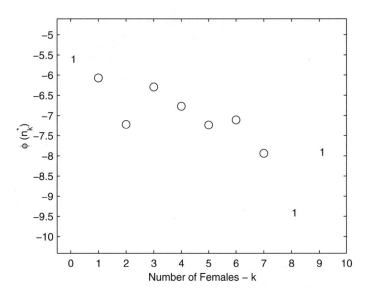

**FIGURE 5.13**
This shows the binomialness plot for the data in Table 5.2. From this it seems reasonable to use the binomial distribution to model the data.

## Box Plots

Box plots (sometimes called box-and-whisker diagrams) have been in use for many years [Tukey, 1977]. As with most visualization techniques, they are used to display the distribution of a sample. Five values from a data set are used to construct the box plot. These are the three sample quartiles $(\hat{q}_{0.25}, \hat{q}_{0.5}, \hat{q}_{0.75})$, the minimum value in the sample and the maximum value.

There are many variations of the box plot, and it is important to note that they are defined differently depending on the software package that is used. Frigge, Hoaglin and Iglewicz [1989] describe a study on how box plots are implemented in some popular statistics programs such as Minitab, S, SAS, SPSS and others. The main difference lies in how outliers and quartiles are defined. Therefore, depending on how the software calculates these, different plots might be obtained [Frigge, Hoaglin and Iglewicz, 1989].

Before we describe the box plot, we need to define some terms. Recall from Chapter 3, that the *interquartile range* (IQR) is the difference between the first and the third sample quartiles. This gives the range of the middle 50% of the data. It is estimated from the following

$$I\hat{Q}R = \hat{q}_{0.75} - \hat{q}_{0.25}.$$  (5.5)

Two limits are also defined: a lower limit (LL) and an upper limit (UL). These are calculated from the estimated IQR as follows

$$LL = \hat{q}_{0.25} - 1.5 \cdot \hat{IQR}$$
$$UL = \hat{q}_{0.75} + 1.5 \cdot \hat{IQR}.$$
(5.6)

The idea is that observations that lie outside these limits are possible outliers. *Outliers* are data points that lie away from the rest of the data. This might mean that the data were incorrectly measured or recorded. On the other hand, it could mean that they represent extreme points that arise naturally according to the distribution. In any event, they are sample points that are suitable for further investigation.

*Adjacent values* are the most extreme observations in the data set that are within the lower and the upper limits. If there are no potential outliers, then the adjacent values are simply the maximum and the minimum data points.

To construct a box plot, we place horizontal lines at each of the three quartiles and draw vertical lines to create a box. We then extend a line from the first quartile to the smallest adjacent value and do the same for the third quartile and largest adjacent value. These lines are sometimes called the whiskers. Finally, any possible outliers are shown as an asterisk or some other plotting symbol. An example of a box plot is shown in Figure 5.14.

Box plots for different samples can be plotted together for visually comparing the corresponding distributions. The MATLAB Statistics Toolbox contains a function called **boxplot** for creating this type of display. It displays one box plot for each column of data. When we want to compare data sets, it is better to display a box plot with notches. These notches represent the uncertainty in the locations of central tendency and provide a rough measure of the significance of the differences between the values. If the notches do not overlap, then there is evidence that the medians are significantly different. The length of the whisker is easily adjusted using optional input arguments to **boxplot**. For more information on this function and to find out what other options are available, type **help boxplot** at the MATLAB command line.

### Example 5.10

In this example, we first generate random variables from a uniform distribution on the interval $(0, 1)$, a standard normal distribution, and an exponential distribution. We will then display the box plots corresponding to each sample using the MATLAB function **boxplot**.

```
% Generate a sample from the uniform distribution.
xunif = rand(100,1);
% Generate sample from the standard normal.
xnorm = randn(100,1);
% Generate a sample from the exponential distribution.
```

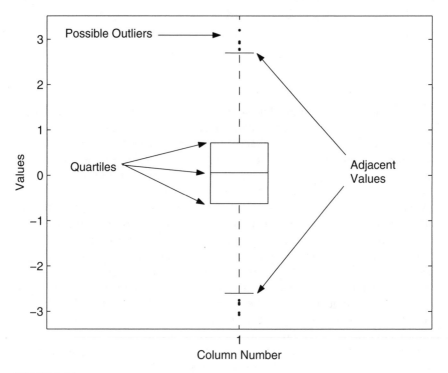

**FIGURE 5.14**
An example of a box plot with possible outliers shown as points.

```
% NOTE: this function is from the Statistics Toolbox.
xexp = exprnd(1,100,1);
boxplot([xunif,xnorm,xexp],1)
```

It can be seen in Figure 5.15 that the box plot readily conveys the shape of the distribution. A symmetric distribution will have whiskers with approximately equal lengths, and the two sides of the box will also be approximately equal. This would be the case for the uniform or normal distribution. A skewed distribution will have one side of the box and whisker longer than the other. This is seen in Figure 5.15 for the exponential distribution. If the interquartile range is small, then the data in the middle are packed around the median. Conversely, if it is large, then the middle 50% of the data are widely dispersed.

❑

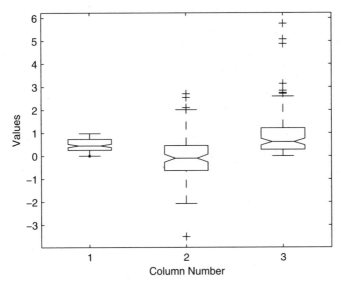

**FIGURE 5.15**
Here we have three box plots. The one on the left is for a sample from the uniform distribution. The data for the middle box plot came from a standard normal distribution, while the data for the box plot on the right came from an exponential. Notice that the shape of each distribution is apparent from the information contained in the box plots.

## 5.3 Exploring Bivariate and Trivariate Data

Using Cartesian coordinates, we can view up to three dimensions. For example, we could view bivariate data as points or trivariate data as a point cloud. We could also view a bivariate function, $z = f(x, y)$ as a surface. Visualizing anything more than three dimensions is very difficult, but we do offer some techniques in the next section. In this section, we present several methods for visualizing 2-D and 3-D data, looking first at bivariate data. Most of the techniques that we discuss are readily available in the basic MATLAB program.

### Scatterplots

Perhaps one of the easiest ways to visualize bivariate data is with the scatterplot. A scatterplot is obtained by displaying the ordered pairs as points using some plotting symbol. This type of plot conveys useful information such as how the data are distributed in the two dimensions and how the two variables are related (e.g., a linear or a nonlinear relationship). Before any modeling, such as regression, is done using bivariate data, the analyst should always look at a scatterplot to see what type of relationship is reasonable. We will explore this further in Chapters 7 and 10.

A scatterplot can be obtained easily in MATLAB using the **plot** command. One simply enters the marker style or plotting symbol as one of the arguments. See the **help** on **plot** for more information on what characters are available. By entering a marker (or line) style, you tell MATLAB that you do *not* want to connect the points with a straight line, which is the default. We have already seen many examples of how to use the **plot** function in this way when we constructed the quantile and q-q plots.

An alternative function for scatterplots that is available with MATLAB is the function called **scatter**. This function takes the input vectors **x** and **y** and plots them as symbols. There are optional arguments that will plot the markers as different colors and sizes. These alternatives are explored in Example 5.11.

### Example 5.11
We first generate a set of bivariate normal random variables using the technique described in Chapter 4. However, it should be noted that we find the matrix **R** in Equation 4.19 using singular value decomposition rather than Cholesky factorization. We then create a scatterplot using the **plot** function and the **scatter** function. The resulting plots are shown in Figure 5.16 and Figure 5.17.

```
% Create a positive definite covariance matrix.
vmat = [2, 1.5; 1.5, 9];
% Create mean at (2,3).
mu = [2 3];
[u,s,v] = svd(vmat);
vsqrt = ( v*(u'.*sqrt(s)))';
% Get standard normal random variables.
td = randn(250,2);
% Use x=z*sigma+mu to transform - see Chapter 4.
data = td*vsqrt+ones(250,1)*mu;
% Create a scatterplot using the plot function.
% Figure 5.16.
plot(data(:,1),data(:,2),'x')
axis equal
% Create a scatterplot using the scatter fumction.
% Figure 5.17.
% Use filled-in markers.
scatter(data(:,1),data(:,2),'filled')
axis equal
box on
```

❑

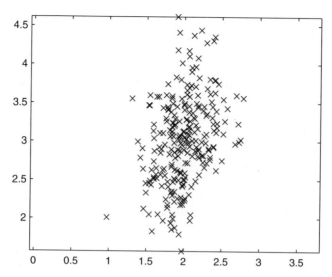

**FIGURE 5.16**
This is a scatterplot of the sample in Example 5.11 using the `plot` function. We can see that the data seem to come from a bivariate normal distribution. Here we use `'x'` as an argument to the `plot` function to plot the symbols as x's.

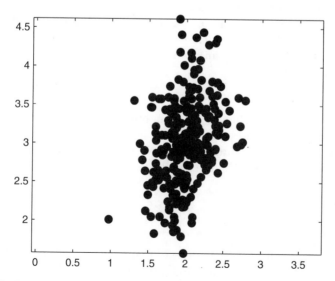

**FIGURE 5.17**
This is a scatterplot of the sample in Example 5.11 using the `scatter` function with filled markers.

## Surface Plots

If we have data that represents a function defined over a bivariate domain, such as $z = f(x, y)$, then we can view our values for $z$ as a surface. MATLAB provides two functions that display a matrix of $z$ values as a surface: **mesh** and **surf**.

The **mesh** function displays the values as points above a rectangular grid in the $x$-$y$ plane and connects adjacent points with straight lines. The mesh lines can be colored using various options, but the default method maps the height of the surface to a color.

The **surf** function is similar to **mesh**, except that the open spaces between the lines are filled in with color, with lines shown in black. Other options available with the **shading** command remove the lines or interpolate the color across the patches. An example of where the ability to display a surface can be used is in visualizing a probability density function (see Chapter 8).

## Example 5.12

In this example, we begin by generating a grid over which we evaluate a bivariate normal density function. We then calculate the $z$ values that correspond to the function evaluated at each $x$ and $y$. We can display this as a surface using **surf**, which is shown in Figure 5.18.

```
% Create a bivariate standard normal.
% First create a grid for the domain.
[x,y] = meshgrid(-3:.1:3,-3:.1:3);
% Evaluate using the bivariate standard normal.
z = (1/(2*pi))*exp(-0.5*(x.^2+y.^2));
% Do the plot as a surface.
surf(x,y,z)
```

❑

Special effects can be achieved by changing color maps and using lighting. For example, lighting and color can help highlight structure or features on functions that have many bumps or a jagged surface. We will see some examples of how to use these techniques in the next section and in the exercises at the end of the chapter.

## Contour Plots

We can also use contour plots to view our surface. Contour plots show lines of constant surface values, similar to topographical maps. Two functions are available in MATLAB for creating 2-D and 3-D contour plots. These are called **contour** and **contour3**.

The **pcolor** function shows the same information that is in a contour plot by mapping the surface height to a set of colors. It is sometimes useful to combine the two on the same plot. MATLAB provides the **contourf** function

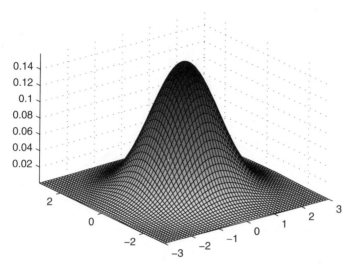

**FIGURE 5.18**
This shows a **surf** plot of a bivariate normal probability density function.

that will create a combination **pcolor** and **contour** plot. The various options that are available for creating contour plots are illustrated in Example 5.13.

### Example 5.13

MATLAB has a function called **peaks** that returns a surface with peaks and depressions that can be used to illustrate contour plots. We show how to use the **peaks** function in this example. The following MATLAB code demonstrates how to create the 2-D contour plot in Figure 5.19.

```
% Get the data for plotting.
[x,y,z] = peaks;
% Create a 2-D contour plot with labels.
% This returns the information for the labels.
c = contour(x,y,z);
% Add the labels to the plot.
clabel(c)
```

A filled contour plot, which is a combination of **pcolor** and **contour**, is given in Figure 5.20. The MATLAB command needed to get this plot is given here.

```
% Create a 2-D filled contour plot.
contourf(x,y,z,15)
```

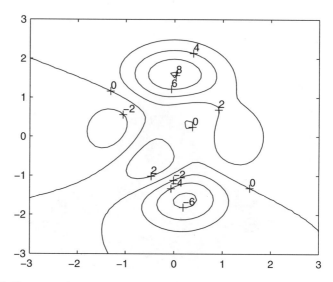

**FIGURE 5.19**
This is a labeled contour plot of the **peaks** function. The labels make it easier to understand the hills and valleys in the surface.

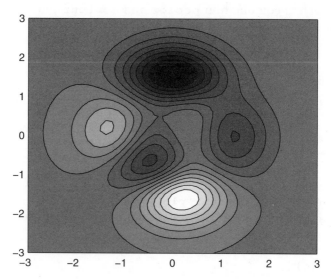

**FIGURE 5.20**
This is a filled contour plot of the **peaks** surface. It is created using the **contourf** function.

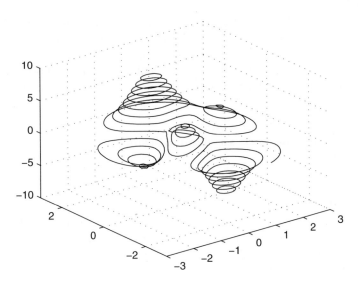

**FIGURE 5.21**
This is a 3-D contour plot of the **peaks** function.

Finally, a 3-D contour plot is easily obtained using the **contour3** function as shown below. The resulting contour plot is shown in Figure 5:21.

```
% Create a 3-D contour plot.
contour3(x,y,z,15)
```

❑

### Bivariate Histogram

In the last section, we described the univariate density histogram as a way of viewing how our data are distributed over the range of the data. We can extend this to any number of dimensions over a partition of the space [Scott, 1992]. However, in this section we restrict our attention to the bivariate histogram given by

$$\hat{f}(\mathbf{x}) = \frac{v_k}{nh_1h_2} \qquad \text{x in } B_k, \tag{5.7}$$

where $v_k$ represents the number of observations falling into the bivariate bin $B_k$ and $h_i$ is the width of the bin for the $x_i$ coordinate axis. Example 5.14 shows how to get the bivariate density histogram in MATLAB.

**Example 5.14**

We generate bivariate standard normal random variables and use them to illustrate how to get the bivariate density histogram. We use the optimal bin width for data generated from a standard bivariate normal given in Scott [1992]. We postpone discussion of the optimal bin width and how to obtain it until Chapter 8. A scatterplot of the data and the resulting histogram are shown in Figure 5.22.

```
% Generate sample that is
% standard normal in each dimension.
n = 1000;
d = 2;
x = randn(n,d);
% Need bin origins.
bin0 = [floor(min(x(:,1))) floor(min(x(:,2)))];
% The bin widths - h - are covered later.
h = 3.504*n^(-0.25)*ones(1,2);
% find the number of bins
nb1 = ceil((max(x(:,1))-bin0(1))/h(1));
nb2 = ceil((max(x(:,2))-bin0(2))/h(2));
% find the mesh
t1 = bin0(1):h(1):(nb1*h(1)+bin0(1));
t2 = bin0(2):h(2):(nb2*h(2)+bin0(2));
[X,Y] = meshgrid(t1,t2);
% Find bin frequencies.
[nr,nc] = size(X);
vu = zeros(nr-1,nc-1);
for i = 1:(nr-1)
    for j = 1:(nc-1)
        xv = [X(i,j) X(i,j+1) X(i+1,j+1) X(i+1,j)];
        yv = [Y(i,j) Y(i,j+1) Y(i+1,j+1) Y(i+1,j)];
        in = inpolygon(x(:,1),x(:,2),xv,yv);
        vu(i,j) = sum(in(:));
    end
end
Z = vu/(n*h(1)*h(2));
% Get some axes that make sense.
[XX,YY] = meshgrid(linspace(-3,3,nb1),...
    linspace(-3,3,nb2));
surf(XX,YY,Z)
```

❑

We displayed the resulting bivariate histogram using the **surf** plot in MATLAB. The matrix **Z** in Example 5.14 contains the bin heights. When MATLAB constructs a **mesh** or **surf** plot, the elements of the **Z** matrix represent heights above the *x-y* plane. The surface is obtained by plotting the

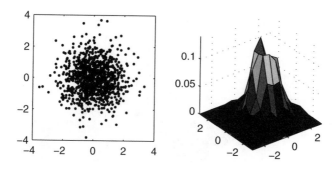

**FIGURE 5.22**
On the left is a scatterplot of the data. A surface plot of the bivariate density histogram is on the right. Compare the estimated density given by the surface with the one shown in Figure 5.18.

points and joining adjacent points with straight lines. Therefore, a **surf** or **mesh** plot of the bivariate histogram bin heights is a linear interpolation between adjacent bins. In essence, it provides a smooth version of a histogram. In the next example, we offer another method for viewing the bivariate histogram.

**Example 5.15**
In this example, we show the bin heights of the bivariate histogram as bars using the MATLAB function **bar3**. The colors are mapped to the column number of the **Z** matrix, not to the heights of the bins. The resulting histogram is shown in Figure 5.23.

```
% The Z matrix is obtained in Example 5.14.
bar3(Z,1)
% Use some Handle Graphics.
set(gca,'YTickLabel',' ','XTickLabel',' ')
set(gca,'YTick',0,'XTick',0)
grid off
```

The following MATLAB code constructs a plot that displays the distribution in a different way. We can use the **scatter** plotting function with arguments

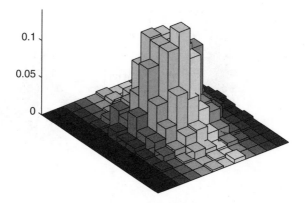

**FIGURE 5.23**
This shows the same bivariate histogram of Figure 5.22, where the heights of the bars are plotted using the MATLAB function **bar3**.

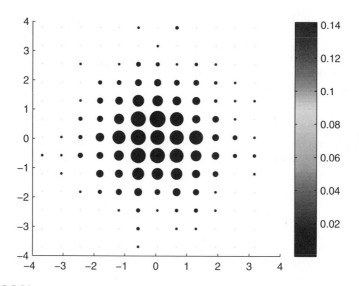

**FIGURE 5.24**
Here is a different display of the bivariate histogram of Example 5.15. The size and color of the markers indicate the heights of the bins.

that relate the marker size and color to the height of the bins. We add the `colorbar` to map the heights of the bins to the color.

```
% Plot the 2-D histogram as a scatterplot with
% heights proportional to marker size.
% Find the bin centers to use in the scatterplot.
n1 = length(t1);
n2 = length(t2);
tt1 = linspace(((t1(1)+t1(2))/2,...
    (t1(n1-1)+t1(n1))/2,nb1);
tt2 = linspace(((t2(1)+t2(2))/2,...
    (t2(n2-1)+t2(n2))/2,nb2);
[xxs,yys] = meshgrid(tt1,tt2);
scatter(xxs(:),yys(:),(Z(:)+eps)*1000,...
    (Z(:)+eps)*1000,'filled')
% Create a colorbar and set the axis
% to the correct scale
h_ax = colorbar;
% Get the current labels.
temp = get(h_ax,'Yticklabel');
[nr,nc] = size(temp);
% Convert from strings to numbers.
newlab = cell(nr,1);
tempcell = cellstr(temp);
% Re-scale and convert back to numbers.
for i=1:nr
    newlab{i}=num2str((str2num(tempcell{i})/1000));
end
set(h_ax,'Yticklabel',newlab)
```

This graphic is given in Figure 5.24. Note that we still see the same bivariate normal distribution. The reader might want to compare this plot with the scatterplot of the sample shown in Figure 5.22.
◻

## 3-D Scatterplot

As with 2-D data, one way we can view trivariate data is with the scatterplot. This is the 3-D analog of the bivariate scatterplot. In this case, the ordered triples $(x, y, z)$ are plotted as points. MATLAB provides a function called `scatter3` that will create a 3-D scatterplot. Analogous to the bivariate case, you can also use the `plot3` function using a symbol for the marker style to obtain a 3-D scatterplot.

A useful MATLAB command when visualizing anything in 3-D is `rotate3d`. Simply type this in at the command line, and you will be able to rotate your graphic using the mouse. There is also a toolbar button that activates the same capability. One reason for looking at scatterplots of the data is to look for interesting structures. The ability to view these structures for 3-D data is dependent on the viewpoint or projection to the screen. When looking at 3-D scatterplots, the analyst should rotate them to search the data for patterns or structure.

## Example 5.16

Three variables were measured on ten insects from each of three species [Hand, et al.,1994]. The variables correspond to the width of the first joint of the first tarsus, the width of the first joint of the second tarsus and the maximal width of the aedeagus. All widths are measured in microns. These data were originally used in cluster analysis [Lindsey, Herzberg, and Watts, 1987]. What we would like to see from the scatterplot is whether the data for each species can be separated from the others. In other words, is there clear separation or clustering between the species using these variables? The 3-D scatterplot for these data is shown in Figure 5.25. This view of the scatterplot indicates that using these variables for pattern recognition or clustering (see Chapter 9) is reasonable.

```
% Load the insect data
load insect
% Create a 3-D scatter plot using a
% different color and marker
% for each class of insect.
% Plot the first class and hold the plot.
plot3(insect(1:10,1),insect(1:10,2),...
    insect(1:10,3),'ro')
hold on
% Plot the second class.
plot3(insect(11:20,1),insect(11:20,2),...
    insect(11:20,3),'gx')
% Plot the third class.
plot3(insect(21:30,1),insect(21:30,2),...
    insect(21:30,3),'b*')
% Be sure to turn the hold off!
hold off
```

❑

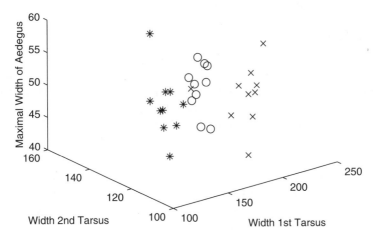

**FIGURE 5.25**
This is a 3-D scatterplot of the `insect` data. Each species is plotted using a different symbol. This plot indicates that we should be able to identify (with reasonable success) the species based on these three variables.

## 5.4 Exploring Multi-Dimensional Data

Several methods have been developed to address the problem of visualizing multi-dimensional data. Here we consider applications where we are trying to explore data that has more than three dimensions $(d > 3)$.

We discuss several ways of statically visualizing multi-dimensional data. These include the scatterplot matrix, slices, 3-D contours, star plots, Andrews curves, and parallel coordinates. We finish this section with a description of projection pursuit exploratory data analysis and the grand tour. The grand tour provides a dynamic display of projections of multi-dimensional data, and projection pursuit looks for structure in 1-D or 2-D projections. It should be noted that some of the methods presented here are not restricted to the case where the dimensionality of our data is greater than 3-D.

### Scatterplot Matrix

In the previous sections, we presented the scatterplot as a way of looking at 2-D and 3-D data. We can extend this to multi-dimensional data by looking at 2-D scatterplots of all possible pairs of variables. This allows one to view pairwise relationships and to look for interesting structures in two dimensions. MATLAB provides a function called **plotmatrix** that will create a scatterplot matrix. Its use is illustrated below.

### Example 5.17

The **iris** data are well-known to statisticians and are often used to illustrate classification, clustering or visualization techniques. The data were collected by Anderson [1935] and were analyzed by Fisher [1936], so the data are often called *Fisher's iris data* by statisticians. The data consist of 150 observations containing four measurements based on the petals and sepals of three species of iris. These three species are: *Iris setosa, Iris virginica* and *Iris versicolor*. We apply the **plotmatrix** function to the iris data set.

```
load iris
% This loads up three matrices, one for each species.
% Get the plotmatrix display of the Iris setosa data.
[H,ax,bigax,P] = plotmatrix(setosa);
axes(bigax),title('Iris Setosa')
```

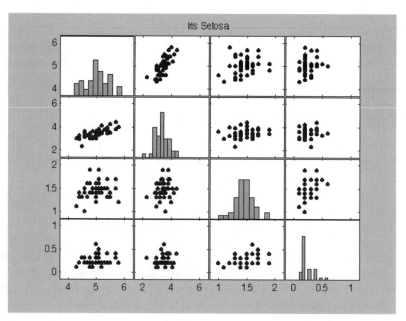

FIGURE 5.26
This is the scatterplot matrix for the *Iris setosa* data using the **plotmatrix** function.

The results are shown in Figure 5.26. Several argument options are available for the **plotmatrix** function. If the first two arguments are matrices, then MATLAB plots one column versus the other column. In our example, we use a single matrix argument, and MATLAB creates scatterplots of all possible pairs of variables. Histograms of each variable or column are shown along the diagonal of the scatterplot matrix. Optional output arguments allow one to add a title or change the plot as shown in the following MATLAB commands. Here we replace the histograms with text that identifies the variable names and display the result in Figure 5.27.

```
% Create the labels as a cell array of strings.
labs = {'Sepal Length','Sepal Width',...
    'Petal Length', 'Petal Width'};
[H,ax,bigax,P] = plotmatrix(virginica);
axes(bigax)
title('Virginica')
% Delete the histograms.
delete(P)
%Put the labels in - the positions might have
% to be adjusted depending on the text.
for i = 1:4
    txtax = axes('Position',get(ax(i,i),'Position'),...
        'units','normalized');
    text(.1, .5,labs{i})
    set(txtax,'xtick',[],'ytick',[],...
        'xgrid','off','ygrid','off','box','on')
end
```
□

## Slices and Isosurfaces

If we have a function defined over a volume, $f(x, y, z)$, then we can view it using the MATLAB **slice** function or the **isosurface** function (available in MATLAB 5.3 and higher). This situation could arise in cases where we have a probability density function defined over a volume. The **slice** capability allows us to view the distribution of our data on slices through a volume. The **isosurface** function allows us to view 3-D contours through our volume. These are illustrated in the following examples.

### Example 5.18

To illustrate the **slice** function, we need $f(x, y, z)$ values that are defined over a 3-D grid or volume. We will use a trivariate normal distribution centered at the origin with covariance equal to the identity matrix. The following MATLAB code displays slices through the $x = 0$, $y = 0$, and $z = 0$ planes, and the resulting display is shown in Figure 5.28. A standard normal bivari-

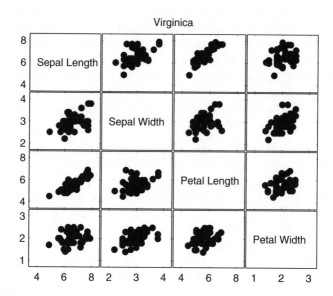

**FIGURE 5.27**
By using MATLAB's Handle Graphics, we can add text for the variable name to the diagonal boxes.

ate density is given in Figure 5.29 to help the reader understand what the `slice` function is showing. The density or height of the surface defined over the volume is mapped to a color. Therefore, in the `slice` plot, you can see that the maximum density or surface height is at the origin with the height decreasing at the edges of the slices. The color at each point is obtained by interpolation into the volume $f(x, y, z)$.

```
% Create a grid for the domain.
[x,y,z] = meshgrid(-3:.1:3,-3:.1:3,-3:.1:3);
[n,d] = size(x(:));
% Evaluate the trivariate standard normal.
a = (2*pi)^(3/2);
arg = (x.^2 + y.^2 + z.^2);
prob = exp((-.5)*arg)/a;
% Slice through the x=0, y=0, z=0 planes.
slice(x,y,z,prob,0,0,0)
xlabel('X Axis'),ylabel('Y Axis'),zlabel('Z Axis')
```

❑

Isosurfaces are a way of viewing contours through a volume. An isosurface is a surface where the function values $f(x, y, z)$ are constant. These are similar to $\alpha$-level contours [Scott, 1992], which are defined by

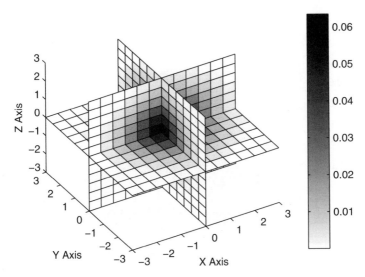

**FIGURE 5.28**
These are slices through the $x = 0, y = 0, z = 0$ planes for a standard trivariate normal distribution. Each of these planes slice through the volume, and the value of the volume (in this case, the height of the trivariate normal density) is represented by the color. The mode at the origin is clearly seen. We can also see that it is symmetric, because the volume is a mirror image in every slice. Finally, note that the ranges for all the axes are consistent with a standard normal distribution.

$$S_\alpha = \{ \mathbf{x} \colon f(\mathbf{x}) = \alpha f_{max} \}; \qquad 0 \le \alpha \le 1, \qquad (5.8)$$

where $\mathbf{x}$ is a $d$-dimensional vector. Generally, the $\alpha$-level contours are nested surfaces.

The MATLAB function `isosurface(X,Y,Z,V,isovalue)` determines the contour from the volume data **V** at the value given by `isovalue`. The arrays in **X**, **Y**, and **Z** define the coordinates for the volume. The outputs from this function are the faces and vertices corresponding to the isosurface and can be passed directly into the `patch` function for displaying.

**Example 5.19**
We illustrate several isosurfaces of 3-D contours for data that is uniformly distributed over the volume defined by a unit cube. We display two contours of different levels in Figures 5.30 and 5.31.

```
% Get some data that will be between 0 and 1.
data = rand(10,10,10);
data = smooth3(data,'gaussian');
```

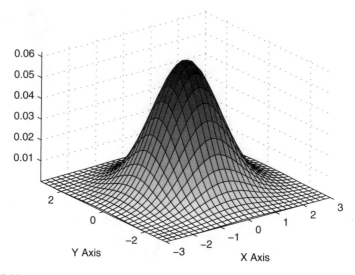

**FIGURE 5.29**
This is the surface plot for a standard normal bivariate distribution. to help the reader understand what is shown in Figure 5.28.

```
% Just in case there are some figure windows
% open - we should start anew.
close all
for i = [0.4 0.6]
    figure
    hpatch=patch(isosurface(data,i),...
        'Facecolor','blue',...
        'Edgecolor','none',...
        'AmbientStrength',.2,...
        'SpecularStrength',.7,...
        'DiffuseStrength',.4);
    isonormals(data,hpatch)
    title(['f(x,y,z) = ' num2str(i)])
    daspect([1,1,1])
    axis tight
    axis off
    view(3)
    camlight right
    camlight left
    lighting phong
    drawnow
end
```

In Figure 5.30, we have the isosurface for $f(x, y, z) = 0.4$. The isosurface for $f(x, y, z) = 0.6$ is given in Figure 5.31. Again, these are surface contours where the value of the volume is the same.
❑

f(x,y,z) = 0.4

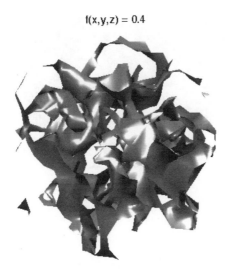

**FIGURE 5.30**
This is the isosurface of Example 5.19 for $f(x, y, z) = 0.4$.

It would be better if we had a context to help us understand what we are viewing with the isosurfaces. This can be done easily in MATLAB using the function called **isocaps**. This function puts caps on the boundaries of the domain and shows the distribution of the volume $f(x, y, z)$ above the isosurface. The color of the cap is mapped to the values $f(x, y, z)$ that are above the given value **isovalue**. Values below the **isovalue** can be shown on the **isocap** via the optional input argument, **enclose**. The following example illustrates this concept by adding isocaps to the surfaces obtained in Example 5.19.

**Example 5.20**
These MATLAB commands show how to add **isocaps** to the isosurfaces in the previous example.

```
for i=[0.4 0.6]
  figure
  hpatch = patch(isosurface(data,i),...
     'Facecolor','blue',...
     'Edgecolor','none',...
```

$f(x,y,z) = 0.6$

**FIGURE 5.31**
This is the isosurface of Example 5.19 for $f(x, y, z) = 0.6$.

```
    'AmbientStrength',.2,...
    'SpecularStrength',.7,...
    'DiffuseStrength',.4);
isonormals(data,hpatch)
patch(isocaps(data,i),...
    'Facecolor','interp',...
    'EdgeColor','none')
colormap hsv
title(['f(x,y,z) = ' num2str(i)])
daspect([1,1,1])
axis tight
axis off
view(3)
camlight right
camlight left
lighting phong
drawnow
end
```

Figure 5.32 shows the **isosurface** of Figure 5.30 with the **isocaps**. It is easier now to see what values are *'inside'* the isosurface or contour. Figure 5.33 shows the **isocaps** added to the **isosurface** corresponding to Figure 5.31.
❑

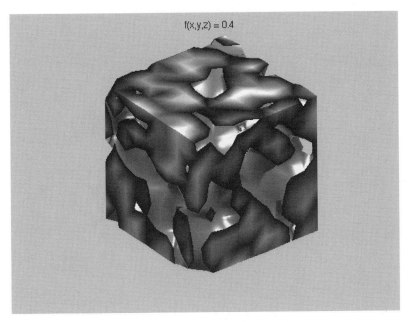

**FIGURE 5.32**
This is the **isosurface** of Figure 5.30 with **isocaps** added. Note that the color of the edges is mapped to the volume. The default is to map all values above $f(x, y, z) = 0.4$ to the color on the **isocaps**. This can be changed by an input argument to **isocaps**.

## Star Plots

Star diagrams were developed by Fienberg [1979] as a way of viewing multi-dimensional observations as a glyph or star. Each observed data point in the sample is plotted as a star, with the value of each measurement shown as a radial line from a common center point. Thus, each measured value for an observation is plotted as a spoke that is proportional to the size of the measured variable with the ends of the spokes connected with line segments to form a star. Star plots are a nice way to view the entire data set over all dimensions, but they are not suitable when there is a large number of observations ($n > 10$) or many dimensions (e.g., $d > 15$).

The next example applies this technique to data obtained from ratings of eight brands of cereal [Chakrapani and Ehrenberg, 1981; Venables and Ripley, 1994]. In our version of the star plot, the first variable is plotted as the spoke at angle $\theta = 0$, and the rest are shown counter-clockwise from there.

## Example 5.21

This example shows the MATLAB code to plot $d$-dimensional observations in a star plot. The **cereal** file contains a matrix where each row corresponds to

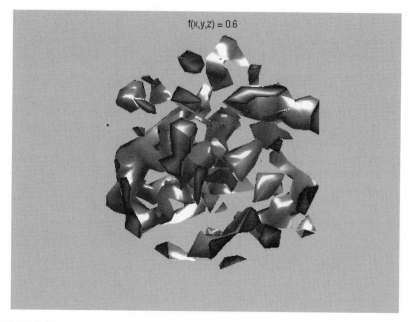

**FIGURE 5.33**
This is the `isosurface` of Figure 5.31 with `isocaps` added. Note that the color of the edges is mapped to the volume.

an observation and each column represents one of the variables or the percent agreement with the following statements about the cereal:

- come back to
- tastes nice
- popular with all the family
- very easy to digest
- nourishing
- natural flavor
- reasonably priced
- a lot of food value
- stays crispy in milk
- helps to keep you fit
- fun for children to eat

The resulting star plot is shown in Figure 5.34.

```
load cereal
% This file contains the labels and
% the matrix of 8 observations.
```

```
clf
n = 8;
p = 11;
% Find number of rows and columns for the stars.
ncol = floor(sqrt(n));
nrow = ceil(n/ncol);
% Re-scale the data.
md = min(cereal(:));
data = 1 + cereal - md;
% Get angles that are linearly spaced.
% Do not use the last point.
theta = linspace(0,2*pi,p+1);
theta(end) = [];
k = 0;
for i = 1:n
    k = k+1;
    % get the observation for plotting
    r = data(k,:);
    [x,y] = pol2cart(theta,r);
    X = x(:);  % make col vectors
    Y = y(:);
    X = [zeros(p,1) X];
    Y = [zeros(p,1) Y];
    x = [x(:); x(1)];
    y = [y(:); y(1)];
    subplot(nrow,ncol,k),
    patch(x,y,'w')
    hold on
    plot(X(1,:),Y(1,:))
    for ii = 2:p
      plot(X(ii,:),Y(ii,:))
    end
    title(labs{k})
    axis off
    hold off
end
```

❏

## Andrews Curves

Andrews curves [Andrews, 1972] were developed as a method for visualizing multi-dimensional data by mapping each observation onto a function. This is similar to star plots in that each observation or sample point is represented by a glyph, except that in this case the glyph is a curve. This function is defined as

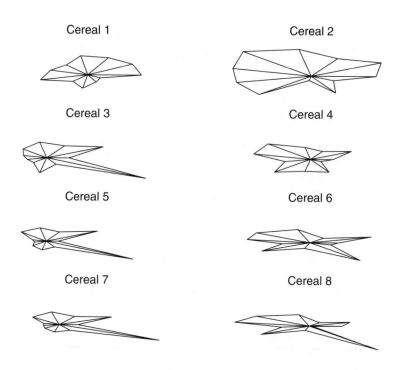

Cereal 1                              Cereal 2

Cereal 3                              Cereal 4

Cereal 5                              Cereal 6

Cereal 7                              Cereal 8

**FIGURE 5.34**
This is the star plot of the `cereal` data.

$$f_{\mathbf{x}}(t) = x_1/\sqrt{2} + x_2\sin t + x_3\cos t + x_4\sin 2t + x_5\cos 2t + \dots, \qquad (5.9)$$

where the range of $t$ is given by $-\pi \le t \le \pi$. Each observation is projected onto a set of orthogonal basis functions represented by sines and cosines and then plotted. Thus, each sample point is now represented by a curve given by Equation 5.9. We illustrate how to get the Andrews curves in Example 5.22.

### Example 5.22

We use a simple example to show how to get Andrews curves. The data we have are the following observations:

$$\mathbf{x}_1 = (2, 6, 4)$$
$$\mathbf{x}_2 = (5, 7, 3)$$
$$\mathbf{x}_3 = (1, 8, 9).$$

Using Equation 5.9, we construct three curves, one corresponding to each data point. The Andrews curves for the data are:

$$f_{x_1}(t) = 2/\sqrt{2} + 6\sin t + 4\cos t$$

$$f_{x_2}(t) = 5/\sqrt{2} + 7\sin t + 3\cos t$$

$$f_{x_3}(t) = 1/\sqrt{2} + 8\sin t + 9\cos t.$$

We can plot these three functions in MATLAB using the following commands. The Andrews curves for these data are shown in Figure 5.35.

```
% Get the domain.
t = linspace(-pi,pi);
% Evaluate function values for each observation.
f1 = 2/sqrt(2)+6*sin(t)+4*cos(t);
f2 = 5/sqrt(2)+7*sin(t)+3*cos(t);
f3 = 1/sqrt(2)+8*sin(t)+9*cos(t);
plot(t,f1,'.',t,f2,'*',t,f3,'o')
legend('F1','F2','F3')
xlabel('t')
```

❑

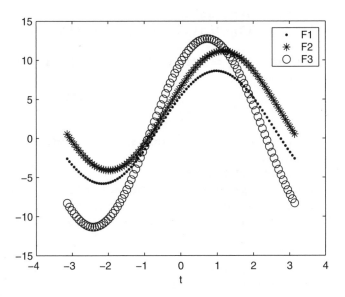

**FIGURE 5.35**
Andrews curves for the three data points in Example 5.22.

It has been shown [Andrews, 1972; Embrechts and Herzberg, 1991] that because of the mathematical properties of the trigonometric functions, the Andrews curves preserve means, distance (up to a constant) and variances. One consequence of this is that Andrews curves showing functions close together suggest that the corresponding data points will also be close together. Thus, one use of Andrews curves is to look for clustering of the data points.

## Example 5.23

We show how to construct Andrews curves for the **iris** data, using only the observations for *Iris setosa* and *Iris virginica* observations. We plot the curves for each species in a different line style to see if there is evidence that we can distinguish between the species using these variables.

```
load iris
% This defines the domain that will be plotted.
theta = (-pi+eps):0.1:(pi-eps);
n = 50;
p = 4;
ysetosa = zeros(n,p);
% There will n curves plotted,
% one for each data point.
yvirginica = zeros(n,p);
% Take dot product of each row with observation.
ang = zeros(length(theta),p);
fstr = '[1/sqrt(2) sin(i) cos(i) sin(2*i)]';
k = 0;
% Evaluate sin and cos functions at each angle theta.
for i = theta
    k = k+1;
    ang(k,:) = eval(fstr);
end
% Now generate a 'y' for each observation.
for i = 1:n
  for j = 1:length(theta)
      % Find dot product with observation.
      ysetosa(i,j)=setosa(i,:)*ang(j,:)';
      yvirginica(i,j)=virginica(i,:)*ang(j,:)';
  end
end
% Do all of the plots.
plot(theta,ysetosa(1,:),'r',...
     theta,yvirginica(1,:),'b-.')
legend('Iris Setosa','Iris Virginica')
hold
for i = 2:n
```

```
   plot(theta,ysetosa(i,:),'r',...
       theta,yvirginica(i,:),'b-.')
end
hold off
title('Andrews Plot')
xlabel('t')
ylabel('Andrews Curve')
```

The curves are shown in Figure 5.36. By plotting the two groups with different line styles, we can gain some insights about whether or not these two species of iris can be distinguished based on these features. From the Andrews curves, we see that the observations exhibit similarity within each class and that they show differences between the classes. Thus, we might get reasonable discrimination using these features.
❏

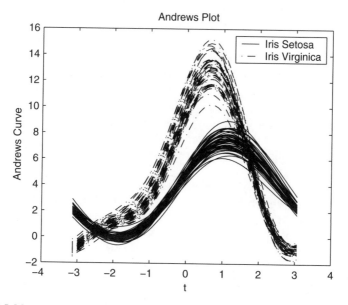

**FIGURE 5.36**
These are the Andrews curves for the *Iris setosa* and *Iris virginica* data. The curves corresponding to each species are plotted with different line styles. Note that the observations within each group show similar curves, and that we seem to be able to separate these two species.

Andrews curves are dependent on the order of the variables. Lower frequency terms exert more influence on the shape of the curves, so re-ordering the variables and viewing the resulting plot might provide insights about the data. By lower frequency terms, we mean those that are first in the sum given in Equation 5.9. Embrechts and Herzberg [1991] also suggest that the data be rescaled so they are centered at the origin and have covariance equal to the identity matrix. Andrews curves can be extended by using orthogonal bases other than sines and cosines. For example, Embrechts and Herzberg [1991] illustrate Andrews curves using Legendre polynomials and Chebychev polynomials.

### Parallel Coordinates

In the Cartesian coordinate system the axes are orthogonal, so the most we can view is three dimensions. If instead we draw the axes parallel to each other, then we can view many axes on the same display. This technique was developed by Wegman [1986] as a way of viewing and analyzing multidimensional data and was introduced by Inselberg [1985] in the context of computational geometry and computer vision. Parallel coordinate techniques were expanded on and described in a statistical setting by Wegman [1990]. Wegman [1990] also gave a rigorous explanation of the properties of parallel coordinates as a projective transformation and illustrated the duality properties between the parallel coordinate representation and the Cartesian orthogonal coordinate representation.

A parallel coordinate plot for $d$-dimensional data is constructed by drawing $d$ lines parallel to each other. We draw $d$ copies of the real line representing the coordinates for $x_1, x_2, ..., x_d$. The lines are the same distance apart and are perpendicular to the Cartesian $y$ axis. Additionally, they all have the same positive orientation as the Cartesian $x$ axis. Some versions of parallel coordinates [Inselberg, 1985] draw the parallel axes perpendicular to the Cartesian $x$ axis.

A point $C = (c_1, ..., c_4)$ is shown in Figure 5.37 with the MATLAB code that generates it given in Example 5.24. We see that the point is a polygonal line with vertices at $(c_i, i-1), i = 1, ..., d$ in Cartesian coordinates on the $x_i$ parallel axis. Thus, a point in Cartesian coordinates is represented in parallel coordinates as a series of connected line segments.

### Example 5.24
We now plot the point $C = (1, 3, 7, 2)$ in parallel coordinates using these MATLAB commands.

```
c = [1 3 7 2];
% Get range of parallel axes.
x = [1 7];
% Plot the 4 parallel axes.
plot(x,zeros(1,2),x,ones(1,2),x,...
    2*ones(1,2),x,3*ones(1,2))
hold on
% Now plot point c as a polygonal line.
```

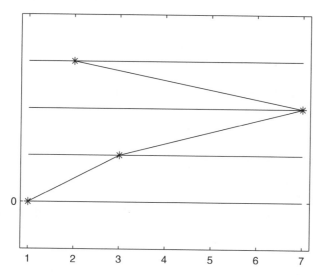

**FIGURE 5.37**
This shows the parallel coordinate representation for the 4-D point (1,3,7,2).

```
plot(c,0:3,c,0:3,'*')
ax = axis;
axis([ax(1) ax(2) -1 4 ])
set(gca,'ytick',0)
hold off
```

❑

If we plot observations in parallel coordinates with colors designating what class they belong to, then the parallel coordinate display can be used to determine whether or not the variables will enable us to separate the classes. This is similar to the Andrews curves in Example 5.23, where we used the Andrews curves to view the separation between two species of iris. The parallel coordinate plot provides graphical representations of multi-dimensional relationships [Wegman, 1990]. The next example shows how parallel coordinates can display the correlation between two variables.

**Example 5.25**
We first generate a set of 20 bivariate normal random variables with correlation given by 1. We plot the data using the function called **csparallel** to show how to recognize various types of correlation in parallel coordinate plots.

```
% Get a covariance matrix with correlation 1.
covmat = [1 1; 1 1];
```

```
% Generate the bivariate normal random variables.
% Note: you could use csmvrnd to get these.
[u,s,v] = svd(covmat);
vsqrt = (v*(u'.*sqrt(s)))';
subdata = randn(20,2);
data = subdata*vsqrt;
% Close any open figure windows.
close all
% Create parallel plot using CS Toolbox function.
csparallel(data)
title('Correlation of 1')
```

This is shown in Figure 5.38. The direct linear relationship between the first variable and the second variable is readily apparent. We can generate data that are correlated differently by changing the covariance matrix. For example, to obtain a random sample for data with a correlation of 0.2, we can use

```
covmat = [4 1.2; 1.2, 9];
```

In Figure 5.39, we show the parallel coordinates plot for data that have a correlation coefficient of -1. Note the different structure that is visible in the parallel coordinates plot.
❑

In the previous example, we showed how parallel coordinates can indicate the relationship between variables. To provide further insight, we illustrate how parallel coordinates can indicate clustering of variables in a dimension. Figure 5.40 shows data that can be separated into clusters in both of the dimensions. This is indicated on the parallel coordinate representation by separation or groups of lines along the $x_1$ and $x_2$ parallel axes. In Figure 5.41, we have data that are separated into clusters in only one dimension, $x_1$, but not in the $x_2$ dimension. This appears in the parallel coordinates plot as a gap in the $x_1$ parallel axis.

As with Andrews curves, the order of the variables makes a difference. Adjacent parallel axes provide some insights about the relationship between consecutive variables. To see other pairwise relationships, we must permute the order of the parallel axes. Wegman [1990] provides a systematic way of finding all permutations such that all adjacencies in the parallel coordinate display will be visited.

Before we proceed to other topics, we provide an example applying parallel coordinates to the **iris** data. In Example 5.26, we illustrate a parallel coordinates plot of the two classes: *Iris setosa* and *Iris virginica*.

## Example 5.26

First we load up the **iris** data. An optional input argument of the **csparallel** function is the line style for the lines. This usage is shown

Correlation of 1

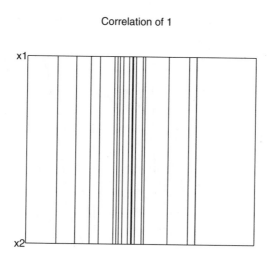

**FIGURE 5.38**
This is a parallel coordinate plot for bivariate data that have a correlation coefficient of 1.

Correlation of −1

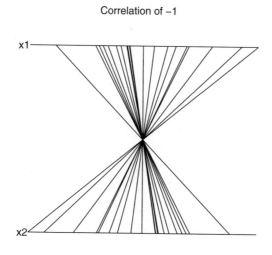

**FIGURE 5.39**
The data shown in this parallel coordinate plot are negatively correlated.

Clustering in Both Dimensions

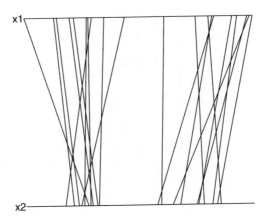

**FIGURE 5.40**
Clustering in two dimensions produces gaps in both parallel axes.

Clustering in $x_1$

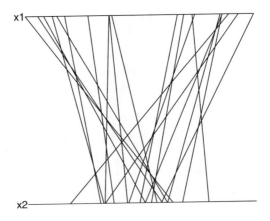

**FIGURE 5.41**
Clustering in only one dimension produces a gap in the corresponding parallel axis.

below, where we plot the *Iris setosa* observations as dot-dash lines and the *Iris virginica* as solid lines. The parallel coordinate plots is given in Figure 5.42.

```
load iris
figure
csparallel(setosa,'-.')
hold on
csparallel(virginica,'-')
hold off
```

From this plot, we see evidence of groups or separation in coordinates $x_2$ and $x_3$.
□

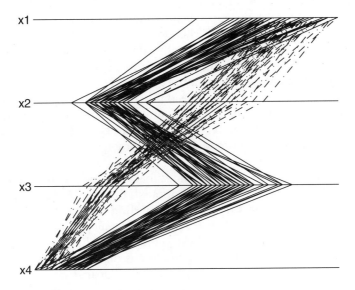

**FIGURE 5.42**
Here we see an example of a parallel coordinate plot for the **iris** data. The *Iris setosa* is shown as dot-dash lines and the *Iris virginica* as solid lines. There is evidence of groups in two of the coordinate axes, indicating that reasonable separation between these species could be made based on these features.

## Projection Pursuit

The Andrews curves and parallel coordinate plots are attempts to visualize all of the data points and all of the dimensions at once. An Andrews curve accomplishes this by mapping a data point to a curve. Parallel coordinate displays accomplish this by mapping each observation to a polygonal line with vertices on parallel axes. Another option is to tackle the problem of visualizing multi-dimensional data by reducing the data to a smaller dimension via a suitable projection. These methods reduce the data to 1-D or 2-D by projecting onto a line or a plane and then displaying each point in some suitable graphic, such as a scatterplot. Once the data are reduced to something that can be easily viewed, then exploring the data for patterns or interesting structure is possible.

One well-known method for reducing dimensionality is *principal component analysis* (PCA) [Jackson, 1991]. This method uses the eigenvector decomposition of the covariance (or the correlation) matrix. The data are then projected onto the eigenvector corresponding to the maximum eigenvalue (sometimes known as the first principal component) to reduce the data to one dimension. In this case, the eigenvector is one that follows the direction of the maximum variation in the data. Therefore, if we project onto the first principal component, then we will be using the direction that accounts for the maximum amount of variation using only one dimension. We illustrate the notion of projecting data onto a line in Figure 5.43.

We could project onto two dimensions using the eigenvectors corresponding to the largest and second largest eigenvalues. This would project onto the plane spanned by these eigenvectors. As we see shortly, PCA can be thought of in terms of projection pursuit, where the interesting structure is the variance of the projected data.

There are an infinite number of planes that we can use to reduce the dimensionality of our data. As we just mentioned, the first two principal components in PCA span one such plane, providing a projection such that the variation in the projected data is maximized over all possible 2-D projections. However, this might not be the best plane for highlighting interesting and informative structure in the data. *Structure* is defined to be departure from normality and includes such things as clusters, linear structures, holes, outliers, etc. Thus, the objective is to find a projection plane that provides a 2-D view of our data such that the structure (or departure from normality) is maximized over all possible 2-D projections.

We can use the Central Limit Theorem to motivate why we are interested in departures from normality. Linear combinations of data (even Bernoulli data) look normal. Since in most of the low-dimensional projections, one observes a Gaussian, if there is something interesting (e.g., clusters, etc.), then it has to be in the few non-normal projections.

Freidman and Tukey [1974] describe projection pursuit as a way of searching for and exploring nonlinear structure in multi-dimensional data by examining many 2-D projections. The idea is that 2-D orthogonal projections of the

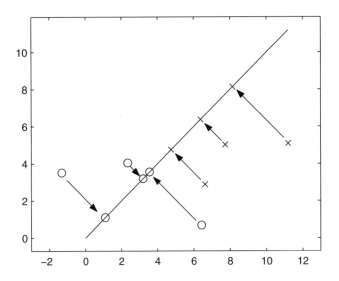

**FIGURE 5.43**
This illustrates the projection of 2-D data onto a line.

data should reveal structure that is in the original data. The projection pursuit technique can also be used to obtain 1-D projections, but we look only at the 2-D case. Extensions to this method are also described in the literature by Friedman [1987], Posse [1995a, 1995b], Huber [1985], and Jones and Sibson [1987]. In our presentation of projection pursuit exploratory data analysis, we follow the method of Posse [1995a, 1995b].

Projection pursuit exploratory data analysis (PPEDA) is accomplished by visiting many projections to find an interesting one, where *interesting* is measured by an index. In most cases, our interest is in non-normality, so the projection pursuit index usually measures the departure from normality. The index we use is known as the *chi-square index* and is developed in Posse [1995a, 1995b]. For completeness, other projection indexes are given in Appendix C, and the interested reader is referred to Posse [1995b] for a simulation analysis of the performance of these indexes.

PPEDA consists of two parts:

1) a projection pursuit index that measures the degree of the structure (or departure from normality), and

2) a method for finding the projection that yields the highest value for the index.

Posse [1995a, 1995b] uses a random search to locate the global optimum of the projection index and combines it with the structure removal of Freidman [1987] to get a sequence of interesting 2-D projections. Each projection found shows a structure that is less important (in terms of the projection index) than the previous one. Before we describe this method for PPEDA, we give a summary of the notation that we use in projection pursuit exploratory data analysis.

*NOTATION - PROJECTION PURSUIT EXPLORATORY DATA ANALYSIS*

**X** is an $n \times d$ matrix, where each row ($\mathbf{X}_i$) corresponds to a $d$-dimensional observation and $n$ is the sample size.

**Z** is the sphered version of **X**.

$\hat{\boldsymbol{\mu}}$ is the $1 \times d$ sample mean:

$$\hat{\boldsymbol{\mu}} = \sum \mathbf{X}_i / n. \tag{5.10}$$

$\hat{\boldsymbol{\Sigma}}$ is the sample covariance matrix:

$$\hat{\boldsymbol{\Sigma}}_{ij} = \frac{1}{n-1} \sum (\mathbf{X}_i - \hat{\boldsymbol{\mu}})(\mathbf{X}_j - \hat{\boldsymbol{\mu}})^T. \tag{5.11}$$

$\alpha, \beta$ are orthonormal ($\alpha^T \alpha = 1 = \beta^T \beta$ and $\alpha^T \beta = 0$) $d$-dimensional vectors that span the projection plane.

$P(\alpha, \beta)$ is the projection plane spanned by $\alpha$ and $\beta$.

$z_i^\alpha, z_i^\beta$ are the sphered observations projected onto the vectors $\alpha$ and $\beta$:

$$\begin{aligned} z_i^\alpha &= z_i^T \alpha \\ z_i^\beta &= z_i^T \beta \end{aligned} \tag{5.12}$$

$(\alpha^*, \beta^*)$ denotes the plane where the index is maximum.

$PI_{\chi^2}(\alpha, \beta)$ denotes the chi-square projection index evaluated using the data projected onto the plane spanned by $\alpha$ and $\beta$.

$\phi_2$ is the standard bivariate normal density.

$c_k$ is the probability evaluated over the $k$-th region using the standard bivariate normal,

$$c_k = \iint_{B_k} \phi_2 dz_1 dz_2. \tag{5.13}$$

$B_k$ is a box in the projection plane.

$I_{B_k}$ is the indicator function for region $B_k$.

$\eta_j = \pi j / 36$, $j = 0, ..., 8$ is the angle by which the data are rotated in the plane before being assigned to regions $B_k$.

$\alpha(\eta_j)$ and $\beta(\eta_j)$ are given by

$$
\begin{aligned}
\alpha(\eta_j) &= \alpha \cos \eta_j - \beta \sin \eta_j \\
\beta(\eta_j) &= \alpha \sin \eta_j + \beta \cos \eta_j
\end{aligned}
\tag{5.14}
$$

$c$ is a scalar that determines the size of the neighborhood around $(\alpha^*, \beta^*)$ that is visited in the search for planes that provide better values for the projection pursuit index.

$v$ is a vector uniformly distributed on the unit $d$-dimensional sphere.

*half* specifies the number of steps without an increase in the projection index, at which time the value of the neighborhood is halved.

$m$ represents the number of searches or random starts to find the best plane.

### Projection Pursuit Index

Posse [1995a, 1995b] developed an index based on the chi-square. The plane is first divided into 48 regions or boxes $B_k$ that are distributed in rings. See Figure 5.44 for an illustration of how the plane is partitioned. All regions have the same angular width of 45 degrees and the inner regions have the same radial width of $(2\log 6)^{1/2} / 5$. This choice for the radial width provides regions with approximately the same probability for the standard bivariate normal distribution. The regions in the outer ring have probability $1/48$. The regions are constructed in this way to account for the radial symmetry of the bivariate normal distribution.

Posse [1995a, 1995b] provides the population version of the projection index. We present only the empirical version here, because that is the one that must be implemented on the computer. The projection index is given by

$$
PI_{\chi^2}(\alpha, \beta) = \frac{1}{9} \sum_{j=1}^{8} \sum_{k=1}^{48} \frac{1}{c_k} \left[ \frac{1}{n} \sum_{i=1}^{n} I_{B_k}(z_i^{\alpha(\eta_j)}, z_i^{\beta(\eta_j)}) - c_k \right]^2 .
\tag{5.15}
$$

The chi-square projection index is not affected by the presence of outliers. This means that an interesting projection obtained using this index will not be one that is interesting solely because of outliers, unlike some of the other indexes (see Appendix C). It is sensitive to distributions that have a hole in the core, and it will also yield projections that contain clusters. The chi-square projection pursuit index is fast and easy to compute, making it appropriate

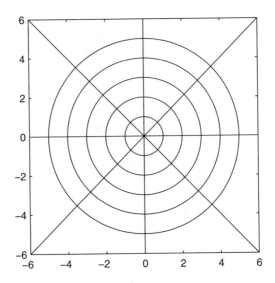

**FIGURE 5.44**
This shows the layout of the regions $B_k$ for the chi-square projection index. [Posse, 1995a]

for large sample sizes. Posse [1995a] provides a formula to approximate the percentiles of the chi-square index so the analyst can assess the significance of the observed value of the projection index.

### Finding the Structure

The second part of PPEDA requires a method for optimizing the projection index over all possible projections onto 2-D planes. Posse [1995a] shows that his optimization method outperforms the steepest-ascent techniques [Friedman and Tukey, 1974]. The Posse algorithm starts by randomly selecting a starting plane, which becomes the current best plane $(\alpha^*, \beta^*)$. The method seeks to improve the current best solution by considering two candidate solutions within its neighborhood. These candidate planes are given by

$$a_1 = \frac{\alpha^* + cv}{\|\alpha^* + cv\|} \qquad b_1 = \frac{\beta^* - (a_1^T \beta^*)a_1}{\|\beta^* - (a_1^T \beta^*)a_1\|}$$

$$a_2 = \frac{\alpha^* - cv}{\|\alpha^* - cv\|} \qquad b_1 = \frac{\beta^* - (a_2^T \beta^*)a_2}{\|\beta^* - (a_2^T \beta^*)a_2\|}. \tag{5.16}$$

In this approach, we start a global search by looking in large neighborhoods of the current best solution plane $(\alpha^*, \beta^*)$ and gradually focus in on a maximum by decreasing the neighborhood by half after a specified number of steps with no improvement in the value of the projection pursuit index. When the neighborhood is small, then the optimization process is terminated.

A summary of the steps for the exploratory projection pursuit algorithm is given here. Details on how to implement these steps are provided in Example 5.27 and in Appendix C. The complete search for the best plane involves repeating steps 2 through 9 of the procedure $m$ times, using $m$ random starting planes. Keep in mind that the best plane $(\alpha^*, \beta^*)$ is the plane where the projected data exhibit the greatest departure from normality.

*PROCEDURE - PROJECTION PURSUIT EXPLORATORY DATA ANALYSIS*

1. Sphere the data using the following transformation

$$\mathbf{Z}_i = \mathbf{\Lambda}^{-1/2}\mathbf{Q}^T(\mathbf{X}_i - \hat{\mathbf{\mu}}) \qquad i = 1, \ldots, n,$$

   where the columns of $\mathbf{Q}$ are the eigenvectors obtained from $\hat{\mathbf{\Sigma}}$, $\mathbf{\Lambda}$ is a diagonal matrix of corresponding eigenvalues, and $\mathbf{X}_i$ is the $i$-th observation.

2. Generate a random starting plane, $(\alpha_0, \beta_0)$. This is the current best plane, $(\alpha^*, \beta^*)$.

3. Evaluate the projection index $PI_{\chi^2}(\alpha_0, \beta_0)$ for the starting plane.

4. Generate two candidate planes $(a_1, b_1)$ and $(a_2, b_2)$ according to Equation 5.16.

5. Evaluate the value of the projection index for these planes, $PI_{\chi^2}(a_1, b_1)$ and $PI_{\chi^2}(a_2, b_2)$.

6. If one of the candidate planes yields a higher value of the projection pursuit index, then that one becomes the current best plane $(\alpha^*, \beta^*)$.

7. Repeat steps 4 through 6 while there are improvements in the projection pursuit index.

8. If the index does not improve for *half* times, then decrease the value of $c$ by half.

9. Repeat steps 4 through 8 until $c$ is some small number set by the analyst.

Note that in PPEDA we are working with sphered or standardized versions of the original data. Some researchers in this area [Huber, 1985] discuss the benefits and the disadvantages of this approach.

### Structure Removal

In PPEDA, we locate a projection that provides a maximum of the projection index. We have no reason to assume that there is only one interesting projection, and there might be other views that reveal insights about our data. To locate other views, Friedman [1987] devised a method called structure removal. The overall procedure is to perform projection pursuit as outlined above, remove the structure found at that projection, and repeat the projection pursuit process to find a projection that yields another maximum value of the projection pursuit index. Proceeding in this manner will provide a sequence of projections providing informative views of the data.

Structure removal in two dimensions is an iterative process. The procedure repeatedly transforms data that are projected to the current solution plane (the one that maximized the projection pursuit index) to standard normal until they stop becoming more normal. We can measure '*more normal*' using the projection pursuit index.

We start with a $d \times d$ matrix $\mathbf{U}^*$, where the first two rows of the matrix are the vectors of the projection obtained from PPEDA. The rest of the rows of $\mathbf{U}^*$ have ones on the diagonal and zero elsewhere. For example, if $d = 4$, then

$$
\mathbf{U}^* = \begin{bmatrix} \alpha_1^* & \alpha_2^* & \alpha_3^* & \alpha_4^* \\ \beta_1^* & \beta_2^* & \beta_3^* & \beta_4^* \\ 0 & 0 & 1 & 0 \\ 0 & 0 & 0 & 1 \end{bmatrix}.
$$

We use the Gram-Schmidt process [Strang, 1988] to make $\mathbf{U}^*$ orthonormal. We denote the orthonormal version as $\mathbf{U}$.

The next step in the structure removal process is to transform the $\mathbf{Z}$ matrix using the following

$$
\mathbf{T} = \mathbf{U}\mathbf{Z}^T. \tag{5.17}
$$

In Equation 5.17, $\mathbf{T}$ is $d \times n$, so each column of the matrix corresponds to a $d$-dimensional observation. With this transformation, the first two dimensions (the first two rows of $\mathbf{T}$) of every transformed observation are the projection onto the plane given by $(\alpha^*, \beta^*)$.

We now remove the structure that is represented by the first two dimensions. We let $\Theta$ be a transformation that transforms the first two rows of $\mathbf{T}$ to a standard normal and the rest remain unchanged. This is where we actually remove the structure, making the data normal in that projection (the first two rows). Letting $\mathbf{T}_1$ and $\mathbf{T}_2$ represent the first two rows of $\mathbf{T}$, we define the transformation as follows

$$\Theta(\mathbf{T}_1) = \Phi^{-1}[F(\mathbf{T}_1)]$$
$$\Theta(\mathbf{T}_2) = \Phi^{-1}[F(\mathbf{T}_2)] \tag{5.18}$$
$$\Theta(\mathbf{T}_i) = \mathbf{T}_i; \qquad i = 3, ..., d \;,$$

where $\Phi^{-1}$ is the inverse of the standard normal cumulative distribution function and $F$ is a function defined below (see Equations 5.19 and 5.20). We see from Equation 5.18, that we will be changing only the first two rows of $\mathbf{T}$.

We now describe the transformation of Equation 5.18 in more detail, working only with $\mathbf{T}_1$ and $\mathbf{T}_2$. First, we note that $\mathbf{T}_1$ can be written as

$$\mathbf{T}_1 = (z_1^{\alpha^*}, ..., z_j^{\alpha^*}, ..., z_n^{\alpha^*}),$$

and $\mathbf{T}_2$ as

$$\mathbf{T}_2 = (z_1^{\beta^*}, ..., z_j^{\beta^*}, ..., z_n^{\beta^*}).$$

Recall that $z_j^{\alpha^*}$ and $z_j^{\beta^*}$ would be coordinates of the $j$-th observation projected onto the plane spanned by $(\alpha^*, \beta^*)$.

Next, we define a rotation about the origin through the angle $\gamma$ as follows

$$\tilde{z}_j^{1(t)} = z_j^{1(t)}\cos\gamma + z_j^{2(t)}\sin\gamma$$
$$\tilde{z}_j^{2(t)} = z_j^{2(t)}\cos\gamma - z_j^{1(t)}\sin\gamma, \tag{5.19}$$

where $\gamma = 0, \pi/4, \pi/8, 3\pi/8$ and $z_j^{1(t)}$ represents the $j$-th element of $\mathbf{T}_1$ at the $t$-th iteration of the process. We now apply the following transformation to the rotated points,

$$z_j^{1(t+1)} = \Phi^{-1}\left\{\frac{r(\tilde{z}_j^{1(t)}) - 0.5}{n}\right\} \qquad z_j^{2(t+1)} = \Phi^{-1}\left\{\frac{r(\tilde{z}_j^{2(t)}) - 0.5}{n}\right\}, \tag{5.20}$$

where $r(\tilde{z}_j^{1(t)})$ represents the rank (position in the ordered list) of $\tilde{z}_j^{1(t)}$.

This transformation replaces each rotated observation by its normal score in the projection. With this procedure, we are deflating the projection index by making the data more normal. It is evident in the procedure given below, that this is an iterative process. Friedman [1987] states that during the first few iterations, the projection index should decrease rapidly. After approximate normality is obtained, the index might oscillate with small changes. Usually, the process takes between 5 to 15 complete iterations to remove the structure.

Once the structure is removed using this process, we must transform the data back using

$$\mathbf{Z'} = \mathbf{U}^T \Theta (\mathbf{U} \mathbf{Z}^T) . \tag{5.21}$$

In other words, we transform back using the transpose of the orthonormal matrix $\mathbf{U}$. From matrix theory [Strang, 1988], we see that all directions orthogonal to the structure (i.e., all rows of $\mathbf{T}$ other than the first two) have not been changed. Whereas, the structure has been Gaussianized and then transformed back.

*PROCEDURE - STRUCTURE REMOVAL*

1. Create the orthonormal matrix $\mathbf{U}$, where the first two rows of $\mathbf{U}$ contain the vectors $\alpha^*, \beta^*$.
2. Transform the data $\mathbf{Z}$ using Equation 5.17 to get $\mathbf{T}$.
3. Using only the first two rows of $\mathbf{T}$, rotate the observations using Equation 5.19.
4. Normalize each rotated point according to Equation 5.20.
5. For angles of rotation $\gamma = 0, \pi/4, \pi/8, 3\pi/8$, repeat steps 3 through 4.
6. Evaluate the projection index using $\mathbf{z}_j^{1(t+1)}$ and $\mathbf{z}_j^{2(t+1)}$, after going through an entire cycle of rotation (Equation 5.19) and normalization (Equation 5.20).
7. Repeat steps 3 through 6 until the projection pursuit index stops changing.
8. Transform the data back using Equation 5.21.

## Example 5.27

We use a synthetic data set to illustrate the MATLAB functions used for PPEDA. The source code for the functions used in this example is given in Appendix C. These data contain two structures, both of which are clusters. So we will search for two planes that maximize the projection pursuit index. First we load the data set that is contained in the file called **ppdata**. This loads a matrix $\mathbf{X}$ containing 400 six-dimensional observations. We also set up the constants we need for the algorithm.

```
% First load up a synthetic data set.
% This has structure
% in two planes - clusters.
% Note that the data is in
% ppdata.mat
load ppdata
```

```
% For m random starts, find the best projection plane
% using N structure removal procedures.
% Two structures:
N = 2;
% Four random starts:
m = 4;
c = tan(80*pi/180);
% Number of steps with no increase.
half = 30;
```

We now set up some arrays to store the results of projection pursuit.

```
% To store the N structures:
astar = zeros(d,N);
bstar = zeros(d,N);
ppmax = zeros(1,N);
```

Next we have to sphere the data.

```
% Sphere the data.
[n,d] = size(X);
muhat = mean(X);
[V,D] = eig(cov(X));
Xc = X-ones(n,1)*muhat;
Z = ((D)^(-1/2)*V'*Xc')';
```

We use the sphered data as input to the function **csppeda**. The outputs from this function are the vectors that span the plane containing the structure and the corresponding value of the projection pursuit index.

```
% Now do the PPEDA.
% Find a structure, remove it,
% and look for another one.
Zt = Z;
for i = 1:N
    [astar(:,i),bstar(:,i),ppmax(i)] =,...
        csppeda(Zt,c,half,m);
    % Now remove the structure.
    Zt = csppstrtrem(Zt,astar(:,i),bstar(:,i));
end
```

Note that each column of **astar** and **bstar** contains the projections for a structure, each one found using *m* random starts of the Posse algorithm. To see the first structure and second structures, we project onto the best planes as follows:

```
% Now project and see the structure.
proj1 = [astar(:,1), bstar(:,1)];
proj2 = [astar(:,2), bstar(:,2)];
Zp1 = Z*proj1;
```

```
Zp2 = Z*proj2;
figure
plot(Zp1(:,1),Zp1(:,2),'k.'),title('Structure 1')
xlabel('\alpha^*'),ylabel('\beta^*')
figure
plot(Zp2(:,1),Zp2(:,2),'k.'),title('Structure 2')
xlabel('\alpha^*'),ylabel('\beta^*')
```

The results are shown in Figure 5.45 and Figure 5.46, where we see that projection pursuit did find two structures. The first structure has a projection pursuit index of 2.67, and the second structure has an index equal to 0.572.
□

## Grand Tour

The grand tour of Asimov [1985] is an interactive visualization technique that enables the analyst to look for interesting structure embedded in multi-dimensional data. The idea is to project the $d$-dimensional data to a plane and to rotate the plane through all possible angles, searching for structure in the data. As with projection pursuit, structure is defined as departure from normality, such as clusters, spirals, linear relationships, etc.

In this procedure, we first determine a plane, project the data onto it, and then view it as a 2-D scatterplot. This process is repeated for a sequence of planes. If the sequence of planes is smooth (in the sense that the orientation of the plane changes slowly), then the result is a movie that shows the data points moving in a continuous manner. Asimov [1985] describes two methods for conducting a grand tour, called the *torus algorithm* and the *random interpolation algorithm*. Neither of these methods is ideal. With the torus method we may end up spending too much time in certain regions, and it is computationally intensive. The random interpolation method is better computationally, but cannot be reversed easily (to recover the projection) unless the set of random numbers used to generate the tour is retained. Thus, this method requires a lot of computer storage. Because of these limitations, we describe the *pseudo grand tour* described in Wegman and Shen [1993].

One of the important aspects of the torus grand tour is the need for a continuous space-filling path through the manifold of planes. This requirement satisfies the condition that the tour will visit *all* possible orientations of the projection plane. Here, we do not follow a space-filling curve, so this will be called a pseudo grand tour. In spite of this, the pseudo grand tour has many benefits:

- It can be calculated easily;
- It does not spend a lot of time in any one region;
- It still visits an ample set of orientations; and
- It is easily reversible.

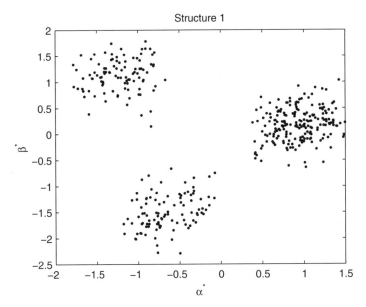

**FIGURE 5.45**
Here we see the first structure that was found using PPEDA. This structure yields a value of 2.67 for the chi-square projection pursuit index.

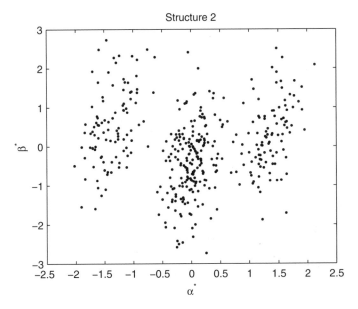

**FIGURE 5.46**
Here is the second structure we found using PPEDA. This structure has a value of 0.572 for the chi-square projection pursuit index.

The fact that the pseudo grand tour is easily reversible enables the analyst to recover the projection for further analysis. Two versions of the pseudo grand tour are available: one that projects onto a line and one that projects onto a plane.

As with projection pursuit, we need unit vectors that comprise the desired projection. In the 1-D case, we require a unit vector $\alpha(t)$ such that

$$\|\alpha(t)\|^2 = \sum_{i=1}^{d} \alpha_i^2(t) = 1$$

for every $t$, where $t$ represents a point in the sequence of projections. For the pseudo grand tour, $\alpha(t)$ must be a continuous function of $t$ and should produce *all* possible orientations of a unit vector.

We obtain the projection of the data using

$$z_i^{\alpha(t)} = \alpha^T(t)\mathbf{x}_i, \tag{5.22}$$

where $\mathbf{x}_i$ is the $i$-th $d$-dimensional data point. To get the movie view of the pseudo grand tour, we plot $z_i^{\alpha(t)}$ on a fixed 1-D coordinate system, re-displaying the projected points as $t$ increases.

The grand tour in two dimensions is similar. We need a second unit vector $\beta(t)$ that is orthonormal to $\alpha(t)$,

$$\|\beta(t)\|^2 = \sum_{i=1}^{d} \beta_i^2(t) = 1 \qquad \alpha^T(t)\beta(t) = 0.$$

We project the data onto the second vector using

$$z_i^{\beta(t)} = \beta^T(t)\mathbf{x}_i. \tag{5.23}$$

To obtain the movie view of the 2-D pseudo grand tour, we display $z_i^{\alpha(t)}$ and $z_i^{\beta(t)}$ in a 2-D scatterplot, replotting the points as $t$ increases.

The basic idea of the grand tour is to project the data onto a 1-D or 2-D space and plot the projected data, repeating this process many times to provide many views of the data. It is important for viewing purposes to make the time steps small to provide a nearly continuous path and to provide smooth motion of the points. The reader should note that the grand tour is an interactive approach to EDA. The analyst must stop the tour when an interesting projection is found.

Asimov [1985] contends that we are viewing more than one or two dimensions because the speed vectors provide further information. For example, the further away a point is from the computer screen, the faster the point rotates. We believe that the extra dimension conveyed by the speed is difficult to understand unless the analyst has experience looking at grand tour movies.

In order to implement the pseudo grand tour, we need a way of obtaining the projection vectors $\boldsymbol{\alpha}(t)$ and $\boldsymbol{\beta}(t)$. First we consider the data vector **x**. If $d$ is odd, then we augment each data point with a zero, to get an even number of elements. In this case,

$$\mathbf{x} = (x_1, \ldots, x_d, 0); \qquad \text{for } d \text{ odd.}$$

This will not affect the projection. So, without loss of generality, we present the method with the understanding that $d$ is even. We take the vector $\boldsymbol{\alpha}(t)$ to be

$$\boldsymbol{\alpha}(t) = \sqrt{2/d}(\sin\omega_1 t, \cos\omega_1 t, \ldots, \sin\omega_{d/2} t, \cos\omega_{d/2} t), \qquad (5.24)$$

and the vector $\boldsymbol{\beta}(t)$ as

$$\boldsymbol{\beta}(t) = \sqrt{2/d}(\cos\omega_1 t, -\sin\omega_1 t, \ldots, \cos\omega_{d/2} t, -\sin\omega_{d/2} t). \qquad (5.25)$$

We choose $\omega_i$ and $\omega_j$ such that the ratio $\omega_i/\omega_j$ is irrational for every $i$ and $j$. Additionally, we must choose these such that no $\omega_i/\omega_j$ is a rational multiple of any other ratio. It is also recommended that the time step $\Delta t$ be a small positive irrational number. One way to obtain irrational values for $\omega_i$ is to let $\omega_i = \sqrt{P_i}$, where $P_i$ is the $i$-th prime number.

The steps for implementing the 2-D pseudo grand tour are given here. The details on how to implement this in MATLAB are given in Example 5.28.

*PROCEDURE - PSEUDO GRAND TOUR*

1. Set each $\omega_i$ to an irrational number.
2. Find vectors $\boldsymbol{\alpha}(t)$ and $\boldsymbol{\beta}(t)$ using Equations 5.24 and 5.25.
3. Project the data onto the plane spanned by these vectors using Equations 5.23 and 5.24.
4. Display the projected points, $z_i^{\alpha(t)}$ and $z_i^{\beta(t)}$, in a 2-D scatterplot.
5. Using $\Delta t$ irrational, increment the time, and repeat steps 2 through 4.

Before we illustrate this in an example, we note that once we stop the tour at an interesting projection, we can easily recover the projection by knowing the time step.

**Example 5.28**

In this example, we use the **iris** data to illustrate the grand tour. First we load up the data and set up some preliminaries.

```
% This is for the iris data.
load iris
% Put data into one matrix.
x = [setosa;virginica;versicolor];
% Set up vector of frequencies.
th = sqrt([2 3]);
% Set up other constants.
[n,d] = size(x);
% This is a small irrational number:
delt = eps*10^14;
% Do the tour for some specified time steps.
maxit = 1000;
cof = sqrt(2/d);
% Set up storage space for projection vectors.
a = zeros(d,1);
b = zeros(d,1);
z = zeros(n,2);
```

We now do some preliminary plotting, just to get the handles we need to use MATLAB's Handle Graphics for plotting. This enables us to update the points that are plotted rather than replotting the entire figure.

```
% Get an initial plot, so the tour can be implemented
% using Handle Graphics.
Hlin1 = plot(z(1:50,1),z(1:50,2),'ro');
set(gcf,'backingstore','off')
set(gca,'Drawmode','fast')
hold on
Hlin2 = plot(z(51:100,1),z(51:100,2),'go');
Hlin3 = plot(z(101:150,1),z(101:150,2),'bo');
hold off
axis equal
axis vis3d
axis off
```

Now we do the actual pseudo grand tour, where we use a maximum number of iterations given by **maxit**.

```
for t = 0:delt:(delt*maxit)
  % Find the transformation vectors.
  for j = 1:d/2
    a(2*(j-1)+1) = cof*sin(th(j)*t);
    a(2*j) = cof*cos(th(j)*t);
    b(2*(j-1)+1) = cof*cos(th(j)*t);
```

```
    b(2*j) = cof*(-sin(th(j)*t));
end
% Project onto the vectors.
z(:,1) = x*a;
z(:,2) = x*b;
set(Hlin1,'xdata',z(1:50,1),'ydata',z(1:50,2))
set(Hlin2,'xdata',z(51:100,1),'ydata',z(51:100,2))
set(Hlin3,'xdata',z(101:150,1),'ydata',z(101:150,2))
drawnow
end
```

❑

## 5.5 MATLAB Code

MATLAB has many functions for visualizing data, both in the main package and in the Statistics Toolbox. Many of these were mentioned in the text and are summarized in Appendix E. Basic MATLAB has functions for scatterplots (**scatter**), histograms (**hist, bar**), and scatterplot matrices (**plotmatrix**). The Statistics Toolbox has functions for constructing q-q plots (**normplot, qqplot, weibplot**), the empirical cumulative distribution function (**cdfplot**), grouped versions of plots (**gscatter, gplotmatrix**), and others. Some other graphing functions in the standard MATLAB package that might be of interest include pie charts (**pie**), stair plots (**stairs**), error bars (**errorbar**), and stem plots (**stem**).

The methods for statistical graphics described in Cleveland's *Visualizing Data* [1993] have been implemented in MATLAB. They are available for download at

http://www.datatool.com/Dataviz_home.htm.

This book contains many useful techniques for visualizing data. Since MATLAB code is available for these methods, we urge the reader to refer to this highly readable text for more information on statistical visualization.

Rousseeuw, Ruts and Tukey [1999] describe a bivariate generalization of the univariate boxplot called a *bagplot*. This type of plot displays the location, spread, correlation, skewness and tails of the data set. Software (MATLAB and S-Plus®) for constructing a bagplot is available for download at

http://win-www.uia.ac.be/u/statis/index.html.

In the Computational Statistics Toolbox, we include several functions that implement some of the algorithms and graphics covered in Chapter 5. These are summarized in Table 5.3.

TABLE 5.3

List of Functions from Chapter 5 Included in the
Computational Statistics Toolbox

| Purpose | MATLAB Function |
|---|---|
| Star Plot | csstars |
| Stem-and-leaf Plot | csstemleaf |
| Parallel Coordinates Plot | csparallel |
| Q-Q Plot | csqqplot |
| Poissonness Plot | cspoissplot |
| Andrews Curves | csandrews |
| Exponential Probability Plot | csexpoplot |
| Binomial Plot | csbinoplot |
| PPEDA | csppeda |
| | csppstrtrem |
| | csppind |

## 5.6 Further Reading

One of the first treatises on graphical exploratory data analysis is John Tukey's *Exploratory Data Analysis* [1977]. In this book, he explains many aspects of EDA, including smoothing techniques, graphical techniques and others. The material in this book is practical and is readily accessible to readers with rudimentary knowledge of data analysis. Another excellent book on this subject is *Graphical Exploratory Data Analysis* [du Toit, Steyn and Stumpf, 1986], which includes several techniques (e.g., Chernoff faces and profiles) that we do not cover. For texts that emphasize the visualization of technical data, see Fortner and Meyer [1997] and Fortner [1995]. The paper by Wegman, Carr and Luo [1993] discusses many of the methods we present, along with others such as stereoscopic displays, generalized nonlinear regression using skeletons and a description of $d$-dimensional grand tour. This paper and Wegman [1990] provide an excellent theoretical treatment of parallel coordinates.

*The Grammar of Graphics* by Wilkinson [1999] describes a foundation for producing graphics for scientific journals, the internet, statistical packages, or any visualization system. It looks at the rules for producing pie charts, bar charts scatterplots, maps, function plots, and many others.

For the reader who is interested in visualization and information design, the three books by Edward Tufte are recommended. His first book, *The Visual Display of Quantitative Information* [Tufte, 1983], shows how to depict numbers. The second in the series is called *Envisioning Information* [Tufte, 1990], and illustrates how to deal with pictures of nouns (e.g., maps, aerial photographs, weather data). The third book is entitled *Visual Explanations* [Tufte, 1997], and it discusses how to illustrate pictures of verbs. These three books also provide many examples of good graphics and bad graphics. We highly recommend the book by Wainer [1997] for any statistician, engineer or data analyst. Wainer discusses the subject of good and bad graphics in a way that is accessible to the general reader.

Other techniques for visualizing multi-dimensional data have been proposed in the literature. One method introduced by Chernoff [1973] represents *d*-dimensional observations by a cartoon face, where features of the face reflect the values of the measurements. The size and shape of the nose, eyes, mouth, outline of the face and eyebrows, etc. would be determined by the value of the measurements. Chernoff faces can be used to determine simple trends in the data, but they are hard to interpret in most cases.

Another graphical EDA method that is often used is called brushing. Brushing [Venables and Ripley, 1994; Cleveland, 1993] is an interactive technique where the user can highlight data points on a scatterplot and the same points are highlighted on all other plots. For example, in a scatterplot matrix, highlighting a point in one plot shows up as highlighted in all of the others. This helps illustrate interesting structure across plots.

High-dimensional data can also be viewed using color histograms or data images. Color histograms are described in Wegman [1990]. Data images are discussed in Minotte and West [1998] and are a special case of color histograms.

For more information on the graphical capabilities of MATLAB, we refer the reader to the MATLAB documentation *Using MATLAB Graphics*. Another excellent resource is the book called *Graphics and GUI's with MATLAB* by Marchand [1999]. These go into more detail on the graphics capabilities in MATLAB that are useful in data analysis such as lighting, use of the camera, animation, etc.

We now describe references that extend the techniques given in this book.

- **_Stem-and-leaf_**: Various versions and extensions of the stem-and-leaf plot are available. We show an ordered stem-and-leaf plot in this book, but ordering is not required. Another version shades the leaves. Most introductory applied statistics books have information on stem-and-leaf plots (e.g., Montgomery, et al. [1998]). Hunter [1988] proposes an enhanced stem-and-leaf called the ***digidot plot***. This combines a stem-and-leaf with a time sequence plot. As data are collected they are plotted as a sequence of connected dots and a stem-and-leaf is created at the same time.

- *Discrete Quantile Plots*: Hoaglin and Tukey [1985] provide similar plots for other discrete distributions. These include the negative binomial, the geometric and the logarithmic series. They also discuss graphical techniques for plotting confidence intervals instead of points. This has the advantage of showing the confidence one has for each count.

- *Box plots*: Other variations of the box plot have been described in the literature. See McGill, Tukey and Larsen [1978] for a discussion of the variable width box plot. With this type of display, the width of the box represents the number of observations in each sample.

- *Scatterplots*: Scatterplot techniques are discussed in Carr, et al. [1987]. The methods presented in this paper are especially pertinent to the situation facing analysts today, where the typical data set that must be analyzed is often very large ($n = 10^3, 10^6, \ldots$). They recommend various forms of binning (including hexagonal binning) and representation of the value by gray scale or symbol area.

- *PPEDA*: Jones and Sibson [1987] describe a steepest-ascent algorithm that starts from either principal components or random starts. Friedman [1987] combines steepest-ascent with a stepping search to look for a region of interest. Crawford [1991] uses genetic algorithms to optimize the projection index.

- *Projection Pursuit*: Other uses for projection pursuit have been proposed. These include projection pursuit probability density estimation [Friedman, Stuetzle, and Schroeder, 1984], projection pursuit regression [Friedman and Stuetzle, 1981], robust estimation [Li and Chen, 1985], and projection pursuit for pattern recognition [Flick, et al., 1990]. A 3-D projection pursuit algorithm is given in Nason [1995]. For a theoretical and comprehensive description of projection pursuit, the reader is directed to Huber [1985]. This invited paper with discussion also presents applications of projection pursuit to computer tomography and to the deconvolution of time series. Another paper that provides applications of projection pursuit is Jones and Sibson [1987]. Not surprisingly, projection pursuit has been combined with the grand tour by Cook, et al. [1995]. Montanari and Lizzani [2001] apply projection pursuit to the variable selection problem. Bolton and Krzanowski [1999] describe the connection between projection pursuit and principal component analysis.

## Exercises

5.1. Generate a sample of 1000 univariate standard normal random variables using **randn**. Construct a frequency histogram, relative frequency histogram, and density histogram. For the density histogram, superimpose the corresponding theoretical probability density function. How well do they match?

5.2. Repeat problem 5.1 for random samples generated from the exponential, gamma, and beta distributions.

5.3. Do a quantile plot of the Tibetan skull data of Example 5.3 using the standard normal quantiles. Is it reasonable to assume the data follow a normal distribution?

5.4. Try the following MATLAB code using the 3-D multivariate normal as defined in Example 5.18. This will create a slice through the volume at an arbitrary angle. Notice that the colors indicate a normal distribution centered at the origin with the covariance matrix equal to the identity matrix.

```
% Draw a slice at an arbitrary angle
hs = surf(linspace(-3,3,20),...
    linspace(-3,3,20),zeros(20));
% Rotate the surface :
rotate(hs,[1,-1,1],30)
% Get the data that will define the
% surface at an arbitrary angle.
xd = get(hs,'XData');
yd = get(hs,'YData');
zd = get(hs,'ZData');
delete(hs)
% Draw slice:
slice(x,y,z,prob,xd,yd,zd)
axis tight
% Now plot this using the peaks surface as the slice.
% Try plotting against the peaks surface
[xd,yd,zd] = peaks;
slice(x,y,z,prob,xd,yd,zd)
axis tight
```

5.5. Repeat Example 5.23 using the data for *Iris virginica* and *Iris versicolor*. Do the Andrews curves indicate separation between the classes? Do you think it will be difficult to separate these classes based on these features?

5.6. Repeat Example 5.4, where you generate random variables such that

(a) $X \sim N(0, 2)$ and $Y \sim N(0, 1)$

(b) $X \sim N(5, 1)$ and $Y \sim N(0, 1)$

How can you tell from the q-q plot that the scale and the location parameters are different?

5.7. Write a MATLAB program that permutes the axes in a parallel coordinates plot. Apply it to the **iris** data.

5.8. Write a MATLAB program that permutes the order of the variables and plots the resulting Andrews curves. Apply it to the **iris** data.

5.9. Implement Andrews curves using a different set of basis functions as suggested in the text.

5.10. Repeat Example 5.16 and use **rotate3d** (or the rotate toolbar button) to rotate about the axes. Do you see any separation of the different types of insects?

5.11. Do a scatterplot matrix of the *Iris versicolor* data.

5.12. Verify that the two vectors used in Equations 5.24 and 5.25 are orthonormal.

5.13. Write a function that implements Example 5.17 for any data set. The user should have the opportunity to input the labels.

5.14. Define a trivariate normal as your volume, $f(x, y, z)$. Use the MATLAB functions **isosurface** and **isocaps** to obtain contours of constant volume or probability (in this case).

5.15. Construct a quantile plot using the **forearm** data, comparing the sample to the quantiles of a normal distribution. Is it reasonable to model the data using the normal distribution?

5.16. The **moths** data represent the number of moths caught in a trap over 24 consecutive nights [Hand, et al., 1994]. Use the stem-and-leaf to explore the shape of the distribution.

5.17. The **biology** data set contains the number of research papers for 1534 biologists [Tripathi and Gupta, 1988; Hand, et al., 1994]. Construct a binomial plot of these data. Analyze your results.

5.18. In the **counting** data set, we have the number of scintillations in 72 second intervals arising from the radioactive decay of polonium [Rutherford and Geiger, 1910; Hand, et al., 1994]. Construct a Poissonness plot. Does this indicate agreement with the Poisson distribution?

5.19. Use the MATLAB Statistics Toolbox function **boxplot** to compare box plots of the features for each species of **iris** data.

5.20. The **thrombos** data set contains measurements of urinary-thromboglobulin excretion in 12 normal and 12 diabetic patients [van Oost, et al.; 1983; Hand, et al., 1994]. Put each of these into a column of a matrix and use the **boxplot** function to compare normal versus diabetic patients.

5.21. To explore the **shading** options in MATLAB, try the following code from the documentation:

```
% The ezsurf function is available in MATLAB 5.3
% and later.
% First get a surface.
ezsurf('sin(sqrt(x^2+y^2))/sqrt(x^2+y^2)',...
    [-6*pi,6*pi])
% Now add some lighting effects:
view(0,75)
shading interp
lightangle(-45,30)
set(findobj('type','surface'),...
    'FaceLighting','phong',...
    'AmbientStrength',0.3,'DiffuseStrength',0.8,...
    'SpecularStrength',0.9,'SpecularExponent',25,...
    'BackFaceLighting','unlit')
axis off
```

5.22. The **bank** data contains two matrices comprised of measurements made on genuine money and forged money. Combine these two matrices into one and use PPEDA to discover any clusters or groups in the data. Compare your results with the known groups in the data.

5.23. Using the data in Example 5.27, do a scatterplot matrix of the original sphered data set. Note the structures in the first four dimensions. Get the first structure and construct another scatterplot matrix of the sphered data after the first structure has been removed. Repeat this process after both structures are removed.

5.24. Load the data sets in **posse**. These contain several data sets from Posse [1995b]. Apply the PPEDA method to these data.

# Chapter 6

## Monte Carlo Methods for Inferential Statistics

### 6.1 Introduction

Methods in inferential statistics are used to draw conclusions about a population and to measure the reliability of these conclusions using information obtained from a random sample. Inferential statistics involves techniques such as estimating population parameters using point estimates, calculating confidence interval estimates for parameters, hypothesis testing, and modeling (e.g., regression and density estimation). To measure the reliability of the inferences that are made, the statistician must understand the distribution of any statistics that are used in the analysis. In situations where we use a well-understood statistic, such as the sample mean, this is easily done analytically. However, in many applications, we do not want to be limited to using such simple statistics or to making simplifying assumptions. The goal of this chapter is to explain how simulation or Monte Carlo methods can be used to make inferences when the traditional or analytical statistical methods fail.

According to Murdoch [2000], the term *Monte Carlo* originally referred to simulations that involved random walks and was first used by Jon von Neumann and S. M. Ulam in the 1940's. Today, the *Monte Carlo method* refers to any simulation that involves the use of random numbers. In the following sections, we show that Monte Carlo simulations (or experiments) are an easy and inexpensive way to understand the phenomena of interest [Gentle, 1998]. To conduct a simulation experiment, you need a model that represents your population or phenomena of interest and a way to generate random numbers (according to your model) using a computer. The data that are generated from your model can then be studied as if they were observations. As we will see, one can use statistics based on the simulated data (means, medians, modes, variance, skewness, etc.) to gain understanding about the population.

In Section 6.2, we give a short overview of methods used in classical inferential statistics, covering such topics as hypothesis testing, power, and confidence intervals. The reader who is familiar with these may skip this section. In Section 6.3, we discuss Monte Carlo simulation methods for hypothesis testing and for evaluating the performance of the tests. The bootstrap method for estimating the bias and variance of estimates is presented in Section 6.4. Finally, Sections 6.5 and 6.6 conclude the chapter with information about available MATLAB code and references on Monte Carlo simulation and the bootstrap.

## 6.2 Classical Inferential Statistics

In this section, we will cover two of the main methods in inferential statistics: hypothesis testing and calculating confidence intervals. With confidence intervals, we are interested in obtaining an interval of real numbers that we expect (with specified confidence) contains the true value of a population parameter. In hypothesis testing, our goal is to make a decision about not rejecting or rejecting some statement about the population based on data from a random sample. We give a brief summary of the concepts in classical inferential statistics, endeavoring to keep the theory to a minimum. There are many books available that contain more information on these topics. We recommend Casella and Berger [1990], Walpole and Myers [1985], Bickel and Doksum [1977], Lindgren [1993], Montgomery, Runger and Hubele [1998], and Mood, Graybill and Boes [1974].

### Hypothesis Testing

In hypothesis testing, we start with a *statistical hypothesis*, which is a conjecture about one or more populations. Some examples of these are:

- A transportation official in the Washington, D.C. area thinks that the mean travel time to work for northern Virginia residents has increased from the average time it took in 1995.
- A medical researcher would like to determine whether aspirin decreases the risk of heart attacks.
- A pharmaceutical company needs to decide whether a new vaccine is superior to the one currently in use.
- An engineer has to determine whether there is a difference in accuracy between two types of instruments.

We generally formulate our statistical hypotheses in two parts. The first is the *null hypothesis* represented by $H_0$, which denotes the hypothesis we would like to test. Usually, we are searching for departures from this statement. Using one of the examples given above, the engineer would have the null hypothesis that there is no difference in the accuracy between the two instruments.

There must be an *alternative hypothesis* such that we would decide in favor of one or the other, and this is denoted by $H_1$. If we reject $H_0$, then this leads to the acceptance of $H_1$. Returning to the engineering example, the alternative hypothesis might be that there is a difference in the instruments or that one is more accurate than the other. When we perform a statistical hypothesis test, we can never know with certainty what hypothesis is true. For ease of exposition, we will use the terms *accept the null hypothesis* and *reject the null hypothesis* for our decisions resulting from statistical hypothesis testing.

To clarify these ideas, let's look at the example of the transportation official who wants to determine whether the average travel time to work has increased from the time it took in 1995. The mean travel time to work for northern Virginia residents in 1995 was 45 minutes. Since he wants to determine whether the mean travel time has increased, the statistical hypotheses are given by:

$$H_0: \quad \mu = 45 \text{ minutes}$$
$$H_1: \quad \mu > 45 \text{ minutes.}$$

The logic behind statistical hypothesis testing is summarized below, with details and definitions given after.

*STEPS OF HYPOTHESIS TESTING*

1. Determine the null and alternative hypotheses, using mathematical expressions if applicable. Usually, this is an expression that involves a characteristic or descriptive measure of a population.
2. Take a random sample from the population of interest.
3. Calculate a statistic from the sample that provides information about the null hypothesis. We use this to make our decision.
4. If the value of the statistic is consistent with the null hypothesis, then do not reject $H_0$.
5. If the value of the statistic is not consistent with the null hypothesis, then reject $H_0$ and accept the alternative hypothesis.

The problem then becomes one of determining when a statistic is consistent with the null hypothesis. Recall from Chapter 3 that a statistic is itself a random variable and has a probability distribution associated with it. So, in order to decide whether or not an observed value of the statistic is consistent with the null hypothesis, we must know the distribution of the statistic when the null hypothesis is true. The statistic used in step 3 is called a *test statistic*.

Let's return to the example of the travel time to work for northern Virginia residents. To perform the analysis, the transportation official takes a random sample of 100 residents in northern Virginia and measures the time it takes them to travel to work. He uses the sample mean to help determine whether there is sufficient evidence to reject the null hypothesis and conclude that the mean travel time has increased. The sample mean that he calculates is 47.2 minutes. This is slightly higher than the mean of 45 minutes for the null hypothesis. However, the sample mean is a random variable and has some variation associated with it. If the variance of the sample mean under the null hypothesis is large, then the observed value of $\bar{x} = 47.2$ minutes might not be inconsistent with $H_0$. This is explained further in Example 6.1.

### Example 6.1

We continue with the transportation example. We need to determine whether or not the value of the statistic obtained from a random sample drawn from the population is consistent with the null hypothesis. Here we have a random sample comprised of $n = 100$ commute times. The sample mean of these observations is $\bar{x} = 47.2$ minutes. If the transportation official assumes that the travel times to work are normally distributed with $\sigma = 15$ minutes (one might know a reasonable value for $\sigma$ based on previous experience with the population), then we know from Chapter 3 that $\bar{x}$ is approximately normally distributed with mean $\mu_X$ and standard deviation $\sigma_{\bar{X}} = \sigma_X / \sqrt{n}$. Standardizing the observed value of the sample mean, we have

$$z_o = \frac{\bar{x} - \mu_0}{\sigma_X / \sqrt{n}} = \frac{\bar{x} - \mu_0}{\sigma_{\bar{X}}} = \frac{47.2 - 45}{15 / \sqrt{100}} = \frac{2.2}{1.5} = 1.47, \qquad (6.1)$$

where $z_o$ is the observed value of the test statistic, and $\mu_0$ is the mean under the null hypothesis. Thus, we have that the value of $\bar{x} = 47.2$ minutes is 1.47 standard deviations away from the mean, if the null hypothesis is really true. (This is why we use $\mu_0$ in Equation 6.1.) We know that approximately 95% of normally distributed random variables fall within two standard deviations either side of the mean. Thus, $\bar{x} = 47.2$ minutes is not inconsistent with the null hypothesis.
□

In hypothesis testing, the rule that governs our decision might be of the form: *if the observed statistic is within some region, then we reject the null hypothesis*. The **critical region** is an interval for the test statistic over which we would reject $H_0$. This is sometimes called the **rejection region**. The **critical value** is that value of the test statistic that divides the domain of the test statistic into a region where $H_0$ will be rejected and one where $H_0$ will be accepted. We need to know the distribution of the test statistic under the null hypothesis to find the critical value(s).

The critical region depends on the distribution of the statistic under the null hypothesis, the alternative hypothesis, and the amount of error we are willing to tolerate. Typically, the critical regions are areas in the tails of the distribution of the test statistic when $H_0$ is true. It could be in the lower tail, the upper tail or both tails, and which one is appropriate depends on the alternative hypothesis. For example:

- If a large value of the test statistic would provide evidence for the alternative hypothesis, then the critical region is in the upper tail of the distribution of the test statistic. This is sometimes referred to as an ***upper tail test***.

- If a small value of test statistic provides evidence for the alternative hypothesis, then the critical region is in the lower tail of the distribution of the test statistic. This is sometimes referred to as a ***lower tail test***.

- If small or large values of the test statistic indicate evidence for the alternative hypothesis, then the critical region is in the lower and upper tails. This is sometimes referred to as a ***two-tail test***.

There are two types of errors that can occur when we make a decision in statistical hypothesis testing. The first is a ***Type I error***, which arises when we reject $H_0$ when it is really true. The other error is called ***Type II error***, and this happens when we fail to detect that $H_0$ is actually false. These errors are summarized in Table 6.1.

TABLE 6.1

Types of Error in Statistical Hypothesis Testing

| Type of Error | Description | Probability of Error |
|---|---|---|
| Type I Error | Rejecting $H_0$ when it is true | $\alpha$ |
| Type II Error | Not rejecting $H_0$ when it is false | $\beta$ |

Recall that we are usually searching for significant evidence that the alternative hypothesis is valid, and we do not want to change from the status quo (i.e., reject $H_0$) unless there is sufficient evidence in the data to lead us in that direction. So, when setting up a hypothesis test we ensure that the probability of wrongly rejecting $H_0$ is controlled. The probability of making a Type I error is denoted by $\alpha$ and is sometimes called the ***significance level*** of the test. The $\alpha$ is set by the analyst, and it represents the maximum probability of Type I error that will be tolerated. Typical values of $\alpha$ are $\alpha = 0.01, 0.05, 0.10$. The critical value is found as the quantile (under the null hypothesis) that gives a significance level of $\alpha$.

The specific procedure for conducting an hypothesis test using these ideas is given below. This is called the ***critical value approach***, because the decision is based on whether the value of the test statistic falls in the rejection region. We will discuss an alternative method later in this section. The concepts of hypothesis testing using the critical value approach are illustrated in Example 6.2.

*PROCEDURE - HYPOTHESIS TESTING (CRITICAL VALUE APPROACH)*

1. Determine the null and alternative hypotheses.
2. Find a test statistic $T$ that will provide evidence that $H_0$ should be accepted or rejected (e.g, a large value of the test statistic indicates $H_0$ should be rejected).
3. Obtain a random sample from the population of interest and compute the observed value of the test statistic $t_o$ using the sample.
4. Using the sampling distribution of the test statistic under the null hypothesis and the significance level, find the critical value(s). That is, find the $t$ such that

   Upper Tail Test: $P_{H_0}(T \le t) = 1 - \alpha$

   Lower Tail Test: $P_{H_0}(T \le t) = \alpha$

   Two-Tail Test: $P_{H_0}(T \le t_1) = \alpha/2$ and $P_{H_0}(T \le t_2) = 1 - \alpha/2$,

   where $P_{H_0}(.)$ denotes the probability under the null hypothesis.

5. If the value of the test statistic $t_o$ falls in the critical region, then reject the null hypothesis.

**Example 6.2**
Here, we illustrate the critical value approach to hypothesis testing using the transportation example. Our test statistic is given by

$$z = \frac{\bar{x} - \mu_0}{\sigma_{\bar{X}}},$$

and we observed a value of $z_o = 1.47$ based on the random sample of $n = 100$ commute times. We want to conduct the hypothesis test at a significance level given by $\alpha = 0.05$. Since our alternative hypothesis is that the commute times have increased, a large value of the test statistic provides evidence for $H_1$. We can find the critical value using the MATLAB Statistics Toolbox as follows:

```
cv = norminv(0.95,0,1);
```

This yields a critical value of 1.645. Thus, if $z_o \geq 1.645$, then we reject $H_0$. Since the observed value of the test statistic is less than the critical value, we do not reject $H_0$. The regions corresponding to this hypothesis test are illustrated in Figure 6.1.
❏

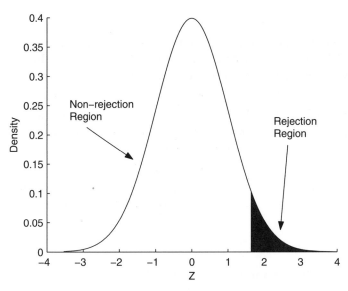

**FIGURE 6.1**
This shows the critical region (shaded region) for the hypothesis test of Examples 6.1 and 6.2. If the observed value of the test statistic falls in the shaded region, then we reject the null hypothesis. Note that this curve reflects the distribution for the test statistic under the null hypothesis.

The probability of making a Type II error is represented by β, and it depends on the sample size, the significance level of the test, and the alternative hypothesis. The last part is important to remember: *the probability that we will not detect a departure from the null hypothesis depends on the distribution of the test statistic under the alternative hypothesis.* Recall that the alternative hypothesis allows for many different possibilities, yielding many distributions under $H_1$. So, we must determine the Type II error for every alternative hypothesis of interest.

A more convenient measure of the performance of a hypothesis test is to determine the probability of not making a Type II error. This is called the *power* of a test. We can consider this to be the probability of rejecting $H_0$ when it is really false. Roughly speaking, one can think of the power as the ability of the hypothesis test to detect a false null hypothesis. The power is given by

$$\text{Power} = 1 - \beta. \tag{6.2}$$

As we see in Example 6.3, the power of the test to detect departures from the null hypothesis depends on the true value of $\mu$.

### Example 6.3

Returning to the transportation example, we illustrate the concepts of Type II error and power. It is important to keep in mind that these values depend on the true mean $\mu$, so we have to calculate the Type II error for different values of $\mu$. First we get a vector of values for $\mu$:

```
% Get several values for the mean under the alternative
% hypothesis. Note that we are getting some values
% below the null hypothesis.
mualt = 40:60;
```

It is actually easier to understand the power when we look at a test statistic based on $\bar{x}$ rather than $z_o$. So, we convert the critical value to its corresponding $\bar{x}$ value:

```
% Note the critical value:
cv = 1.645;
% Note the standard deviation for x-bar:
sig = 1.5;
% It's easier to use the non-standardized version,
% so convert:
ct = cv*1.5 + 45;
```

We find the area under the curve to the left of the critical value (the non rejection region) for each of these values of the true mean. That would be the probability of not rejecting the null hypothesis.

```
% Get a vector of critical values that is
% the same size as mualt.
ctv = ct*ones(size(mualt));
% Now get the probabilities to the left of this value.
% These are the probabilities of the Type II error.
beta = normcdf(ctv,mualt,sig);
```

Note that the variable **beta** contains the probability of Type II error (the area to the left of the critical value **ctv** under a normal curve with mean **mualt** and standard deviation **sig**) for every $\mu$. To get the power, simply subtract all of the values for **beta** from one.

```
% To get the power: 1-beta
```

```
pow = 1 - beta;
```

We plot the power against the true value of the population mean in Figure 6.2. Note that as $\mu > \mu_0$, the power (or the likelihood that we can detect the alternative hypothesis) increases.

```
plot(mualt,pow);
xlabel('True Mean \mu')
ylabel('Power')
axis([40 60 0 1.1])
```

We leave it as an exercise for the reader to plot the probability of making a Type II error.
❏

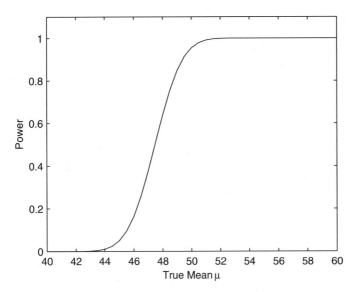

**FIGURE 6.2**
This shows the power (or probability of not making a Type II error) as a function of the true value of the population mean $\mu$. Note that as the true mean gets larger, then the likelihood of not making a Type II error increases.

There is an alternative approach to hypothesis testing, which uses a quantity called a *p*-value. A ***p-value*** is defined as the probability of observing a value of the test statistic as extreme as or more extreme than the one that is observed, when the null hypothesis $H_0$ is true. The word *extreme* refers to the direction of the alternative hypothesis. For example, if a small value of the test statistic (a lower tail test) indicates evidence for the alternative hypothesis, then the *p*-value is calculated as

$$p - \text{value} = P_{H_0}(T \le t_o),$$

where $t_o$ is the observed value of the test statistic $T$, and $P_{H_0}(.)$ denotes the probability under the null hypothesis. The $p$-value is sometimes referred to as the *observed significance level*.

In the $p$-value approach, a small value indicates evidence for the alternative hypothesis and would lead to rejection of $H_0$. Here small refers to a $p$-value that is less than or equal to $\alpha$. The steps for performing hypothesis testing using the $p$-value approach are given below and are illustrated in Example 6.4.

*PROCEDURE - HYPOTHESIS TESTING (P-VALUE APPROACH)*

1. Determine the null and alternative hypotheses.
2. Find a test statistic $T$ that will provide evidence about $H_0$.
3. Obtain a random sample from the population of interest and compute the value of the test statistic $t_o$ from the sample.
4. Calculate the $p$-value:

   <u>Lower Tail Test</u>: $p - \text{value} = P_{H_0}(T \le t_o)$

   <u>Upper Tail Test</u>: $p - \text{value} = P_{H_0}(T \ge t_o)$

5. If the $p$-value $\le \alpha$, then reject the null hypothesis.

For a two-tail test, the $p$-value is determined similarly.

## Example 6.4

In this example, we repeat the hypothesis test of Example 6.2 using the $p$-value approach. First we set some of the values we need:

```
mu = 45;
sig = 1.5;
xbar = 47.2;
% Get the observed value of test statistic.
zobs = (xbar - mu)/sig;
```

The $p$-value is the area under the curve greater than the value for **zobs**. We can find it using the following command:

```
pval = 1-normcdf(zobs,0,1);
```

We get a *p*-value of 0.071. If we are doing the hypothesis test at the 0.05 significance level, then we would not reject the null hypothesis. This is consistent with the results we had previously.
□

Note that in each approach, knowledge of the distribution of *T* under the null hypothesis $H_0$ is needed. How to tackle situations where we do not know the distribution of our statistic is the focus of the rest of the chapter.

## Confidence Intervals

In Chapter 3, we discussed several examples of estimators for population parameters such as the mean, the variance, moments, and others. We call these *point estimates.* It is unlikely that a point estimate obtained from a random sample will exactly equal the true value of the population parameter. Thus, it might be more useful to have an interval of numbers that we expect will contain the value of the parameter. This type of estimate is called an *interval estimate.* An understanding of confidence intervals is needed for the bootstrap methods covered in Section 6.4.

Let $\theta$ represent a population parameter that we wish to estimate, and let *T* denote a statistic that we will use as a point estimate for $\theta$. The observed value of the statistic is denoted as $\hat{\theta}$. An interval estimate for $\theta$ will be of the form

$$\hat{\theta}_{Lo} < \theta < \hat{\theta}_{Up}, \tag{6.3}$$

where $\hat{\theta}_{Lo}$ and $\hat{\theta}_{Up}$ depend on the observed value $\hat{\theta}$ and the distribution of the statistic *T*.

If we know the sampling distribution of *T*, then we are able to determine values for $\hat{\theta}_{Lo}$ and $\hat{\theta}_{Up}$ such that

$$P(\hat{\theta}_{Lo} < \theta < \hat{\theta}_{Up}) = 1 - \alpha, \tag{6.4}$$

where $0 < \alpha < 1$. Equation 6.4 indicates that we have a probability of $1 - \alpha$ that we will select a random sample that produces an interval that contains $\theta$. This interval (Equation 6.3) is called a $(1 - \alpha) \cdot 100\%$ confidence interval. The philosophy underlying confidence intervals is the following. Suppose we repeatedly take samples of size *n* from the population and compute the random interval given by Equation 6.3. Then the relative frequency of the intervals that contain the parameter $\theta$ would approach $(1 - \alpha) \cdot 100\%$. It should be noted that one-sided confidence intervals can be defined similarly [Mood, Graybill and Boes, 1974].

To illustrate these concepts, we use Equation 6.4 to get a confidence interval for the population mean $\mu$. Recall from Chapter 3 that we know the distribution for $\overline{X}$. We define $z^{(\alpha/2)}$ as the *z* value that has an area under the standard normal curve of size $\alpha/2$ to the left of it. In other words, we use $z^{(\alpha/2)}$ to denote that value such that

$$P(Z < z^{(\alpha/2)}) = \alpha/2.$$

Thus, the area between $z^{(\alpha/2)}$ and $z^{(1-\alpha/2)}$ is $1-\alpha$. This is shown in Figure 6.3.

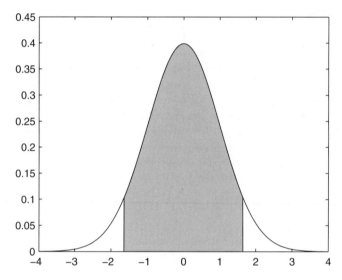

**FIGURE 6.3**
The left vertical line corresponds to $z^{(\alpha/2)}$, and the right vertical line is at $z^{(1-\alpha/2)}$. So, the non-shaded areas in the tails each have an area of $\alpha/2$, and the shaded area in the middle is $1-\alpha$.

We can see from this that the shaded area has probability $1-\alpha$, and

$$P(z^{(\alpha/2)} < Z < z^{(1-\alpha/2)}) = 1-\alpha, \tag{6.5}$$

where

$$Z = \frac{\overline{X} - \mu}{\sigma/\sqrt{n}}. \tag{6.6}$$

If we substitute this into Equation 6.5, then we have

$$P\left(z^{(\alpha/2)} < \frac{\overline{X} - \mu}{\sigma/\sqrt{n}} < z^{(1-\alpha/2)}\right) = 1 - \alpha. \tag{6.7}$$

Rearranging the inequalities in Equation 6.7, we obtain

$$P\left(\overline{X} - z^{(1-\alpha/2)}\frac{\sigma}{\sqrt{n}} < \mu < \overline{X} - z^{(\alpha/2)}\frac{\sigma}{\sqrt{n}}\right) = 1 - \alpha. \tag{6.8}$$

Comparing Equations 6.8 and 6.4, we see that

$$\hat{\theta}_{Lo} = \overline{X} - z^{(1-\alpha/2)}\frac{\sigma}{\sqrt{n}} \qquad \hat{\theta}_{Up} = \overline{X} - z^{(\alpha/2)}\frac{\sigma}{\sqrt{n}}.$$

## Example 6.5

We provide an example of finding a 95% confidence interval, using the transportation application of before. Recall that $n = 100$, $\bar{x} = 47.2$ minutes, and the standard deviation of the travel time to work is $\sigma = 15$ minutes. Since we want a 95% confidence interval, $\alpha = 0.05$.

```
mu = 45;
sig = 15;
n = 100;
alpha = 0.05;
xbar = 47.2;
```

We can get the endpoints for a 95% confidence interval as follows:

```
% Get the 95% confidence interval.
% Get the value for z_alpha/2.
zlo = norminv(1-alpha/2,0,1);
zhi = norminv(alpha/2,0,1);
thetalo = xbar - zlo*sig/sqrt(n);
thetaup = xbar - zhi*sig/sqrt(n);
```

We get a value of $\hat{\theta}_{Lo} = 44.26$ and $\hat{\theta}_{Up} = 50.14$.
□

We return to confidence intervals in Section 6.4 and Chapter 7, where we discuss bootstrap methods for obtaining them. First, however, we look at Monte Carlo methods for hypothesis testing.

## 6.3 Monte Carlo Methods for Inferential Statistics

The sampling distribution is known for many statistics. However, these are typically derived using assumptions about the underlying population under study or for large sample sizes. In many cases, we do not know the sampling distribution for the statistic, or we cannot be sure that the assumptions are satisfied. We can address these cases using Monte Carlo simulation methods, which is the topic of this section. Some of the uses of Monte Carlo simulation for inferential statistics are the following:

- Performing inference when the distribution of the test statistic is not known analytically,
- Assessing the performance of inferential methods when parametric assumptions are violated,
- Testing the null and alternative hypotheses under various conditions,
- Evaluating the performance (e.g., power) of inferential methods,
- Comparing the quality of estimators.

In this section, we cover situations in inferential statistics where we do know something about the distribution of the population our sample came from or we are willing to make assumptions about the distribution. In Section 6.4, we discuss bootstrap methods that can be used when no assumptions are made about the underlying distribution of the population.

### Basic Monte Carlo Procedure

The fundamental idea behind *Monte Carlo simulation* for inferential statistics is that insights regarding the characteristics of a statistic can be gained by repeatedly drawing random samples from the same population of interest and observing the behavior of the statistic over the samples. In other words, we estimate the distribution of the statistic by randomly sampling from the population and recording the value of the statistic for each sample. The observed values of the statistic for these samples are used to estimate the distribution.

The first step is to decide on a pseudo-population that the analyst assumes represents the real population in all relevant aspects. We use the word *pseudo* here to emphasize the fact that we obtain our samples using a computer and pseudo random numbers. For example, we might assume that the underlying population is exponentially distributed if the random variable represents the time before a part fails, or we could assume the random variable comes from a normal distribution if we are measuring IQ scores. The pseudo-population must be something we can sample from using the computer. In this text, we consider this type of Monte Carlo simulation to be a parametric technique, because we sample from a known or assumed distribution.

The basic Monte Carlo procedure is outlined here. Later, we provide procedures illustrating some specific uses of Monte Carlo simulation as applied to statistical hypothesis testing.

*PROCEDURE - BASIC MONTE CARLO SIMULATION*

1. Determine the pseudo-population or model that represents the true population of interest.
2. Use a sampling procedure to sample from the pseudo-population.
3. Calculate a value for the statistic of interest and store it.
4. Repeat steps 2 and 3 for $M$ trials.
5. Use the $M$ values found in step 4 to study the distribution of the statistic.

It is important to keep in mind, that when sampling from the pseudo-population, the analyst should ensure that all relevant characteristics reflect the statistical situation. For example, the same sample size and sampling strategy should be used when trying to understand the performance of a statistic. This means that the distribution for the statistic obtained via Monte Carlo simulation is valid only for the conditions of the·sampling procedure and the assumptions about the pseudo-population.

Note that in the last step of the Monte Carlo simulation procedure, the analyst can use the estimated distribution of the statistic to study characteristics of interest. For example, one could use this information to estimate the skewness, bias, standard deviation, kurtosis and many other characteristics.

## Monte Carlo Hypothesis Testing

Recall that in statistical hypothesis testing, we have a test statistic that provides evidence that the null hypothesis should be rejected or not. Once we observe the value of the test statistic, we decide whether or not that particular value is consistent with the null hypothesis. To make that decision, we must know the distribution of the statistic when the null hypothesis is true. Estimating the distribution of the test statistic under the null hypothesis is one of the goals of Monte Carlo hypothesis testing. We discuss and illustrate the Monte Carlo method as applied to the critical value and $p$-value approaches to hypothesis testing.

Recall that in the critical value approach to hypothesis testing, we are given a significance level $\alpha$. We then use this significance level to find the appropriate critical region in the distribution of the test statistic when the null hypothesis is true. Using the Monte Carlo method, we determine the critical value using the estimated distribution of the test statistic. The basic procedure is to randomly sample many times from the pseudo-population representing the null hypothesis, calculate the value of the test statistic at each trial, and use these values to estimate the distribution of the test statistic.

*PROCEDURE - MONTE CARLO HYPOTHESIS TESTING (CRITICAL VALUE)*

1. Using an available random sample of size $n$ from the population of interest, calculate the observed value of the test statistic, $t_o$.

2. Decide on a pseudo-population that reflects the characteristics of the true population under the null hypothesis.

3. Obtain a random sample of size $n$ from the pseudo-population.

4. Calculate the value of the test statistic using the random sample in step 3 and record it.

5. Repeat steps 3 and 4 for $M$ trials. We now have values $t_1, ..., t_M$, that serve as an estimate of the distribution of the test statistic, $T$, when the null hypothesis is true.

6. Obtain the critical value for the given significance level $\alpha$:

   Lower Tail Test: get the $\alpha$-th sample quantile, $\hat{q}_\alpha$, from the $t_1, ..., t_M$.

   Upper Tail Test: get the $(1 - \alpha)$-th sample quantile, $\hat{q}_{1-\alpha}$, from the $t_1, ..., t_M$.

   Two-Tail Test: get the sample quantiles $\hat{q}_{\alpha/2}$ and $\hat{q}_{1-\alpha/2}$ from the $t_1, ..., t_M$.

7. If $t_o$ falls in the critical region, then reject the null hypothesis.

The critical values in step 6 can be obtained using the estimate of a sample quantile that we discussed in Chapter 3. The function **csquantiles** from the Computational Statistics Toolbox is also available to find these values.

In the examples given below, we apply the Monte Carlo method to a familiar hypothesis testing situation where we are testing an hypothesis about the population mean. As we saw earlier, we can use analytical approaches for this type of test. We use this simple application in the hope that the reader will better understand the ideas of Monte Carlo hypothesis testing and then easily apply them to more complicated problems.

**Example 6.6**

This toy example illustrates the concepts of Monte Carlo hypothesis testing. The **mcdata** data set contains 25 observations. We are interested in using these data to test the following null and alternative hypotheses:

$$H_0: \quad \mu = 454$$
$$H_1: \quad \mu < 454.$$

We will perform our hypothesis test using simulation to get the critical values. We decide to use the following as our test statistic

$$z = \frac{\bar{x} - 454}{\sigma / \sqrt{n}}.$$

First, we take care of some preliminaries.

```
% Load up the data.
load mcdata
n = length(mcdata);
% Population sigma is known.
sigma = 7.8;
sigxbar = sigma/sqrt(n);
% Get the observed value of the test statistic.
Tobs = (mean(mcdata)-454)/sigxbar;
```

The observed value of the test statistic is $t_o = -2.56$. The next step is to decide on a model for the population that generated our data. We suspect that the normal distribution with $\sigma = 7.8$ is a good model, and we check this assumption using a normal probability plot. The resulting plot in Figure 6.4 shows that we can use the normal distribution as the pseudo-population.

```
% This command generates the normal probability plot.
% It is a function in the MATLAB Statistics Toolbox.
normplot(mcdata)
```

We are now ready to implement the Monte Carlo simulation. We use 1000 trials in this example. At each trial, we randomly sample from the distribution of the test statistic under the null hypothesis (the normal distribution with $\mu = 454$ and $\sigma = 7.8$) and record the value of the test statistic.

```
M = 1000;% Number of Monte Carlo trials
% Storage for test statistics from the MC trials.
Tm = zeros(1,M);
% Start the simulation.
for i = 1:M
    % Generate a random sample under H_0
    % where n is the sample size.
    xs = sigma*randn(1,n) + 454;
    Tm(i) = (mean(xs) - 454)/sigxbar;
end
```

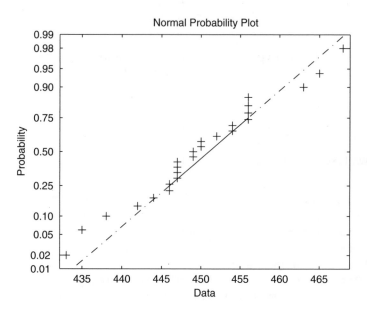

**FIGURE 6.4**
This normal probability plot for the **mcdata** data shows that assuming a normal distribution for the data is reasonable.

Now that we have the estimated distribution of the test statistic contained in the variable **Tm**, we can use that to estimate the critical value for a lower tail test.

```
% Get the critical value for alpha.
% This is a lower-tail test, so it is the
% alpha quantile.
alpha = 0.05;
cv = csquantiles(Tm,alpha);
```

We get an estimated critical value of -1.75. Since the observed value of our test statistic is $t_o = -2.56$, which is less than the estimated critical value, we reject $H_0$.
❑

The procedure for Monte Carlo hypothesis testing using the *p*-value approach is similar. Instead of finding the critical value from the simulated distribution of the test statistic, we use it to estimate the *p*-value.

PROCEDURE - MONTE CARLO HYPOTHESIS TESTING (P-VALUE)

1. For a random sample of size $n$ to be used in a statistical hypothesis test, calculate the observed value of the test statistic, $t_o$.
2. Decide on a pseudo-population that reflects the characteristics of the population under the null hypothesis.
3. Obtain a random sample of size $n$ from the pseudo-population.
4. Calculate the value of the test statistic using the random sample in step 3 and record it as $t_i$.
5. Repeat steps 3 and 4 for $M$ trials. We now have values $t_1, ..., t_M$, that serve as an estimate of the distribution of the test statistic, $T$, when the null hypothesis is true.
6. Estimate the $p$-value using the distribution found in step 5, using the following.
   Lower Tail Test:

   $$\hat{p}\text{-value} = \frac{\#(t_i \leq t_o)}{M}; \qquad i = 1, ..., M$$

   Upper Tail Test:

   $$\hat{p}\text{-value} = \frac{\#(t_i \geq t_o)}{M}; \qquad i = 1, ..., M$$

7. If $\hat{p}$-value $\leq \alpha$, then reject the null hypothesis.

## Example 6.7

We return to the situation in Example 6.6 and apply Monte Carlo simulation to the $p$-value approach to hypothesis testing. Just to change things a bit, we use the sample mean as our test statistic.

```
% Let's change the test statistic to xbar.
Tobs = mean(mcdata);
% Number of Monte Carlo trials.
M = 1000;
% Start the simulation.
Tm = zeros(1,M);
for i = 1:M
    % Generate a random sample under H_0.
    xs = sigma*randn(1,n) + 454;
    Tm(i) = mean(xs);
end
```

We find the estimated $p$-value by counting the number of observations in **Tm** that are below the value of the observed value of the test statistic and dividing by $M$.

```
% Get the p-value. This is a lower tail test.
% Find all of the values from the simulation that are
% below the observed value of the test statistic.
ind = find(Tm <= Tobs);
pvalhat = length(ind)/M;
```

We have an estimated $p$-value given by 0.007. If the significance level of our test is $\alpha = 0.05$, then we would reject the null hypothesis.
❏

## Monte Carlo Assessment of Hypothesis Testing

Monte Carlo simulation can be used to evaluate the performance of an inference model or hypothesis test in terms of the Type I error and the Type II error. For some statistics, such as the sample mean, these errors can be determined analytically. However, what if we have an inference test where the assumptions of the standard methods might be violated or the analytical methods cannot be applied? For instance, suppose we choose the critical value by using a normal approximation (when our test statistic is *not* normally distributed), and we need to assess the results of doing that? In these situations, we can use Monte Carlo simulation to estimate the Type I and the Type II error.

We first outline the procedure for estimating the Type I error. Because the Type I error occurs when we reject the null hypothesis test when it is true, we must sample from the pseudo-population that represents $H_0$.

*PROCEDURE - MONTE CARLO ASSESSMENT OF TYPE I ERROR*

1. Determine the pseudo-population when the null hypothesis is *true.*

2. Generate a random sample of size $n$ from this pseudo-population.

3. Perform the hypothesis test using the critical value.

4. Determine whether a Type I error has been committed. In other words, was the null hypothesis rejected? We know that it should not be rejected because we are sampling from the distribution according to the null hypothesis. Record the result for this trial as,

$$I_i = \begin{cases} 1; & \text{Type I error is made} \\ 0; & \text{Type I error is not made.} \end{cases}$$

5. Repeat steps 2 through 4 for $M$ trials.

6. The probability of making a Type I error is

$$\hat{\alpha} = \frac{1}{M}\sum_{i=1}^{M} I_i. \tag{6.9}$$

Note that in step 6, this is the same as calculating the proportion of times the null hypothesis is falsely rejected out of $M$ trials. This provides an estimate of the significance level of the test for a given critical value.

The procedure is similar for estimating the Type II error of a hypothesis test. However, this error is determined by sampling from the distribution when the null hypothesis is false. There are many possibilities for the Type II error, and the analyst should investigate the Type II error for those alternative hypotheses that are of interest.

*PROCEDURE - MONTE CARLO ASSESSMENT OF TYPE II ERROR*

1. Determine a pseudo-population of interest where the null hypothesis is *false*.

2. Generate a random sample of size $n$ from this pseudo-population.

3. Perform the hypothesis test using the significance level $\alpha$ and corresponding critical value.

4. Note whether a Type II error has been committed; i.e., was the null hypothesis *not* rejected? Record the result for this trial as,

$$I_i = \begin{cases} 1; & \text{Type II error is made} \\ 0; & \text{Type II error is not made.} \end{cases}$$

5. Repeat steps 2 through 4 for $M$ trials.

6. The probability of making a Type II error is

$$\hat{\beta} = \frac{1}{M}\sum_{i=1}^{M} I_i. \tag{6.10}$$

The Type II error rate is estimated using the proportion of times the null hypothesis is not rejected (when it should be) out of $M$ trials.

## Example 6.8

For the hypothesis test in Example 6.6, we had a critical value (from theory) of -1.645. We can estimate the significance level of the test using the following steps:

```
M = 1000;
alpha = 0.05;
% Get the critical value, using z as test statistic.
cv = norminv(alpha,0,1);
% Start the simulation.
Im = 0;
for i = 1:M
 % Generate a random sample under H_0.
 xs = sigma*randn(1,n) + 454;
 Tm = (mean(xs)-454)/sigxbar;
 if Tm <= cv    % then reject H_0
   Im = Im +1;
 end
end
alphahat = Im/M;
```

A critical value of -1.645 in this situation corresponds to a desired probability of Type I error of 0.05. From this simulation, we get an estimated value of 0.045, which is very close to the theoretical value. We now check the Type II error in this test. Note that we now have to sample from the alternative hypotheses of interest.

```
% Now check the probability of Type II error.
% Get some alternative hypotheses:
mualt = 445:458;
betahat = zeros(size(mualt));
for j = 1:length(mualt)
   Im = 0;
   % Get the true mean.
   mu = mualt(j);
   for i = 1:M
      % Generate a sample from H_1.
      xs = sigma*randn(1,n) + mu;
      Tm = (mean(xs)-454)/sigxbar;
      if Tm > cv    % Then did not reject H_0.
         Im = Im +1;
      end
   end
   betahat(j) = Im/M;
end
% Get the estimated power.
powhat = 1-betahat;
```

We plot the estimated power as a function of $\mu$ in Figure 6.5. As expected, as the true value for $\mu$ gets closer to 454 (the mean under the null hypothesis), the power of the test decreases.

❑

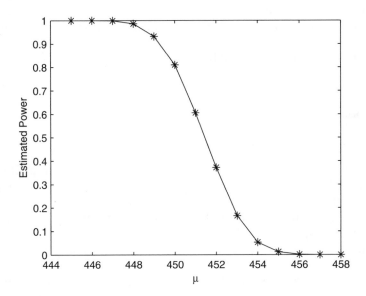

**FIGURE 6.5**
Here is the curve for the estimated power corresponding to the hypothesis test of Example 6.8.

An important point to keep in mind about the Monte Carlo simulations discussed in this section is that the experiment is applicable only for the situation that has been simulated. For example, when we assess the Type II error in Example 6.8, it is appropriate only for those alternative hypotheses, sample size and critical value. What would be the probability of Type II error, if some other departure from the null hypothesis is used in the simulation? In other cases, we might need to know whether the distribution of the statistic changes with sample size or skewness in the population or some other characteristic of interest. These variations are easily investigated using multiple Monte Carlo experiments.

One quantity that the researcher must determine is the number of trials that are needed in Monte Carlo simulations. This often depends on the computing assets that are available. If time and computer resources are not an issue, then $M$ should be made as large as possible. Hope [1968] showed that results from a Monte Carlo simulation are unbiased for any $M$, under the assumption that the programming is correct.

Mooney [1997] states that there is no general theory that governs the number of trials in Monte Carlo simulation. However, he recommends the following general guidelines. The researcher should first use a small number of trials and ensure that the program is working properly. Once the code has been checked, the simulation or experiments can be run for very large $M$.

Most simulations would have $M > 1000$, but $M$ between 10,000 and 25,000 is not uncommon. One important guideline for determining the number of trials, is the purpose of the simulation. If the tail of the distribution is of interest (e.g., estimating Type I error, getting $p$-values, etc.), then more trials are needed to ensure that there will be a good estimate of that area.

## 6.4 Bootstrap Methods

The treatment of the bootstrap methods described here comes from Efron and Tibshirani [1993]. The interested reader is referred to that text for more information on the underlying theory behind the bootstrap. There does not seem to be a consistent terminology in the literature for what techniques are considered bootstrap methods. Some refer to the resampling techniques of the previous section as bootstrap methods. Here, we use *bootstrap* to refer to Monte Carlo simulations that treat the original sample as the pseudo-population or as an estimate of the population. Thus, in the steps where we randomly sample from the pseudo-population, we now resample from the original sample.

In this section, we discuss the general bootstrap methodology, followed by some applications of the bootstrap. These include bootstrap estimates of the standard error, bootstrap estimates of bias, and bootstrap confidence intervals.

### General Bootstrap Methodology

The bootstrap is a method of Monte Carlo simulation where no parametric assumptions are made about the underlying population that generated the random sample. Instead, we use the sample as an estimate of the population. This estimate is called the empirical distribution $\hat{F}$ where each $x_i$ has probability mass $1/n$. Thus, each $x_i$ has the same likelihood of being selected in a new sample taken from $\hat{F}$.

When we use $\hat{F}$ as our pseudo-population, then we resample *with replacement* from the original sample $\mathbf{x} = (x_1, \ldots, x_n)$. We denote the new sample obtained in this manner by $\mathbf{x}^* = (x_1^*, \ldots, x_n^*)$. Since we are sampling with replacement from the original sample, there is a possibility that some points $x_i$ will appear more than once in $\mathbf{x}^*$ or maybe not at all. We are looking at the univariate situation, but the bootstrap concepts can also be applied in the $d$-dimensional case.

A small example serves to illustrate these ideas. Let's say that our random sample consists of the four numbers $\mathbf{x} = (5, 8, 3, 2)$. The following are possible samples $\mathbf{x}^*$, when we sample with replacement from $\mathbf{x}$:

$$\mathbf{x}^{*1} = (x_4, x_4, x_2, x_1) = (2, 2, 8, 5)$$

$$\mathbf{x}^{*2} = (x_4, x_2, x_3, x_4) = (2, 8, 3, 2).$$

We use the notation $\mathbf{x}^{*b}$, $b = 1, ..., B$ for the $b$-th bootstrap data set.

In many situations, the analyst is interested in estimating some parameter $\theta$ by calculating a statistic from the random sample. We denote this estimate by

$$\hat{\theta} = T = t(x_1, ..., x_n). \tag{6.11}$$

We might also like to determine the standard error in the estimate $\hat{\theta}$ and the bias. The bootstrap method can provide an estimate of this when analytical methods fail. The method is also suitable for situations when the estimator $\hat{\theta} = t(x)$ is complicated.

To get estimates of bias or standard error of a statistic, we obtain $B$ bootstrap samples by sampling with replacement from the original sample. For every bootstrap sample, we calculate the same statistic to obtain the **bootstrap replications** of $\hat{\theta}$, as follows

$$\hat{\theta}^{*b} = t(\mathbf{x}^{*b}); \qquad b = 1, ..., B. \tag{6.12}$$

These $B$ bootstrap replicates provide us with an estimate of the distribution of $\hat{\theta}$. This is similar to what we did in the previous section, except that we are not making any assumptions about the distribution for the original sample. Once we have the bootstrap replicates in Equation 6.12, we can use them to understand the distribution of the estimate.

The steps for the basic bootstrap methodology are given here, with detailed procedures for finding specific characteristics of $\hat{\theta}$ provided later. The issue of how large to make $B$ is addressed with each application of the bootstrap.

*PROCEDURE - BASIC BOOTSTRAP*

1. Given a random sample, $\mathbf{x} = (x_1, ..., x_n)$, calculate $\hat{\theta}$.
2. Sample with replacement from the original sample to get $\mathbf{x}^{*b} = (x_1^{*b}, ..., x_n^{*b})$.
3. Calculate the same statistic using the bootstrap sample in step 2 to get, $\hat{\theta}^{*b}$.
4. Repeat steps 2 through 3, $B$ times.
5. Use this estimate of the distribution of $\hat{\theta}$ (i.e., the bootstrap replicates) to obtain the desired characteristic (e.g., standard error, bias or confidence interval).

Efron and Tibshirani [1993] discuss a method called the **parametric boot-strap**. In this case, the data analyst makes an assumption about the distribution that generated the original sample. Parameters for that distribution are estimated from the sample, and resampling (in step 2) is done using the assumed distribution and the estimated parameters. The parametric bootstrap is closer to the Monte Carlo methods described in the previous section.

For instance, say we have reason to believe that the data come from an exponential distribution with parameter $\lambda$. We need to estimate the variance and use

$$\hat{\theta} = \frac{1}{n} \sum_{i=1}^{n} (x_i - \bar{x})^2 \tag{6.13}$$

as the estimator. We can use the parametric bootstrap as outlined above to understand the behavior of $\hat{\theta}$. Since we assume an exponential distribution for the data, we estimate the parameter $\lambda$ from the sample to get $\hat{\lambda}$. We then resample from an exponential distribution with parameter $\hat{\lambda}$ to get the bootstrap samples. The reader is asked to implement the parametric bootstrap in the exercises.

## Bootstrap Estimate of Standard Error

When our goal is to estimate the standard error of $\hat{\theta}$ using the bootstrap method, we proceed as outlined in the previous procedure. Once we have the estimated distribution for $\hat{\theta}$, we use it to estimate the standard error for $\hat{\theta}$. This estimate is given by

$$\hat{SE}_B(\hat{\theta}) = \left\{ \frac{1}{B-1} \sum_{b=1}^{B} (\hat{\theta}^{*b} - \bar{\hat{\theta}}^{*})^2 \right\}^{\frac{1}{2}}, \tag{6.14}$$

where

$$\bar{\hat{\theta}}^{*} = \frac{1}{B} \sum_{b=1}^{B} \hat{\theta}^{*b}. \tag{6.15}$$

Note that Equation 6.14 is just the sample standard deviation of the bootstrap replicates, and Equation 6.15 is the sample mean of the bootstrap replicates.

Efron and Tibshirani [1993] show that the number of bootstrap replicates $B$ should be between 50 and 200 when estimating the standard error of a statistic. Often the choice of $B$ is dictated by the computational complexity of $\hat{\theta}$, the sample size $n$, and the computer resources that are available. Even using a small value of $B$, say $B = 25$, the analyst will gain information about the variability of $\hat{\theta}$. In most cases, taking more than 200 bootstrap replicates to estimate the standard error is unnecessary.

The procedure for finding the bootstrap estimate of the standard error is given here and is illustrated in Example 6.9

*PROCEDURE - BOOTSTRAP ESTIMATE OF THE STANDARD ERROR*

1. Given a random sample, $\mathbf{x} = (x_1, ..., x_n)$, calculate the statistic $\hat{\theta}$.
2. Sample with replacement from the original sample to get $\mathbf{x}^{*b} = (x_1^{*b}, ..., x_n^{*b})$.
3. Calculate the same statistic using the sample in step 2 to get the bootstrap replicates, $\hat{\theta}^{*b}$.
4. Repeat steps 2 through 3, $B$ times.
5. Estimate the standard error of $\hat{\theta}$ using Equations 6.14 and 6.15.

## Example 6.9

The lengths of the forearm (in inches) of 140 adult males are contained in the file **forearm** [Hand, et al., 1994]. We use these data to estimate the skewness of the population. We then estimate the standard error of this statistic using the bootstrap method. First we load the data and calculate the skewness.

```
load forearm
% Sample with replacement from this.
% First get the sample size.
n = length(forearm);
B = 100;% number of bootstrap replicates
% Get the value of the statistic of interest.
theta = skewness(forearm);
```

The estimated skewness in the **forearm** data is -0.11. To implement the bootstrap, we use the MATLAB Statistics Toolbox function **unidrnd** to sample with replacement from the original sample. The corresponding function from the Computational Statistics Toolbox can also be used. The output from this function will be indices from 1 to $n$ that point to what observations have been selected for the bootstrap sample.

```
% Use unidrnd to get the indices to the resamples.
% Note that each column corresponds to indices
% for a bootstrap resample.
inds = unidrnd(n,n,B);
% Extract these from the data.
xboot = forearm(inds);
% We can get the skewness for each column using the
% MATLAB Statistics Toolbox function skewness.
```

```
thetab = skewness(xboot);
seb = std(thetab);
```

From this we get an estimated standard error in the skewness of 0.14. Efron and Tibshirani [1993] recommend that one look at histograms of the bootstrap replicates as a useful tool for understanding the distribution of $\hat{\theta}$. We show the histogram in Figure 6.6.

The MATLAB Statistics Toolbox has a function called **bootstrp** that returns the bootstrap replicates. We now show how to get the bootstrap estimate of standard error using this function.

```
% Now show how to do it with MATLAB Statistics Toolbox
% function: bootstrp.
Bmat = bootstrp(B,'skewness',forearm);
% What we get back are the bootstrap replicates.
% Get an estimate of the standard error.
sebmat = std(Bmat);
```

Note that one of the arguments to **bootstrp** is a string representing the function that calculates the statistics. From this, we get an estimated standard error of 0.12.
□

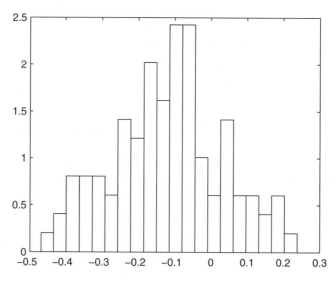

**FIGURE 6.6**
This is a histogram for the bootstrap replicates in Example 6.9. This shows the estimated distribution of the sample skewness of the **forearm** data.

## Bootstrap Estimate of Bias

The standard error of an estimate is one measure of its performance. Bias is another quantity that measures the statistical accuracy of an estimate. From Chapter 3, the bias is defined as the difference between the expected value of the statistic and the parameter,

$$\text{bias}(T) = E[T] - \theta. \tag{6.16}$$

The expectation in Equation 6.16 is taken with respect to the true distribution $F$. To get the bootstrap estimate of bias, we use the empirical distribution $\hat{F}$ as before. We resample from the empirical distribution and calculate the statistic using each bootstrap resample, yielding the bootstrap replicates $\hat{\theta}^{*b}$. We use these to estimate the bias from the following:

$$\hat{\text{bias}}_B = \overline{\hat{\theta}^*} - \hat{\theta}, \tag{6.17}$$

where $\overline{\hat{\theta}^*}$ is given by the mean of the bootstrap replicates (Equation 6.15).

Presumably, one is interested in the bias in order to correct for it. The bias-corrected estimator is given by

$$\widehat{\theta} = \hat{\theta} - \hat{\text{bias}}_B. \tag{6.18}$$

Using Equation 6.17 in Equation 6.18, we have

$$\widehat{\theta} = 2\hat{\theta} - \overline{\hat{\theta}^*}. \tag{6.19}$$

More bootstrap samples are needed to estimate the bias, than are required to estimate the standard error. Efron and Tibshirani [1993] recommend that $B \geq 400$.

It is useful to have an estimate of the bias for $\hat{\theta}$, but caution should be used when correcting for the bias. Equation 6.19 will hopefully yield a less biased estimate, but it could turn out that $\widehat{\theta}$ will have a larger variation or standard error. It is recommended that if the estimated bias is small relative to the estimate of standard error (both of which can be estimated using the bootstrap method), then the analyst should not correct for the bias [Efron and Tibshirani, 1993]. However, if this is not the case, then perhaps some other, less biased, estimator should be used to estimate the parameter $\theta$.

*PROCEDURE - BOOTSTRAP ESTIMATE OF THE BIAS*

1. Given a random sample, $\mathbf{x} = (x_1, ..., x_n)$, calculate the statistic $\hat{\theta}$.

2. Sample with replacement from the original sample to get
   $\mathbf{x}^{*b} = (x_1^{*b}, ..., x_n^{*b})$.

3. Calculate the same statistic using the sample in step 2 to get the
   bootstrap replicates, $\hat{\theta}^{*b}$.

4. Repeat steps 2 through 3, $B$ times.

5. Using the bootstrap replicates, calculate $\overline{\hat{\theta}^*}$.

6. Estimate the bias of $\hat{\theta}$ using Equation 6.17.

## Example 6.10

We return to the `forearm` data of Example 6.9, where now we want to estimate the bias in the sample skewness. We use the same bootstrap replicates
as before, so all we have to do is to calculate the bias using Equation 6.17.

```
% Use the same replicates from before.
% Evaluate the mean using Equation 6.15.
meanb = mean(thetab);
% Now estimate the bias using Equation 6.17.
biasb = meanb - theta;
```

We have an estimated bias of -0.011. Note that this is small relative to the standard error.
□

In the next chapter, we discuss another method for estimating the bias and
the standard error of a statistic called the jackknife. The jackknife method is
related to the bootstrap. However, since it is based on the reuse or partitioning of the original sample rather than resampling, we do not include it here.

## Bootstrap Confidence Intervals

There are several ways of constructing confidence intervals using the bootstrap. We discuss three of them here: the standard interval, the bootstrap-*t*
interval and the percentile method. Because it uses the jackknife procedure,
an improved bootstrap confidence interval called the $BC_a$ will be presented
in the next chapter.

### Bootstrap Standard Confidence Interval

The *bootstrap standard confidence interval* is based on the parametric form
of the confidence interval that was discussed in Section 6.2. We showed that
the $(1 - \alpha) \cdot 100\%$ confidence interval for the mean can be found using

$$P\left(\overline{X} - z^{(1-\alpha/2)}\frac{\sigma}{\sqrt{n}} < \mu < \overline{X} - z^{(\alpha/2)}\frac{\sigma}{\sqrt{n}}\right) = 1 - \alpha. \qquad (6.20)$$

Similar to this, the bootstrap standard confidence interval is given by

$$(\hat{\theta} - z^{(1-\alpha/2)}SE_{\hat{\theta}}, \hat{\theta} - z^{(\alpha/2)}SE_{\hat{\theta}}), \qquad (6.21)$$

where $SE_{\hat{\theta}}$ is the standard error for the statistic $\hat{\theta}$ obtained using the bootstrap [Mooney and Duval, 1993]. The confidence interval in Equation 6.21 can be used when the distribution for $\hat{\theta}$ is normally distributed or the normality assumption is plausible. This is easily coded in MATLAB using previous results and is left as an exercise for the reader.

### Bootstrap-t Confidence Interval
The second type of confidence interval using the bootstrap is called the **bootstrap-t**. We first generate $B$ bootstrap samples, and for each bootstrap sample the following quantity is computed:

$$z^{*b} = \frac{\hat{\theta}^{*b} - \hat{\theta}}{\hat{SE}^{*b}}. \qquad (6.22)$$

As before, $\hat{\theta}^{*b}$ is the bootstrap replicate of $\hat{\theta}$, but $\hat{SE}^{*b}$ is the estimated standard error of $\hat{\theta}^{*b}$ for that bootstrap sample. If a formula exists for the standard error of $\hat{\theta}^{*b}$, then we can use that to determine the denominator of Equation 6.22. For instance, if $\hat{\theta}$ is the mean, then we can calculate the standard error as explained in Chapter 3. However, in most situations where we have to resort to using the bootstrap, these formulas are not available. One option is to use the bootstrap method of finding the standard error, keeping in mind that you are estimating the standard error of $\hat{\theta}^{*b}$ using the bootstrap sample $x^{*b}$. In other words, one resamples with replacement from the bootstrap sample $x^{*b}$ to get an estimate of $\hat{SE}^{*b}$.

Once we have the $B$ bootstrapped $z^{*b}$ values from Equation 6.22, the next step is to estimate the quantiles needed for the endpoints of the interval. The $\alpha/2$-th quantile, denoted by $\hat{t}^{(\alpha/2)}$ of the $z^{*b}$, is estimated by

$$\alpha/2 = \frac{\#(z^{*b} \leq \hat{t}^{(\alpha/2)})}{B}. \qquad (6.23)$$

This says that the estimated quantile is the $\hat{t}^{(\alpha/2)}$ such that $100 \cdot \alpha/2$ % of the points $z^{*b}$ are less than this number. For example, if $B = 100$ and $\alpha/2 = 0.05$, then $\hat{t}^{(0.05)}$ could be estimated as the fifth largest value of the $z^{*b}$ $(B \cdot \alpha/2 = 100 \cdot 0.05 = 5)$. One could also use the quantile estimates discussed previously in Chapter 3 or some other suitable estimate.

We are now ready to calculate the bootstrap-$t$ confidence interval. This is given by

$$(\hat{\theta} - \hat{t}^{(1-\alpha/2)} \cdot \hat{SE}_{\hat{\theta}}, \hat{\theta} - \hat{t}^{(\alpha/2)} \cdot \hat{SE}_{\hat{\theta}}), \tag{6.24}$$

where $\hat{SE}$ is an estimate of the standard error of $\hat{\theta}$. The bootstrap-$t$ interval is suitable for location statistics such as the mean or quantiles. However, its accuracy for more general situations is questionable [Efron and Tibshirani, 1993]. The next method based on the bootstrap percentiles is more reliable.

*PROCEDURE - BOOTSTRAP-T CONFIDENCE INTERVAL*

1. Given a random sample, $\mathbf{x} = (x_1, ..., x_n)$, calculate $\hat{\theta}$.
2. Sample with replacement from the original sample to get $\mathbf{x}^{*b} = (x_1^{*b}, ..., x_n^{*b})$.
3. Calculate the same statistic using the sample in step 2 to get $\hat{\theta}^{*b}$.
4. Use the bootstrap sample $\mathbf{x}^{*b}$ to get the standard error of $\hat{\theta}^{*b}$. This can be calculated using a formula or estimated by the bootstrap.
5. Calculate $z^{*b}$ using the information found in steps 3 and 4.
6. Repeat steps 2 through 5, $B$ times, where $B \geq 1000$.
7. Order the $z^{*b}$ from smallest to largest. Find the quantiles $\hat{t}^{(1-\alpha/2)}$ and $\hat{t}^{(\alpha/2)}$.
8. Estimate the standard error $\hat{SE}_{\hat{\theta}}$ of $\hat{\theta}$ using the $B$ bootstrap replicates of $\hat{\theta}^{*b}$ (from step 3).
9. Use Equation 6.24 to get the confidence interval.

The number of bootstrap replicates that are needed is quite large for confidence intervals. It is recommended that $B$ should be 1000 or more. If no formula exists for calculating the standard error of $\hat{\theta}^{*b}$, then the bootstrap method can be used. This means that there are two levels of bootstrapping: one for finding the $\hat{SE}^{*b}$ and one for finding the $z^{*b}$, which can greatly increase the computational burden. For example, say that $B = 1000$ and we use 50 bootstrap replicates to find $\hat{SE}^{*b}$, then this results in a total of 50,000 resamples.

**Example 6.11**

Say we are interested in estimating the variance of the **forearm** data, and we decide to use the following statistic,

$$\hat{\sigma}^2 = \frac{1}{n} \sum_{i=1}^{n} (X_i - \bar{X})^2,$$

which is the sample second central moment. We write our own simple function called **mom** (included in the Computational Statistics Toolbox) to estimate this.

```
% This function will calculate the sample 2nd
% central moment for a given sample vector x.
function mr = mom(x)
n = length(x);
mu = mean(x);
mr = (1/n)*sum((x-mu).^2);
```

We use this function as an input argument to **bootstrp** to get the bootstrap-*t* confidence interval. The MATLAB code given below also shows how to get the bootstrap estimate of standard error for each bootstrap sample. First we load the data and get the observed value of the statistic.

```
load forearm
n = length(forearm);
alpha = 0.1;
B = 1000;
thetahat = mom(forearm);
```

Now we get the bootstrap replicates using the function **bootstrp**. One of the optional output arguments from this function is a matrix of indices for the resamples. As shown below, each column of the output **bootsam** contains the indices to a bootstrap sample. We loop through all of the bootstrap samples to estimate the standard error of the bootstrap replicate using that resample.

```
% Get the bootstrap replicates and samples.
[bootreps, bootsam] = bootstrp(B,'mom',forearm);
% Set up some storage space for the SE's.
sehats = zeros(size(bootreps));
% Each column of bootsam contains indices
% to a bootstrap sample.
for i = 1:B
    % Extract the sample from the data.
    xstar = forearm(bootsam(:,i));
    bvals(i) = mom(xstar);
    % Do bootstrap using that sample to estimate SE.
    sehats(i) = std(bootstrp(25,'mom',xstar));
end
zvals = (bootreps - thetahat)./sehats;
```

Then we get the estimate of the standard error that we need for the endpoints of the interval.

```
% Estimate the SE using the bootstrap.
SE = std(bootreps);
```

Now we get the quantiles that we need for the interval given in Equation 6.24 and calculate the interval.

```
% Get the quantiles.
k = B*alpha/2;
szval = sort(zvals);
tlo = szval(k);
thi = szval(B-k);
% Get the endpoints of the interval.
blo = thetahat - thi*SE;
bhi = thetahat - tlo*SE;
```

The bootstrap-*t* interval for the variance of the **forearm** data is $(1.00, 1.57)$.
❑


### Bootstrap Percentile Interval

An improved bootstrap confidence interval is based on the quantiles of the distribution of the bootstrap replicates. This technique has the benefit of being more stable than the bootstrap-*t*, and it also enjoys better theoretical coverage properties [Efron and Tibshirani, 1993]. The *bootstrap percentile confidence interval* is

$$(\hat{\theta}_B^{*(\alpha/2)}, \hat{\theta}_B^{*(1-\alpha/2)}),\qquad(6.25)$$

where $\hat{\theta}_B^{*(\alpha/2)}$ is the $\alpha/2$ quantile in the bootstrap distribution of $\hat{\theta}^*$. For example, if $\alpha/2 = 0.025$ and $B = 1000$, then $\hat{\theta}_B^{*(0.025)}$ is the $\hat{\theta}^{*b}$ in the 25th position of the ordered bootstrap replicates. Similarly, $\hat{\theta}_B^{*(0.975)}$ is the replicate in position 975. As discussed previously, some other suitable estimate for the quantile can be used.

The procedure is the same as the general bootstrap method, making it easy to understand and to implement. We outline the steps below.

*PROCEDURE - BOOTSTRAP PERCENTILE INTERVAL*

1. Given a random sample, $\mathbf{x} = (x_1, ..., x_n)$, calculate $\hat{\theta}$.
2. Sample with replacement from the original sample to get $\mathbf{x}^{*b} = (x_1^{*b}, ..., x_n^{*b})$.
3. Calculate the same statistic using the sample in step 2 to get the bootstrap replicates, $\hat{\theta}^{*b}$.
4. Repeat steps 2 through 3, $B$ times, where $B \geq 1000$.
5. Order the $\hat{\theta}^{*b}$ from smallest to largest.
6. Calculate $B \cdot \alpha/2$ and $B \cdot (1 - \alpha/2)$.

7. The lower endpoint of the interval is given by the bootstrap replicate that is in the $B \cdot \alpha/2$-th position of the ordered $\hat{\theta}^{*b}$, and the upper endpoint is given by the bootstrap replicate that is in the $B \cdot (1 - \alpha/2)$-th position of the same ordered list. Alternatively, using quantile notation, the lower endpoint is the estimated quantile $\hat{q}_{\alpha/2}$ and the upper endpoint is the estimated quantile $\hat{q}_{1-\alpha/2}$, where the estimates are taken from the bootstrap replicates.

## Example 6.12

Let's find the bootstrap percentile interval for the same **forearm** data. The confidence interval is easily found from the bootstrap replicates, as shown below.

```
% Use Statistics Toolbox function
% to get the bootstrap replicates.
bvals = bootstrp(B,'mom',forearm);
% Find the upper and lower endpoints
k = B*alpha/2;
sbval = sort(bvals);
blo = sbval(k);
bhi = sbval(B-k);
```

This interval is given by $(1.03, 1.45)$, which is slightly narrower than the bootstrap-$t$ interval from Example 6.11.
□

So far, we discussed three types of bootstrap confidence intervals. The standard interval is the easiest and assumes that $\hat{\theta}$ is normally distributed. The bootstrap-$t$ interval estimates the standardized version of $\hat{\theta}$ from the data, avoiding the normality assumptions used in the standard interval. The percentile interval is simple to calculate and obtains the endpoints directly from the bootstrap estimate of the distribution for $\hat{\theta}$. It has another advantage in that it is range-preserving. This means that if the parameter $\theta$ can take on values in a certain range, then the confidence interval will reflect that. This is not always the case with the other intervals.

According to Efron and Tibshirani [1993], the bootstrap-$t$ interval has good coverage probabilities, but does not perform well in practice. The bootstrap percentile interval is more dependable in most situations, but does not enjoy the good coverage property of the bootstrap-$t$ interval. There is another bootstrap confidence interval, called the $BC_a$ interval, that has both good coverage and is dependable. This interval is described in the next chapter.

The bootstrap estimates of bias and standard error are also random variables, and they have their own error associated with them. So, how accurate are they? In the next chapter, we discuss how one can use the jackknife method to evaluate the error in the bootstrap estimates.

As with any method, the bootstrap is not appropriate in every situation. When analytical methods are available to understand the uncertainty associated with an estimate, then those are more efficient than the bootstrap. In what situations should the analyst use caution in applying the bootstrap? One important assumption that underlies the theory of the bootstrap is the notion that the empirical distribution function is representative of the true population distribution. If this is not the case, then the bootstrap will not yield reliable results. For example, this can happen when the sample size is small or the sample was not gathered using appropriate random sampling techniques. Chernick [1999] describes other examples from the literature where the bootstrap should not be used. We also address a situation in Chapter 7 where the bootstrap fails. This can happen when the statistic is non-smooth, such as the median.

## 6.5 MATLAB Code

We include several functions with the Computational Statistics Toolbox that implement some of the bootstrap techniques discussed in this chapter. These are listed in Table 6.2. Like **bootstrp**, these functions have an input argument that specifies a MATLAB function that calculates the statistic.

TABLE 6.2

List of MATLAB Functions for Chapter 6

| Purpose | MATLAB Function |
|---|---|
| General bootstrap: resampling, estimates of standard error and bias | csboot bootstrp |
| Constructing bootstrap confidence Intervals | csbootint csbooperint csbootbca |

As we saw in the examples, the MATLAB Statistics Toolbox has a function called **bootstrp** that will return the bootstrap replicates from the input argument **bootfun** (e.g., **mean, std, var,** etc.). It takes an input data set, finds the bootstrap resamples, applies the **bootfun** to the resamples, and stores the replicate in the first row of the output argument. The user can get two outputs from the function: the bootstrap replicates and the indices that correspond to the points selected in the resample.

There is a Bootstrap MATLAB Toolbox written by Zoubir and Iskander at the Curtin University of Technology. It is available for download at

**www.atri.curtin.edu.au/csp**. It requires the MATLAB Statistics Toolbox and has a postscript version of the reference manual.

Other software exists for Monte Carlo simulation as applied to statistics. The Efron and Tibshirani [1993] book has a description of S code for implementing the bootstrap. This code, written by the authors, can be downloaded from the statistics archive at Carnegie-Mellon University that was mentioned in Chapter 1. Another software package that has some of these capabilities is called Resampling Stats® [Simon, 1999], and information on this can be found at **www.resample.com**. Routines are available from Resampling Stats for MATLAB [Kaplan, 1999] and Excel.

---

## 6.6 Further Reading

Mooney [1997] describes Monte Carlo simulation for inferential statistics that is written in a way that is accessible to most data analysts. It has some excellent examples of using Monte Carlo simulation for hypothesis testing using multiple experiments, assessing the behavior of an estimator, and exploring the distribution of a statistic using graphical techniques. The text by Gentle [1998] has a chapter on performing Monte Carlo studies in statistics. He discusses how simulation can be considered as a scientific experiment and should be held to the same high standards. Hoaglin and Andrews [1975] provide guidelines and standards for reporting the results from computations. Efron and Tibshirani [1991] explain several computational techniques, written at a level accessible to most readers. Other articles describing Monte Carlo inferential methods can be found in Joeckel [1991], Hope [1968], Besag and Diggle [1977], Diggle and Gratton [ 1984], Efron [1979], Efron and Gong [1983], and Teichroew [1965].

There has been a lot of work in the literature on bootstrap methods. Perhaps the most comprehensive and easy to understand treatment of the topic can be found in Efron and Tibshirani [1993]. Efron's [1982] earlier monogram on resampling techniques describes the jackknife, the bootstrap and cross-validation. A more recent book by Chernick [1999] gives an updated description of results in this area, and it also has an extensive bibliography (over 1,600 references!) on the bootstrap. Hall [1992] describes the connection between Edgeworth expansions and the bootstrap. A volume of papers on the bootstrap was edited by LePage and Billard [1992], where many applications of the bootstrap are explored. Politis, Romano, and Wolf [1999] present subsampling as an alternative to the bootstrap. A subset of articles that present the theoretical justification for the bootstrap are Efron [1981, 1985, 1987]. The paper by Boos and Zhang [2000] looks at a way to ease the computational burden of Monte Carlo estimation of the power of tests that uses resampling methods. For a nice discussion on the coverage of the bootstrap percentile confidence interval, see Polansky [1999].

___

## Exercises

6.1. Repeat Example 6.1 where the population standard deviation for the travel times to work is $\sigma_X = 5$ minutes. Is $\bar{x} = 47.2$ minutes still consistent with the null hypothesis?

6.2. Using the information in Example 6.3, plot the probability of Type II error as a function of $\mu$. How does this compare with Figure 6.2?

6.3. Would you reject the null hypothesis in Example 6.4 if $\alpha = 0.10$?

6.4. Using the same value for the sample mean, repeat Example 6.3 for different sample sizes of $n = 50, 100, 200$. What happens to the curve showing the power as a function of the true mean as the sample size changes?

6.5. Repeat Example 6.6 using a two-tail test. In other words, test for the alternative hypothesis that the mean is not equal to 454.

6.6. Repeat Example 6.8 for larger $M$. Does the estimated Type I error get closer to the true value?

6.7. Write MATLAB code that implements the parametric bootstrap. Test it using the **forearm** data. Assume that the normal distribution is a reasonable model for the data. Use your code to get a bootstrap estimate of the standard error and the bias of the coefficient of skewness and the coefficient of kurtosis. Get a bootstrap percentile interval for the sample central second moment using your parametric bootstrap approach.

6.8. Write MATLAB code that will get the bootstrap standard confidence interval. Use it with the **forearm** data to get a confidence interval for the sample central second moment. Compare this interval with the ones obtained in the examples and in the previous problem.

6.9. Use your program from problem 6.8 and the **forearm** data to get a bootstrap confidence interval for the mean. Compare this to the theoretical one.

6.10. The **remiss** data set contains the remission times for 42 leukemia patients. Some of the patients were treated with the drug called 6-mercaptopurine (**mp**), and the rest were part of the control group (**control**). Use the techniques from Chapter 5 to help determine a suitable model (e.g., Weibull, exponential, etc.) for each group. Devise a Monte Carlo hypothesis test to test for the equality of means between the two groups [Hand, et al., 1994; Gehan, 1965]. Use the $p$-value approach.

6.11. Load the **lawpop** data set [Efron and Tibshirani, 1993]. These data contain the average scores on the LSAT (**lsat**) and the corresponding average undergraduate grade point average (**gpa**) for the 1973 freshman class at 82 law schools. Note that these data constitute the entire population. The data contained in **law** comprise a random sample of 15 of these classes. Obtain the true population variances for the **lsat** and the **gpa**. Use the sample in **law** to estimate the population variance using the sample central second moment. Get bootstrap estimates of the standard error and the bias in your estimate of the variance. Make some comparisons between the known population variance and the estimated variance.

6.12. Using the **lawpop** data, devise a test statistic to test for the significance of the correlation between the LSAT scores and the corresponding grade point averages. Get a random sample from the population, and use that sample to test your hypothesis. Do a Monte Carlo simulation of the Type I and Type II error of the test you devise.

6.13. In 1961, 16 states owned the retail liquor stores. In 26 others, the stores were owned by private citizens. The data contained in **whisky** reflect the price (in dollars) of a fifth of whisky from these 42 states. Note that this represents the population, not a sample. Use the **whisky** data to get an appropriate bootstrap confidence interval for the median price of whisky at the state owned stores and the median price of whisky at the privately owned stores. First get the random sample from each of the populations, and then use the bootstrap with that sample to get the confidence intervals. Do a Monte Carlo study where you compare the confidence intervals for different sample sizes. Compare the intervals with the known population medians [Hand, et al., 1994].

6.14. The **quakes** data [Hand, et al., 1994] give the time in days between successive earthquakes. Use the bootstrap to get an appropriate confidence interval for the average time between earthquakes.

# Chapter 7

## Data Partitioning

---

## 7.1 Introduction

In this book, data partitioning refers to procedures where some observations from the sample are removed as part of the analysis. These techniques are used for the following purposes:

- To evaluate the accuracy of the model or classification scheme;
- To decide what is a reasonable model for the data;
- To find a smoothing parameter in density estimation;
- To estimate the bias and error in parameter estimation;
- And many others.

We start off with an example to motivate the reader. We have a sample where we measured the average atmospheric temperature and the corresponding amount of steam used per month [Draper and Smith, 1981]. Our goal in the analysis is to model the relationship between these variables. Once we have a model, we can use it to predict how much steam is needed for a given average monthly temperature. The model can also be used to gain understanding about the structure of the relationship between the two variables.

The problem then is deciding what model to use. To start off, one should always look at a scatterplot (or scatterplot matrix) of the data as discussed in Chapter 5. The scatterplot for these data is shown in Figure 7.1 and is examined in Example 7.3. We see from the plot that as the temperature increases, the amount of steam used per month decreases. It appears that using a line (i.e., a first degree polynomial) to model the relationship between the variables is not unreasonable. However, other models might provide a better fit. For example, a cubic or some higher degree polynomial might be a better model for the relationship between average temperature and steam usage.

So, how can we decide which model is better? To make that decision, we need to assess the accuracy of the various models. We could then choose the model that has the best accuracy or lowest error. In this chapter, we use the prediction error (see Equation 7.5) to measure the accuracy. One way to assess the error would be to observe new data (average temperature and corresponding monthly steam usage) and then determine what is the predicted monthly steam usage for the new observed average temperatures. We can compare this prediction with the true steam used and calculate the error. We do this for all of the proposed models and pick the model with the smallest error. The problem with this approach is that it is sometimes impossible to obtain new data, so all we have available to evaluate our models (or our statistics) is the original data set. In this chapter, we consider two methods that allow us to use the data already in hand for the evaluation of the models. These are cross-validation and the jackknife.

Cross-validation is typically used to determine the classification error rate for pattern recognition applications or the prediction error when building models. In Chapter 9, we will see two applications of cross-validation where it is used to select the best classification tree and to estimate the misclassification rate. In this chapter, we show how cross-validation can be used to assess the prediction accuracy in a regression problem.

In the previous chapter, we covered the bootstrap method for estimating the bias and standard error of statistics. The jackknife procedure has a similar purpose and was developed prior to the bootstrap [Quenouille,1949]. The connection between the methods is well known and is discussed in the literature [Efron and Tibshirani, 1993; Efron, 1982; Hall, 1992]. We include the jackknife procedure here, because it is more a data partitioning method than a simulation method such as the bootstrap. We return to the bootstrap at the end of this chapter, where we present another method of constructing bootstrap confidence intervals using the jackknife. In the last section, we show how the jackknife method can be used to assess the error in our bootstrap estimates.

## 7.2 Cross-Validation

Often, one of the jobs of a statistician or engineer is to create models using sample data, usually for the purpose of making predictions. For example, given a data set that contains the drying time and the tensile strength of batches of cement, can we model the relationship between these two variables? We would like to be able to predict the tensile strength of the cement for a given drying time that we will observe in the future. We must then decide what model best describes the relationship between the variables and estimate its accuracy.

Unfortunately, in many cases the naive researcher will build a model based on the data set and then use that same data to assess the performance of the model. The problem with this is that the model is being evaluated or tested with data it has already seen. Therefore, that procedure will yield an overly optimistic (i.e., low) prediction error (see Equation 7.5). Cross-validation is a technique that can be used to address this problem by iteratively partitioning the sample into two sets of data. One is used for building the model, and the other is used to test it.

We introduce cross-validation in a linear regression application, where we are interested in estimating the expected prediction error. We use linear regression to illustrate the cross-validation concept, because it is a topic that most engineers and data analysts should be familiar with. However, before we describe the details of cross-validation, we briefly review the concepts in linear regression. We will return to this topic in Chapter 10, where we discuss methods of nonlinear regression.

Say we have a set of data, $(X_i, Y_i)$, where $X_i$ denotes a **predictor variable** and $Y_i$ represents the corresponding **response variable**. We are interested in modeling the dependency of $Y$ on $X$. The easiest example of linear regression is in situations where we can fit a straight line between $X$ and $Y$. In Figure 7.1, we show a scatterplot of 25 observed $(X_i, Y_i)$ pairs [Draper and Smith, 1981]. The $X$ variable represents the average atmospheric temperature measured in degrees Fahrenheit, and the $Y$ variable corresponds to the pounds of steam used per month. The scatterplot indicates that a straight line is a reasonable model for the relationship between these variables. We will use these data to illustrate linear regression.

The linear, first-order model is given by

$$Y = \beta_0 + \beta_1 X + \varepsilon, \tag{7.1}$$

where $\beta_0$ and $\beta_1$ are parameters that must be estimated from the data, and $\varepsilon$ represents the error in the measurements. It should be noted that the word **linear** refers to the linearity of the parameters $\beta_i$. The **order** (or **degree**) of the model refers to the highest power of the predictor variable $X$. We know from elementary algebra that $\beta_1$ is the slope and $\beta_0$ is the $y$-intercept. As another example, we represent the linear, second-order model by

$$Y = \beta_0 + \beta_1 X + \beta_2 X^2 + \varepsilon. \tag{7.2}$$

To get the model, we need to estimate the parameters $\beta_0$ and $\beta_1$. Thus, the estimate of our model given by Equation 7.1 is

$$\hat{Y} = \hat{\beta}_0 + \hat{\beta}_1 X, \tag{7.3}$$

where $\hat{Y}$ denotes the predicted value of $Y$ for some value of $X$, and $\hat{\beta}_0$ and $\hat{\beta}_1$ are the estimated parameters. We do not go into the derivation of the estimators, since it can be found in most introductory statistics textbooks.

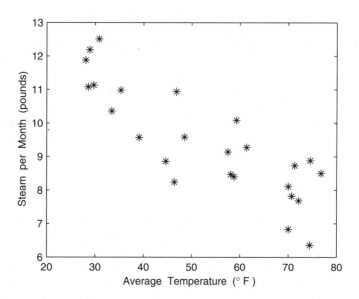

**FIGURE 7.1**
Scatterplot of a data set where we are interested in modeling the relationship between average temperature (the predictor variable) and the amount of steam used per month (the response variable). The scatterplot indicates that modeling the relationship with a straight line is reasonable.

Assume that we have a sample of observed predictor variables with corresponding responses. We denote these by $(X_i, Y_i)$, $i = 1, ..., n$. The least squares fit is obtained by finding the values of the parameters that minimize the sum of the squared errors

$$RSE = \sum_{i=1}^{n} \varepsilon^2 = \sum_{i=1}^{n} (Y_i - (\beta_0 + \beta_1 X_i))^2, \qquad (7.4)$$

where $RSE$ denotes the **residual squared error**.

Estimates of the parameters $\hat{\beta}_0$ and $\hat{\beta}_1$ are easily obtained in MATLAB using the function **polyfit**, and other methods available in MATLAB will be explored in Chapter 10. We use the function **polyfit** in Example 7.1 to model the linear relationship between the atmospheric temperature and the amount of steam used per month (see Figure 7.1).

**Example 7.1**
In this example, we show how to use the MATLAB function **polyfit** to fit a line to the **steam** data. The **polyfit** function takes three arguments: the

observed **x** values, the observed **y** values and the degree of the polynomial that we want to fit to the data. The following commands fit a polynomial of degree one to the steam data.

```
% Loads the vectors x and y.
load steam
% Fit a first degree polynomial to the data.
[p,s] = polyfit(x,y,1);
```

The output argument **p** is a vector of coefficients of the polynomial in decreasing order. So, in this case, the first element of **p** is the estimated slope $\hat{\beta}_1$ and the second element is the estimated $y$-intercept $\hat{\beta}_0$. The resulting model is

$$\hat{\beta}_0 = 13.62 \qquad \hat{\beta}_1 = -0.08.$$

The predictions that would be obtained from the model (i.e., points on the line given by the estimated parameters) are shown in Figure 7.2, and we see that it seems to be a reasonable fit.
❑

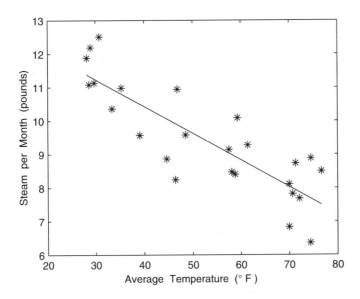

**FIGURE 7.2**
This figure shows a scatterplot of the **steam** data along with the line obtained using **polyfit**. The estimate of the slope is $\hat{\beta}_1 = -0.08$, and the estimate of the $y$-intercept is $\hat{\beta}_0 = 13.62$.

The *prediction error* is defined as

$$PE = E[(Y - \hat{Y})^2],$$ (7.5)

where the expectation is with respect to the true population. To estimate the error given by Equation 7.5, we need to test our model (obtained from **poly-fit**) using an independent set of data that we denote by $(x_i', y_i')$. This means that we would take an observed $(x_i', y_i')$ and obtain the estimate of $\hat{y}_i'$ using our model:

$$\hat{y}_i' = \hat{\beta}_0 + \hat{\beta}_1 x_i'.$$ (7.6)

We then compare $\hat{y}_i'$ with the true value of $y_i'$. Obtaining the outputs or $\hat{y}_i'$ from the model is easily done in MATLAB using the **polyval** function as shown in Example 7.2.

Say we have $m$ independent observations $(x_i', y_i')$ that we can use to test the model. We estimate the prediction error (Equation 7.5) using

$$\hat{PE} = \frac{1}{m} \sum_{i=1}^{m} (y_i' - \hat{y}_i')^2.$$ (7.7)

Equation 7.7 measures the average squared error between the predicted response obtained from the model and the true measured response. It should be noted that other measures of error can be used, such as the absolute difference between the observed and predicted responses.

## Example 7.2
We now show how to estimate the prediction error using Equation 7.7. We first choose some points from the **steam** data set and put them aside to use as an independent test sample. The rest of the observations are then used to obtain the model.

```
load steam
% Get the set that will be used to
% estimate the line.
indtest = 2:2:20; % Just pick some points.
xtest = x(indtest);
ytest = y(indtest);
% Now get the observations that will be
% used to fit the model.
xtrain = x;
ytrain = y;
% Remove the test observations.
```

```
xtrain(indtest) = [];
ytrain(indtest) = [];
```

The next step is to fit a first degree polynomial:

```
% Fit a first degree polynomial (the model)
% to the data.
[p,s] = polyfit(xtrain,ytrain,1);
```

We can use the MATLAB function **polyval** to get the predictions at the *x* values in the testing set and compare these to the observed *y* values in the testing set.

```
% Now get the predictions using the model and the
% testing data that was set aside.
yhat = polyval(p,xtest);
% The residuals are the difference between the true
% and the predicted values.
r = (ytest - yhat);
```

Finally, the estimate of the prediction error (Equation 7.7) is obtained as follows:

```
pe = mean(r.^2);
```

The estimated prediction error is $\hat{PE} = 0.91$. The reader is asked to explore this further in the exercises.
□

What we just illustrated in Example 7.2 was a situation where we partitioned the data into one set for building the model and one for estimating the prediction error. This is perhaps not the best use of the data, because we have all of the data available for evaluating the error in the model. We could repeat the above procedure, repeatedly partitioning the data into *many* training and testing sets. This is the fundamental idea underlying cross-validation.

The most general form of this procedure is called *K*-fold cross-validation. The basic concept is to split the data into *K* partitions of approximately equal size. One partition is reserved for testing, and the rest of the data are used for fitting the model. The test set is used to calculate the squared error $(y_i - \hat{y}_i)^2$. Note that the prediction $\hat{y}_i$ is from the model obtained using the current training set (one without the *i*-th observation in it). This procedure is repeated until all *K* partitions have been used as a test set. Note that we have *n* squared errors because each observation will be a member of one testing set. The average of these errors is the estimated expected prediction error.

In most situations, where the size of the data set is relatively small, the analyst can set $K = n$, so the size of the testing set is one. Since this requires fitting the model *n* times, this can be computationally expensive if *n* is large. We note, however, that there are efficient ways of doing this [Gentle 1998; Hjorth,

1994]. We outline the steps for cross-validation below and demonstrate this approach in Example 7.3.

*PROCEDURE - CROSS-VALIDATION*

1. Partition the data set into $K$ partitions. For simplicity, we assume that $n = r \cdot K$, so there are $r$ observations in each set.
2. Leave out one of the partitions for testing purposes.
3. Use the remaining $n - r$ data points for training (e.g., fit the model, build the classifier, estimate the probability density function).
4. Use the test set with the model and determine the squared error between the observed and predicted response: $(y_i - \hat{y}_i)^2$.
5. Repeat steps 2 through 4 until all $K$ partitions have been used as a test set.
6. Determine the average of the $n$ errors.

Note that the error mentioned in step 4 depends on the application and the goal of the analysis [Hjorth, 1994]. For example, in pattern recognition applications, this might be the cost of misclassifying a case. In the following example, we apply the cross-validation technique to help decide what type of model should be used for the **steam** data.

## Example 7.3

In this example, we apply cross-validation to the modeling problem of Example 7.1. We fit linear, quadratic (degree 2) and cubic (degree 3) models to the data and compare their accuracy using the estimates of prediction error obtained from cross-validation.

```
% Set up the array to store the prediction errors.
n = length(x);
r1 = zeros(1,n);% store error - linear fit
r2 = zeros(1,n);% store error - quadratic fit
r3 = zeros(1,n);% store error - cubic fit
% Loop through all of the data. Remove one point at a
% time as the test point.
for i = 1:n
    xtest = x(i);% Get the test point.
    ytest = y(i);
    xtrain = x;% Get the points to build model.
    ytrain = y;
    xtrain(i) = [];% Remove test point.
    ytrain(i) = [];
    % Fit a first degree polynomial to the data.
    [p1,s] = polyfit(xtrain,ytrain,1);
```

```
% Fit a quadratic to the data.
[p2,s] = polyfit(xtrain,ytrain,2);
% Fit a cubic to the data
[p3,s] = polyfit(xtrain,ytrain,3);
% Get the errors
r1(i) = (ytest - polyval(p1,xtest)).^2;
r2(i) = (ytest - polyval(p2,xtest)).^2;
r3(i) = (ytest - polyval(p3,xtest)).^2;
end
```

We obtain the estimated prediction error of both models as follows,

```
% Get the prediction error for each one.
pe1 = mean(r1);
pe2 = mean(r2);
pe3 = mean(r3);
```

From this, we see that the estimated prediction error for the linear model is 0.86; the corresponding error for the quadratic model is 0.88; and the error for the cubic model is 0.95. Thus, between these three models, the first-degree polynomial is the best in terms of minimum expected prediction error.
□

## 7.3 Jackknife

The jackknife is a data partitioning method like cross-validation, but the goal of the jackknife is more in keeping with that of the bootstrap. The jackknife method is used to estimate the bias and the standard error of statistics.

Let's say that we have a random sample of size $n$, and we denote our estimator of a parameter $\theta$ as

$$\hat{\theta} = T = t(x_1, x_2, ..., x_n). \tag{7.8}$$

So, $\hat{\theta}$ might be the mean, the variance, the correlation coefficient or some other statistic of interest. Recall from Chapters 3 and 6 that $T$ is also a random variable, and it has some error associated with it. We would like to get an estimate of the bias and the standard error of the estimate $T$, so we can assess the accuracy of the results.

When we cannot determine the bias and the standard error using analytical techniques, then methods such as the bootstrap or the jackknife may be used. The jackknife is similar to the bootstrap in that no parametric assumptions are made about the underlying population that generated the data, and the variation in the estimate is investigated by looking at the sample data.

The jackknife method is similar to cross-validation in that we leave out one observation $x_i$ from our sample to form a *jackknife sample* as follows

$$x_1, \ldots, x_{i-1}, x_{i+1}, \ldots, x_n \ .$$

This says that the $i$-th jackknife sample is the original sample with the $i$-th data point removed. We calculate the value of the estimate using this reduced jackknife sample to obtain the $i$-th *jackknife replicate*. This is given by

$$T^{(-i)} = t(x_1, \ldots, x_{i-1}, x_{i+1}, \ldots, x_n) \ .$$

This means that we leave out one point at a time and use the rest of the sample to calculate our statistic. We continue to do this for the entire sample, leaving out one observation at a time, and the end result is a sequence of $n$ jackknife replications of the statistic.

The estimate of the bias of $T$ obtained from the jackknife technique is given by [Efron and Tibshirani, 1993]

$$\widehat{\text{Bias}}_{Jack}(T) = (n-1)(\overline{T^{(J)}} - T), \tag{7.9}$$

where

$$\overline{T^{(J)}} = \sum_{i=1}^{n} T^{(-i)}/n \ . \tag{7.10}$$

We see from Equation 7.10 that $\overline{T^{(J)}}$ is simply the average of the jackknife replications of $T$.

The estimated standard error using the jackknife is defined as follows

$$\widehat{SE}_{Jack}(T) = \left[ \frac{n-1}{n} \sum_{i=1}^{n} (T^{(-i)} - \overline{T^{(J)}})^2 \right]^{1/2}. \tag{7.11}$$

Equation 7.11 is essentially the sample standard deviation of the jackknife replications with a factor $(n-1)/n$ in front of the summation instead of $1/(n-1)$. Efron and Tibshirani [1993] show that this factor ensures that the jackknife estimate of the standard error of the sample mean, $\widehat{SE}_{Jack}(\bar{x})$, is an unbiased estimate.

PROCEDURE - JACKKNIFE

1. Leave out an observation.
2. Calculate the value of the statistic using the remaining sample points to obtain $T^{(-i)}$.
3. Repeat steps 1 and 2, leaving out one point at a time, until all $n$ $T^{(-i)}$ are recorded.
4. Calculate the jackknife estimate of the bias of $T$ using Equation 7.9.
5. Calculate the jackknife estimate of the standard error of $T$ using Equation 7.11.

The following two examples show how this is used to obtain jackknife estimates of the bias and standard error for an estimate of the correlation coefficient.

## Example 7.4

In this example, we use a data set that has been examined in Efron and Tibshirani [1993]. Note that these data are also discussed in the exercises for Chapter 6. These data consist of measurements collected on the freshman class of 82 law schools in 1973. The average score for the entering class on a national law test (**lsat**) and the average undergraduate grade point average (**gpa**) were recorded. A random sample of size $n = 15$ was taken from the population. We would like to use these sample data to estimate the correlation coefficient $\rho$ between the test scores (**lsat**) and the grade point average (**gpa**). We start off by finding the statistic of interest.

```
% Loads up a matrix - law.
load law
% Estimate the desired statistic from the sample.
lsat = law(:,1);
gpa = law(:,2);
tmp = corrcoef(gpa,lsat);
% Recall from Chapter 3 that the corrcoef function
% returns a matrix of correlation coefficients. We
% want the one in the off-diagonal position.
T = tmp(1,2);
```

We get an estimated correlation coefficient of $\hat{\rho} = 0.78$, and we would like to get an estimate of the bias and the standard error of this statistic. The following MATLAB code implements the jackknife procedure for estimating these quantities.

```
% Set up memory for jackknife replicates.
n = length(gpa);
reps = zeros(1,n);
for i = 1:n
```

```
% Store as temporary vector:
gpat = gpa;
lsatt = lsat;
% Leave i-th point out:
gpat(i) = [];
lsatt(i) = [];
% Get correlation coefficient:
% In this example, we want off-diagonal element.
tmp = corrcoef(gpat,lsatt);
reps(i) = tmp(1,2);
end
mureps = mean(reps);
sehat = sqrt((n-1)/n*sum((reps-mureps).^2));
% Get the estimate of the bias:
biashat = (n-1)*(mureps-T);
```

Our estimate of the standard error of the sample correlation coefficient is

$$\hat{SE}_{Jack}(\hat{\rho}) = 0.14,$$

and our estimate of the bias is

$$\hat{Bias}_{Jack}(\hat{\rho}) = -0.0065.$$

This data set will be explored further in the exercises.
□

## Example 7.5

We provide a MATLAB function called **csjack** that implements the jack-knife procedure. This will work with any MATLAB function that takes the random sample as the argument and returns a statistic. This function can be one that comes with MATLAB, such as **mean** or **var**, or it can be one written by the user. We illustrate its use with a user-written function called **corr** that returns the single correlation coefficient between two univariate random variables.

```
function r = corr(data)
% This function returns the single correlation
% coefficient between two variables.
tmp = corrcoef(data);
r = tmp(1,2);
```

The data used in this example are taken from Hand, et al. [1994]. They were originally from Anscombe [1973], where they were created to illustrate the point that even though an observed value of a statistic is the same for data sets ($\rho = 0.82$), that does not tell the entire story. He also used them to show the importance of looking at scatterplots, because it is obvious from the plots that the relationships between the variables are not similar. The scatterplots are shown in Figure 7.3.

```
% Here is another example.
% We have 4 data sets with essentially the same
% correlation coefficient.
% The scatterplots look very different.
% When this file is loaded, you get four sets
% of x and y variables.
load anscombe
% Do the scatterplots.
subplot(2,2,1),plot(x1,y1,'k*');
subplot(2,2,2),plot(x2,y2,'k*');
subplot(2,2,3),plot(x3,y3,'k*');
subplot(2,2,4),plot(x4,y4,'k*');
```

We now determine the jackknife estimate of bias and standard error for $\hat{\rho}$ using `csjack`.

```
% Note that 'corr' is something we wrote.
[b1,se1,jv1] = csjack([x1,y1],'corr');
[b2,se2,jv2] = csjack([x2,y2],'corr');
[b3,se3,jv3] = csjack([x3,y3],'corr');
[b4,se4,jv4] = csjack([x4,y4],'corr');
```

The jackknife estimates of bias are:

```
b1 = -0.0052
b2 =  0.0008
b3 =  0.1514
b4 =  NaN
```

The jackknife estimates of the standard error are:

```
se1 = 0.1054
se2 = 0.1026
se3 = 0.1730
se4 = NaN
```

Note that the jackknife procedure does not work for the fourth data set, because when we leave out the last data point, the correlation coefficient is undefined for the remaining points.
□


The jackknife method is also described in the literature using pseudo-values. The *jackknife pseudo-values* are given by

$$\widehat{T}_i = nT - (n-1)T^{(-i)} \qquad i = 1, ..., n, \qquad (7.12)$$

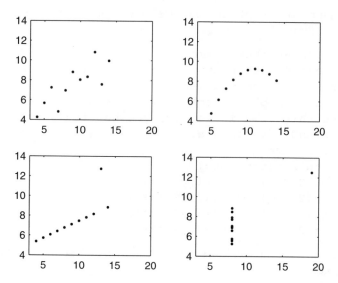

**FIGURE 7.3**
This shows the scatterplots of the four data sets discussed in Example 7.5. These data were created to show the importance of looking at scatterplots [Anscombe, 1973]. All data sets have the same estimated correlation coefficient of $\hat{\rho} = 0.82$, but it is obvious that the relationship between the variables is very different.

where $T^{(-i)}$ is the value of the statistic computed on the sample with the $i$-th data point removed.

We take the average of the pseudo-values given by

$$J(T) = \sum_{i=1}^{n} \widehat{T}_i / n,$$ (7.13)

and use this to get the jackknife estimate of the standard error, as follows

$$\hat{SE}_{JackP}(T) = \left[ \frac{1}{n(n-1)} \sum_{i=1}^{n} (\widehat{T}_i - J(T))^2 \right]^{1/2}.$$ (7.14)

*PROCEDURE - PSEUDO-VALUE JACKKNIFE*

1. Leave out an observation.
2. Calculate the value of the statistic using the remaining sample points to obtain $T^{(-i)}$.

3. Calculate the pseudo-value $\widehat{T}_i$ using Equation 7.12.

4. Repeat steps 2 and 3 for the remaining data points, yielding $n$ values of $\widehat{T}_i$.

5. Determine the jackknife estimate of the standard error of $T$ using Equation 7.14.

## Example 7.6

We now repeat Example 7.4 using the jackknife pseudo-value approach and compare estimates of the standard error of the correlation coefficient for these data. The following MATLAB code implements the pseudo-value procedure.

```
% Loads up a matrix.
load law
lsat = law(:,1);
gpa = law(:,2);
% Get the statistic from the original sample
tmp = corrcoef(gpa,lsat);
T = tmp(1,2);
% Set up memory for jackknife replicates
n = length(gpa);
reps = zeros(1,n);
for i = 1:n
    % store as temporary vector
    gpat = gpa;
    lsatt = lsat;
    % leave i-th point out
    gpat(i) = [];
    lsatt(i) = [];
    % get correlation coefficient
    tmp = corrcoef(gpat,lsatt);
    % In this example, is off-diagonal element.
    % Get the jackknife pseudo-value for the i-th point.
    reps(i) = n*T-(n-1)*tmp(1,2);
end
JT = mean(reps);
sehatpv = sqrt(1/(n*(n-1))*sum((reps - JT).^2));
```

We obtain an estimated standard error of $\widehat{SE}_{JackP}(\hat{\rho}) = 0.14$, which is the same result we had before.

❑

Efron and Tibshirani [1993] describe a situation where the jackknife procedure does not work and suggest that the bootstrap be used instead. These are applications where the statistic is not smooth. An example of this type of statistic is the median. Here *smoothness* refers to statistics where small changes in the data set produce small changes in the value of the statistic. We illustrate this situation in the next example.

### Example 7.7

Researchers collected data on the weight gain of rats that were fed four different diets based on the amount of protein (high and low) and the source of the protein (beef and cereal) [Snedecor and Cochran, 1967; Hand, et al., 1994]. We will use the data collected on the rats who were fed a low protein diet of cereal. The sorted data are

```
x = [58, 67, 74, 74, 80, 89, 95, 97, 98, 107];
```

The median of this data set is $\hat{q}_{0.5} = 84.5$. To see how the median changes with small changes of $x$, we increment the fourth observation $x = 74$ by one. The change in the median is zero, because it is still at $\hat{q}_{0.5} = 84.5$. In fact, the median does not change until we increment the fourth observation by 7, at which time the median becomes $\hat{q}_{0.5} = 85$. Let's see what happens when we use the jackknife approach to get an estimate of the standard error in the median.

```
% Set up memory for jackknife replicates.
n = length(x);
reps = zeros(1,n);
for i = 1:n
   % Store as temporary vector.
   xt = x;
   % Leave i-th point out.
   xt(i) = [];
   % Get the median.
   reps(i) = median(xt);
end
mureps = mean(reps);
sehat = sqrt((n-1)/n*sum((reps-mureps).^2));
```

The jackknife replicates are:

$$89 \quad 89 \quad 89 \quad 89 \quad 89 \quad 81 \quad 81 \quad 81 \quad 81 \quad 81.$$

These give an estimated standard error of the median of $\hat{SE}_{Jack}(\hat{q}_{0.5}) = 12$. Because the median is not a smooth statistic, we have only a few distinct values of the statistic in the jackknife replicates. To understand this further, we now estimate the standard error using the bootstrap.

```
% Now get the estimate of standard error using
% the bootstrap.
[bhat,seboot,bvals]=csboot(x','median',500);
```

This yields an estimate of the standard error of the median of $\hat{SE}_{Boot}(\hat{q}_{0.5}) = 7.1$. In the exercises, the reader is asked to see what happens when the statistic is the mean and should find that the jackknife and bootstrap estimates of the standard error of the mean are similar.
❑

It can be shown [Efron & Tibshirani, 1993] that the jackknife estimate of the standard error of the median does not converge to the true standard error as $n \rightarrow \infty$. For the data set of Example 7.7, we had only two distinct values of the median in the jackknife replicates. This gives a poor estimate of the standard error of the median. On the other hand, the bootstrap produces data sets that are not as similar to the original data, so it yields reasonable results. The delete-$d$ jackknife [Efron and Tibshirani, 1993; Shao and Tu, 1995] deletes $d$ observations at a time instead of only one. This method addresses the problem of inconsistency with non-smooth statistics.

## 7.4 Better Bootstrap Confidence Intervals

In Chapter 6, we discussed three types of confidence intervals based on the bootstrap: the bootstrap standard interval, the bootstrap-$t$ interval and the bootstrap percentile interval. Each of them is applicable under more general assumptions and is superior in some sense (e.g., coverage performance, range-preserving, etc.) to the previous one. The bootstrap confidence interval that we present in this section is an improvement on the bootstrap percentile interval. This is called the $BC_a$ *interval*, which stands for bias-corrected and accelerated.

Recall that the upper and lower endpoints of the $(1 - \alpha) \cdot 100\%$ bootstrap percentile confidence interval are given by

$$\text{Percentile Interval: } (\hat{\theta}_{Lo}, \hat{\theta}_{Hi}) = (\hat{\theta}_B^{*(\alpha/2)}, \hat{\theta}_B^{*(1-\alpha/2)}). \tag{7.15}$$

Say we have $B = 100$ bootstrap replications of our statistic, which we denote as $\hat{\theta}^{*b}$, $b = 1, \ldots, 100$. To find the percentile interval, we sort the bootstrap replicates in ascending order. If we want a 90% confidence interval, then one way to obtain $\hat{\theta}_{Lo}$ is to use the bootstrap replicate in the 5th position of the ordered list. Similarly, $\hat{\theta}_{Hi}$ is the bootstrap replicate in the 95th position. As discussed in Chapter 6, the endpoints could also be obtained using other quantile estimates.

The $BC_a$ interval adjusts the endpoints of the interval based on two parameters, $\hat{a}$ and $\hat{z}_0$. The $(1 - \alpha) \cdot 100\%$ confidence interval using the $BC_a$ method is

$$BC_a \text{ Interval: } (\hat{\theta}_{Lo}, \hat{\theta}_{Hi}) = (\hat{\theta}_B^{*(\alpha_1)}, \hat{\theta}_B^{*(\alpha_2)}), \tag{7.16}$$

where

$$\alpha_1 = \Phi\left(\hat{z}_0 + \frac{\hat{z}_0 + z^{(\alpha/2)}}{1 - \hat{a}(\hat{z}_0 + z^{(\alpha/2)})}\right)$$

$$\alpha_2 = \Phi\left(\hat{z}_0 + \frac{\hat{z}_0 + z^{(1-\alpha/2)}}{1 - \hat{a}(\hat{z}_0 + z^{(1-\alpha/2)})}\right). \tag{7.17}$$

Let's look a little closer at $\alpha_1$ and $\alpha_2$ given in Equation 7.17. Since $\Phi$ denotes the standard normal cumulative distribution function, we know that $0 \le \alpha_1 \le 1$ and $0 \le \alpha_2 \le 1$. So we see from Equation 7.16 and 7.17 that instead of basing the endpoints of the interval on the confidence level of $1 - \alpha$, they are adjusted using information from the distribution of bootstrap replicates.

We discuss, shortly, how to obtain the acceleration $\hat{a}$ and the bias $\hat{z}_0$. However, before we do, we want to remind the reader of the definition of $z^{(\alpha/2)}$. This denotes the $\alpha/2$-th quantile of the standard normal distribution. It is the value of $z$ that has an area to the left of size $\alpha/2$. As an example, for $\alpha/2 = 0.05$, we have $z^{(\alpha/2)} = z^{(0.05)} = -1.645$, because $\Phi(-1.645) = 0.05$.

We can see from Equation 7.17 that if $\hat{a}$ and $\hat{z}_0$ are both equal to zero, then the $BC_a$ is the same as the bootstrap percentile interval. For example,

$$\alpha_1 = \Phi\left\{0 + \frac{0 + z^{(\alpha/2)}}{1 - 0(0 + z^{(\alpha/2)})}\right\} = \Phi(z^{(\alpha/2)}) = \alpha/2,$$

with a similar result for $\alpha_2$. Thus, when we do not account for the bias $\hat{z}_0$ and the acceleration $\hat{a}$, then Equation 7.16 reduces to the bootstrap percentile interval (Equation 7.15).

We now turn our attention to how we determine the parameters $\hat{a}$ and $\hat{z}_0$. The bias-correction is given by $\hat{z}_0$, and it is based on the proportion of bootstrap replicates $\hat{\theta}^{*b}$ that are less than the statistic $\hat{\theta}$ calculated from the original sample. It is given by

$$\hat{z}_0 = \Phi^{-1}\left(\frac{\#(\hat{\theta}^{*b} < \hat{\theta})}{B}\right), \tag{7.18}$$

where $\Phi^{-1}$ denotes the inverse of the standard normal cumulative distribution function.

The acceleration parameter $\hat{a}$ is obtained using the jackknife procedure as follows,

$$\hat{a} = \frac{\displaystyle\sum_{i=1}^{n}\left\{\overline{\hat{\theta}^{(J)}} - \hat{\theta}^{(-i)}\right\}^{3}}{6\left\{\displaystyle\sum_{i=1}^{n}\left(\overline{\hat{\theta}^{(J)}} - \hat{\theta}^{(-i)}\right)^{2}\right\}^{3/2}}, \qquad (7.19)$$

where $\hat{\theta}^{(-i)}$ is the value of the statistic using the sample with the $i$-th data point removed (the $i$-th jackknife sample) and

$$\overline{\hat{\theta}^{(J)}} = \frac{1}{n}\sum_{i=1}^{n}\hat{\theta}^{(-i)}. \qquad (7.20)$$

According to Efron and Tibshirani [1993], $\hat{z}_0$ is a measure of the difference between the median of the bootstrap replicates and $\hat{\theta}$ in normal units. If half of the bootstrap replicates are less than or equal to $\hat{\theta}$, then there is no median bias and $\hat{z}_0$ is zero. The parameter $\hat{a}$ measures the rate acceleration of the standard error of $\hat{\theta}$. For more information on the theoretical justification for these corrections, see Efron and Tibshirani [1993] and Efron [1987].

*PROCEDURE - BC$_a$ INTERVAL*

1. Given a random sample, $\mathbf{x} = (x_1, ..., x_n)$, calculate the statistic of interest $\hat{\theta}$.
2. Sample with replacement from the original sample to get the bootstrap sample

$$\mathbf{x}^{*b} = (x_1^{*b}, ..., x_n^{*b}).$$

3. Calculate the same statistic as in step 1 using the sample found in step 2. This yields a bootstrap replicate $\hat{\theta}^{*b}$.
4. Repeat steps 2 through 3, $B$ times, where $B \geq 1000$.
5. Calculate the bias correction (Equation 7.18) and the acceleration factor (Equation 7.19).
6. Determine the adjustments for the interval endpoints using Equation 7.17.
7. The lower endpoint of the confidence interval is the $\alpha_1$ quantile $\hat{q}_{\alpha_1}$ of the bootstrap replicates, and the upper endpoint of the confidence interval is the $\alpha_2$ quantile $\hat{q}_{\alpha_2}$ of the bootstrap replicates.

## Example 7.8

We use an example from Efron and Tibshirani [1993] to illustrate the $BC_a$ interval. Here we have a set of measurements of 26 neurologically impaired children who took a test of spatial perception called test A. We are interested in finding a 90% confidence interval for the variance of a random score on test A. We use the following estimate for the variance

$$\hat{\theta} = \frac{1}{n}\sum_{i=1}^{n}(x_i - \bar{x})^2,$$

where $x_i$ represents one of the test scores. This is a biased estimator of the variance, and when we calculate this statistic from the sample we get a value of $\hat{\theta} = 171.5$. We provide a function called **csbootbca** that will determine the $BC_a$ interval. Because it is somewhat lengthy, we do not include the MATLAB code here, but the reader can view it in Appendix D. However, before we can use the function **csbootbca**, we have to write an M-file function that will return the estimate of the second sample central moment using only the sample as an input. It should be noted that MATLAB Statistics Toolbox has a function (**moment**) that will return the sample central moments of any order. We do not use this with the **csbootbca** function, because the function specified as an input argument to **csbootbca** can only use the sample as an input. Note that the function **mom** is the same function used in Chapter 6. We can get the bootstrap $BC_a$ interval with the following command.

```
% First load the data.
load spatial
% Now find the BC-a bootstrap interval.
alpha = 0.10;
B = 2000;
% Use the function we wrote to get the
% 2nd sample central moment - 'mom'.
[blo,bhi,bvals,z0,ahat] = ...
    csbootbca(spatial','mom',B,alpha);
```

From this function, we get a bias correction of $\hat{z}_0 = 0.16$ and an acceleration factor of $\hat{a} = 0.061$. The endpoints of the interval from **csbootbca** are (115.97, 258.54). In the exercises, the reader is asked to compare this to the bootstrap-$t$ interval and the bootstrap percentile interval.

☐

## 7.5 Jackknife-After-Bootstrap

In Chapter 6, we presented the bootstrap method for estimating the statistical accuracy of estimates. However, the bootstrap estimates of standard error and bias are also estimates, so they too have error associated with them. This error arises from two sources, one of which is the usual sampling variability because we are working with the sample instead of the population. The other variability comes from the fact that we are working with a finite number $B$ of bootstrap samples.

We now turn our attention to estimating this variability using the jackknife-after-bootstrap technique. The characteristics of the problem are the same as in Chapter 6. We have a random sample $\mathbf{x} = (x_1, ..., x_n)$, from which we calculate our statistic $\hat{\theta}$. We estimate the distribution of $\hat{\theta}$ by creating $B$ bootstrap replicates $\hat{\theta}^{*b}$. Once we have the bootstrap replicates, we estimate some feature of the distribution of $\hat{\theta}$ by calculating the corresponding feature of the distribution of bootstrap replicates. We will denote this feature or bootstrap estimate as $\hat{\gamma}_B$. As we saw before, $\hat{\gamma}_B$ could be the bootstrap estimate of the standard error, the bootstrap estimate of a quantile, the bootstrap estimate of bias or some other quantity.

To obtain the jackknife-after-bootstrap estimate of the variability of $\hat{\gamma}_B$, we leave out one data point $x_i$ at a time and calculate $\hat{\gamma}_B^{(-i)}$ using the bootstrap method on the remaining $n-1$ data points. We continue in this way until we have the $n$ values of $\hat{\gamma}_B^{(-i)}$. We estimate the variance of $\hat{\gamma}_B$ using the $\hat{\gamma}_B^{(-i)}$ values, as follows

$$\hat{\mathrm{var}}_{Jack}(\hat{\gamma}_B) = \frac{n-1}{n} \sum_{i=1}^{n} (\hat{\gamma}_B^{(-i)} - \overline{\hat{\gamma}_B})^2 , \qquad (7.21)$$

where

$$\overline{\hat{\gamma}_B} = \frac{1}{n} \sum_{i=1}^{n} \hat{\gamma}_B^{(-i)} .$$

Note that this is just the jackknife estimate for the variance of a statistic, where the statistic that we have to calculate for each jackknife replicate is a bootstrap estimate.

This can be computationally intensive, because we would need a new set of bootstrap samples when we leave out each data point $x_i$. There is a short-cut method for obtaining $\hat{\mathrm{var}}_{Jack}(\hat{\gamma}_B)$ where we use the original $B$ bootstrap samples. There will be some bootstrap samples where the $i$-th data point does not appear. Efron and Tibshirani [1993] show that if $n \geq 10$ and $B \geq 20$, then the probability is low that every bootstrap sample contains a given point $x_i$. We estimate the value of $\hat{\gamma}_B^{(-i)}$ by taking the bootstrap replicates for samples that do not contain the data point $x_i$. These steps are outlined below.

*PROCEDURE - JACKKNIFE-AFTER-BOOTSTRAP*

1. Given a random sample $\mathbf{x} = (x_1, \ldots, x_n)$, calculate a statistic of interest $\hat{\theta}$.

2. Sample with replacement from the original sample to get a bootstrap sample $\mathbf{x}^{*b} = (x_1^*, \ldots, x_n^*)$.

3. Using the sample obtained in step 2, calculate the same statistic that was determined in step one and denote by $\hat{\theta}^{*b}$.

4. Repeat steps 2 through 3, $B$ times to estimate the distribution of $\hat{\theta}$.

5. Estimate the desired feature of the distribution of $\hat{\theta}$ (e.g., standard error, bias, etc.) by calculating the corresponding feature of the distribution of $\hat{\theta}^{*b}$. Denote this bootstrap estimated feature as $\hat{\gamma}_B$.

6. Now get the error in $\hat{\gamma}_B$. For $i = 1, \ldots, n$, find all samples $\mathbf{x}^{*b} = (x_1^*, \ldots, x_n^*)$ that do not contain the point $x_i$. These are the bootstrap samples that can be used to calculate $\hat{\gamma}_B^{(-i)}$.

7. Calculate the estimate of the variance of $\hat{\gamma}_B$ using Equation 7.21.

**Example 7.9**

In this example, we show how to implement the jackknife-after-bootstrap procedure. For simplicity, we will use the MATLAB Statistics Toolbox function called **bootstrp**, because it returns the indices for each bootstrap sample and the corresponding bootstrap replicate $\hat{\theta}^{*b}$. We return now to the **law** data where our statistic is the sample correlation coefficient. Recall that we wanted to estimate the standard error of the correlation coefficient, so $\hat{\gamma}_B$ will be the bootstrap estimate of the standard error.

```
% Use the law data.
load law
lsat = law(:,1);
gpa = law(:,2);

% Use the example in MATLAB documentation.
B = 1000;
[bootstat,bootsam] = bootstrp(B,'corrcoef',lsat,gpa);
```

The output argument **bootstat** contains the $B$ bootstrap replicates of the statistic we are interested in, and the columns of **bootsam** contains the indices to the data points that were in each bootstrap sample. We can loop through all of the data points and find the columns of **bootsam** that do not contain that point. We then find the corresponding bootstrap replicates.

```
% Find the jackknife-after-bootstrap.
n = length(gpa);
% Set up storage space.
jreps = zeros(1,n);
% Loop through all points,
% Find the columns in bootsam that
% do not have that point in it.
for i = 1:n
    % Note that the columns of bootsam are
    % the indices to the samples.
    % Find all columns with the point.
    [I,J] = find(bootsam==i);
    % Find all columns without the point.
    jacksam = setxor(J,1:B);
    % Find the correlation coefficient for
    % each of the bootstrap samples that
    % do not have the point in them.
    bootrep = bootstat(jacksam,2);
    % In this case it is col 2 that we need.
    % Calculate the feature (gamma_b) we want.
    jreps(i) = std(bootrep);
end
% Estimate the error in gamma_b.
varjack = (n-1)/n*sum((jreps-mean(jreps)).^2);
% The original bootstrap estimate of error is:
gamma = std(bootstat(:,2));
```

We see that the estimate of the standard error of the correlation coefficient for this simulation is $\hat{\gamma}_B = \hat{SE}_{Boot}(\hat{\rho}) = 0.14$, and our estimated standard error in this bootstrap estimate is $\hat{SE}_{Jack}(\hat{\gamma}_B) = 0.088$.
❑

Efron and Tibshirani [1993] point out that the jackknife-after-bootstrap works well when the number of bootstrap replicates $B$ is large. Otherwise, it overestimates the variance of $\hat{\gamma}_B$.

## 7.6 MATLAB Code

To our knowledge, MATLAB does not have M-files for either cross-validation or the jackknife. As described earlier, we provide a function (**csjack**) that will implement the jackknife procedure for estimating the bias and standard error in an estimate. We also provide a function called `csjackboot` that will implement the jackknife-after-bootstrap. These functions are summarized in Table 7.1.

The cross-validation method is application specific, so users must write their own code for each situation. For example, we showed in this chapter how to use cross-validation to help choose a model in regression by estimating the prediction error. In Chapter 9, we illustrate two examples of cross-validation: 1) to choose the right size classification tree and 2) to assess the misclassification error. We also describe a procedure in Chapter 10 for using $K$-fold cross-validation to choose the right size regression tree.

**TABLE 7.1**

List of Functions from Chapter 7 Included in the Computational Statistics Toolbox.

| Purpose | MATLAB Function |
|---|---|
| Implements the jackknife and returns the jackknife estimate of standard error and bias. | `csjack` |
| Returns the bootstrap $BC_a$ confidence interval. | `csbootbca` |
| Implements the jackknife-after-bootstrap and returns the jackknife estimate of the error in the bootstrap. | `csjackboot` |

## 7.7 Further Reading

There are very few books available where the cross-validation technique is the main topic, although Hjorth [1994] comes the closest. In that book, he discusses the cross-validation technique and the bootstrap and describes their use in model selection. Other sources on the theory and use of cross-validation are Efron [1982, 1983, 1986] and Efron and Tibshirani [1991, 1993]. Cross-validation is usually presented along with the corresponding applications. For example, to see how cross-validation can be used to select the smoothing parameter in probability density estimation, see Scott [1992]. Breiman, et al. [1984] and Webb [1999] describe how cross-validation is used to choose the right size classification tree.

The initial jackknife method was proposed by Quenouille [1949, 1956] to estimate the bias of an estimate. This was later extended by Tukey [1958] to estimate the variance using the pseudo-value approach. Efron [1982] is an excellent resource that discusses the underlying theory and the connection between the jackknife, the bootstrap and cross-validation. A more recent text by Shao and Tu [1995] provides a guide to using the jackknife and other resampling plans. Many practical examples are included. They also present the theoretical properties of the jackknife and the bootstrap, examining them in an asymptotic framework. Efron and Tibshirani [1993] show the connection between the bootstrap and the jackknife through a geometrical representation. For a reference on the jackknife that is accessible to readers at the undergraduate level, we recommend Mooney and Duval [1993]. This text also gives a description of the delete-$d$ jackknife procedure.

The use of jackknife-after-bootstrap to evaluate the error in the bootstrap is discussed in Efron and Tibshirani [1993] and Efron [1992]. Applying another level of bootstrapping to estimate this error is given in Loh [1987], Tibshirani [1988], and Hall and Martin [1988]. For other references on this topic, see Chernick [1999].

## Exercises

7.1. The `insulate` data set [Hand, et al., 1994] contains observations corresponding to the average outside temperature in degrees Celsius and the amount of weekly gas consumption measured in 1000 cubic feet. Do a scatterplot of the data corresponding to the measurements taken before insulation was installed. What is a good model for this? Use cross-validation with $K = 1$ to estimate the prediction error for your model. Use cross-validation with $K = 4$. Does your error change significantly? Repeat the process for the data taken after insulation was installed.

7.2. Using the same procedure as in Example 7.2, use a quadratic (degree is 2) and a cubic (degree is 3) polynomial to build the model. What is the estimated prediction error from these models? Which one seems best: linear, quadratic or cubic?

7.3. The `peanuts` data set [Hand, et al., 1994; Draper and Smith, 1981] contain measurements of the alfatoxin ($X$) and the corresponding percentage of non-contaminated peanuts in the batch ($Y$). Do a scatterplot of these data. What is a good model for these data? Use cross-validation to choose the best model.

7.4. Generate $n = 25$ random variables from a standard normal distribution that will serve as the random sample. Determine the jackknife estimate of the standard error for $\bar{x}$, and calculate the bootstrap estimate of the standard error. Compare these to the theoretical value of the standard error (see Chapter 3).

7.5. Using a sample size of $n = 15$, generate random variables from a uniform (0,1) distribution. Determine the jackknife estimate of the standard error for $\bar{x}$, and calculate the bootstrap estimate of the standard error for the same statistic. Let's say we decide to use $s/\sqrt{n}$ as an estimate of the standard error for $\bar{x}$. How does this compare to the other estimates?

7.6. Use Monte Carlo simulation to compare the performance of the bootstrap and the jackknife methods for estimating the standard error and bias of the sample second central moment. For every Monte Carlo trial, generate 100 standard normal random variables and calculate the bootstrap and jackknife estimates of the standard error and bias. Show the distribution of the bootstrap estimates (of bias and standard error) and the jackknife estimates (of bias and standard error) in a histogram or a box plot. Make some comparisons of the two methods.

7.7. Repeat problem 7.4 and use Monte Carlo simulation to compare the bootstrap and jackknife estimates of bias for the sample coefficient of skewness statistic and the sample coefficient of kurtosis (see Chapter 3).

7.8. Using the **law** data set in Example 7.4, find the jackknife replicates of the median. How many different values are there? What is the jackknife estimate of the standard error of the median? Use the bootstrap method to get an estimate of the standard error of the median. Compare the two estimates of the standard error of the median.

7.9. For the data in Example 7.7, use the bootstrap and the jackknife to estimate the standard error of the mean. Compare the two estimates.

7.10. Using the data in Example 7.8, find the bootstrap-$t$ interval and the bootstrap percentile interval. Compare these to the $BC_a$ interval found in Example 7.8.

# Chapter 8

## Probability Density Estimation

## 8.1 Introduction

We discussed several techniques for graphical exploratory data analysis in Chapter 5. One purpose of these exploratory techniques is to obtain information and insights about the distribution of the underlying population. For instance, we would like to know if the distribution is multi-modal, skewed, symmetric, etc. Another way to gain understanding about the distribution of the data is to estimate the probability density function from the random sample, possibly using a nonparametric probability density estimation technique.

Estimating probability density functions is required in many areas of computational statistics. One of these is in the modeling and simulation of physical phenomena. We often have measurements from our process, and we would like to use those measurements to determine the probability distribution so we can generate random variables for a Monte Carlo simulation (Chapter 6). Another application where probability density estimation is used is in statistical pattern recognition (Chapter 9). In supervised learning, which is one approach to pattern recognition, we have measurements where each one is labeled with a class membership tag. We could use the measurements for each class to estimate the class-conditional probability density functions, which are then used in a Bayesian classifier. In other applications, we might need to determine the probability that a random variable will fall within some interval, so we would need to evaluate the cumulative distribution function. If we have an estimate of the probability density function, then we can easily estimate the required probability by integrating under the estimated curve. Finally, in Chapter 10, we show how to use density estimation techniques for nonparametric regression.

In this chapter, we cover semi-parametric and nonparametric techniques for probability density estimation. By these, we mean techniques where we make few or no assumptions about what functional form the probability density takes. This is in contrast to a parametric method, where the density is estimated by assuming a distribution and then estimating the parameters.

We present three main methods of semi-parametric and nonparametric density estimation and their variants: histograms, kernel density estimates, and finite mixtures.

In the remainder of this section, we cover some ways to measure the error in functions as background to what follows. Then, in Section 8.2, we present various histogram based methods for probability density estimation. There we cover optimal bin widths for univariate and multivariate histograms, the frequency polygons, and averaged shifted histograms. Section 8.3 contains a discussion of kernel density estimation, both univariate and multivariate. In Section 8.4, we describe methods that model the probability density as a finite (less than $n$) sum of component densities. As usual, we conclude with descriptions of available MATLAB code and references to the topics covered in the chapter.

Before we can describe the various density estimation methods, we need to provide a little background on measuring the error in functions. We briefly present two ways to measure the error between the true function and the estimate of the function. These are called the mean integrated squared error (MISE) and the mean integrated absolute error (MIAE). Much of the underlying theory for choosing optimal parameters for probability density estimation is based on these concepts.

We start off by describing the mean squared error at a given point in the domain of the function. We can find the *mean squared error* **(MSE)** of the estimate $\hat{f}(x)$ at a point $x$ from the following

$$\text{MSE}[\hat{f}(x)] = E[(\hat{f}(x) - f(x))^2]. \tag{8.1}$$

Alternatively, we can determine the error over the domain for $x$ by integrating. This gives us the *integrated squared error* **(ISE)**:

$$\text{ISE} = \int (\hat{f}(x) - f(x))^2 dx. \tag{8.2}$$

The ISE is a random variable that depends on the true function $f(x)$, the estimator $\hat{f}(x)$, and the particular random sample that was used to obtain the estimate. Therefore, it makes sense to look at the expected value of the ISE or *mean integrated squared error*, which is given by

$$\text{MISE} = E\left[\int (\hat{f}(x) - f(x))^2 dx\right]. \tag{8.3}$$

To obtain the *mean integrated absolute error*, we simply replace the integrand with the absolute difference between the estimate and the true function. Thus, we have

$$\text{MIAE} = E\left[\int |\hat{f}(x) - f(x)| \, dx\right]. \tag{8.4}$$

These concepts are easily extended to the multivariate case.

---

## 8.2 Histograms

Histograms were introduced in Chapter 5 as a graphical way of summarizing or describing a data set. A histogram visually conveys how a data set is distributed, reveals modes and bumps, and provides information about relative frequencies of observations. Histograms are easy to create and are computationally feasible. Thus, they are well suited for summarizing large data sets. We revisit histograms here and examine optimal bin widths and where to start the bins. We also offer several extensions of the histogram, such as the frequency polygon and the averaged shifted histogram.

### 1-D Histograms

Most introductory statistics textbooks expose students to the frequency histogram and the relative frequency histogram. The problem with these is that the total area represented by the bins does not sum to 1. Thus, these are not valid probability density estimates. The reader is referred to Chapter 5 for more information on this and an example illustrating the difference between a frequency histogram and a density histogram. Since our goal is to estimate a *bona fide* probability density, we want to have a function $\hat{f}(x)$ that is nonnegative and satisfies the constraint that

$$\int \hat{f}(x) \, dx = 1. \tag{8.5}$$

The histogram is calculated using a random sample $X_1, X_2, \ldots, X_n$. The analyst must choose an origin $t_0$ for the bins and a bin width $h$. These two parameters define the mesh over which the histogram is constructed. In what follows, we will see that it is the bin width that determines the smoothness of the histogram. Small values of $h$ produce histograms with a lot of variation, while larger bin widths yield smoother histograms. This phenomenon is illustrated in Figure 8.1, where we show histograms with different bin widths. For this reason, the bin width $h$ is sometimes referred to as the *smoothing parameter*.

Let $B_k = [t_k, t_{k+1})$ denote the $k$-th bin, where $t_{k+1} - t_k = h$, for all $k$. We represent the number of observations that fall into the $k$-th bin by $v_k$. The 1-D histogram at a point $x$ is defined as

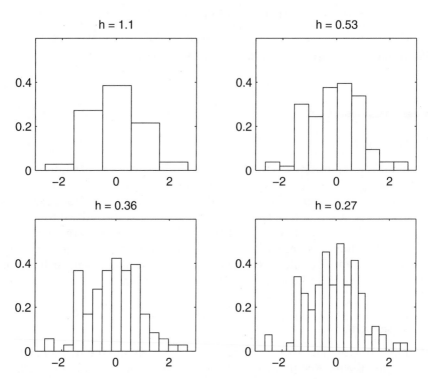

**FIGURE 8.1**
These are histograms for normally distributed random variables. Notice that for the larger bin widths, we have only one bump as expected. As the smoothing parameter gets smaller, the histogram displays more variation and spurious bumps appear in the histogram estimate.

$$\hat{f}_{Hist}(x) = \frac{v_k}{nh} = \frac{1}{nh}\sum_{i=1}^{n}I_{B_k}(X_i); \qquad x \text{ in } B_k, \qquad (8.6)$$

where $I_{B_k}(X_i)$ is the indicator function

$$I_{B_k}(X_i) = \begin{cases} 1, & X_i \text{ in } B_k \\ 0, & X_i \text{ not in } B_k. \end{cases}$$

This means that if we need to estimate the value of the probability density for a given $x$, then we obtain the value $\hat{f}_{Hist}(x)$ by taking the number of observations in the data set that fall into the same bin as $x$ and multiplying by $1/(nh)$.

## Example 8.1

In this example, we illustrate MATLAB code that calculates the estimated value $\hat{f}_{Hist}(x)$ for a given $x$. We first generate random variables from a standard normal distribution.

```
n = 1000;
x = randn(n,1);
```

We then compute the histogram using MATLAB's **hist** function, using the default value of 10 bins. The issue of the bin width (or alternatively the number of bins) will be addressed shortly.

```
% Get the histogram-default is 10 bins.
[vk,bc] = hist(x);
% Get the bin width.
h = bc(2)- bc(1);
```

We can now obtain our histogram estimate at a point using the following code. Note that we have to adjust the output from **hist** to ensure that our estimate is a *bona fide* density. Let's get the estimate of our function at a point $x_0 = 0$.

```
% Now return an estimate at a point xo.
xo = 0;
% Find all of the bin centers less than xo.
ind = find(bc < xo);
% xo should be between these two bin centers.
b1 = bc(ind(end));
b2 = bc(ind(end)+1);
% Put it in the closer bin.
if (xo-b1) < (b2-xo)    % then put it in the 1st bin
    fhat = vk(ind(end))/(n*h);
else
    fhat = vk(ind(end)+1)/(n*h);
end
```

Our result is **fhat = 0.3477**. The true value for the standard normal evaluated at 0 is $1/\sqrt{2\pi} = 0.3989$, so we see that our estimate is close, but not equal to the true value.
❑

We now look at how we can choose the bin width $h$. Using some assumptions, Scott [1992] provides the following upper bound for the MSE (Equation 8.1) of $\hat{f}_{Hist}(x)$:

$$\text{MSE}(\hat{f}_{Hist}(x)) \le \frac{f(\xi_k)}{nh} + \gamma_k^2 h^2; \qquad x \text{ in } B_k, \qquad (8.7)$$

where

$$hf(\xi_k) = \int_{B_k} f(t)dt; \qquad \text{for some } \xi_k \text{ in } B_k . \qquad (8.8)$$

This is based on the assumption that the probability density function $f(x)$ is Lipschitz continuous over the bin interval $B_k$. A function is **Lipschitz continuous** if there is a positive constant $\gamma_k$ such that

$$|f(x) - f(y)| < \gamma_k |x - y|; \qquad \text{for all } x, y \text{ in } B_k . \qquad (8.9)$$

The first term in Equation 8.7 is an upper bound for the variance of the density estimate, and the second term is an upper bound for the squared bias of the density estimate. This upper bound shows what happens to the density estimate when the bin width $h$ is varied.

We can try to minimize the MSE by varying the bin width $h$. We could set $h$ very small to reduce the bias, but this also increases the variance. The increased variance in our density estimate is evident in Figure 8.1, where we see more spikes as the bin width gets smaller. Equation 8.7 shows a common problem in some density estimation methods: the trade-off between variance and bias as $h$ is changed. Most of the optimal bin widths presented here are obtained by trying to minimize the squared error.

A rule for bin width selection that is often presented in introductory statistics texts is called Sturges' Rule. In reality, it is a rule that provides the number of bins in the histogram, and is given by the following formula.

*STURGES' RULE (HISTOGRAM)*

$$k = 1 + \log_2 n .$$

Here $k$ is the number of bins. The bin width $h$ is obtained by taking the range of the sample data and dividing it into the requisite number of bins, $k$.

Some improved values for the bin width $h$ can be obtained by assuming the existence of two derivatives of the probability density function $f(x)$. We include the following results (without proof), because they are the basis for many of the univariate bin width rules presented in this chapter. The interested reader is referred to Scott [1992] for more details. Most of what we present here follows his treatment of the subject.

Equation 8.7 provides a measure of the squared error at a point $x$. If we want to measure the error in our estimate for the entire function, then we can integrate over all values of $x$. Let's assume $f(x)$ has an absolutely continuous and a square-integrable first derivative. If we let $n$ get very large ($n \to \infty$), then the asymptotic MISE is

$$\text{AMISE}_{Hist}(h) \ = \ \frac{1}{nh} + \frac{1}{12}h^2 R(f'),\qquad(8.10)$$

where $R(g) \equiv \int g^2(x)dx$ is used as a measure of the roughness of the function, and $f'$ is the first derivative of $f(x)$. The first term of Equation 8.10 indicates the asymptotic integrated variance, and the second term refers to the asymptotic integrated squared bias. These are obtained as approximations to the integrated squared bias and integrated variance [Scott, 1992]. Note, however, that the form of Equation 8.10 is similar to the upper bound for the MSE in Equation 8.7 and indicates the same trade-off between bias and variance, as the smoothing parameter $h$ changes.

The optimal bin width $h^*_{Hist}$ for the histogram is obtained by minimizing the AMISE (Equation 8.10), so it is the $h$ that yields the smallest MISE as $n$ gets large. This is given by

$$h^*_{Hist} \ = \ \left(\frac{6}{nR(f')}\right)^{1/3}.\qquad(8.11)$$

For the case of data that is normally distributed, we have a roughness of

$$R(f') \ = \ \frac{1}{4\sigma^3\sqrt{\pi}}.$$

Using this in Equation 8.11, we obtain the following expression for the optimal bin width for normal data.

*NORMAL REFERENCE RULE - 1-D HISTOGRAM*

$$h^*_{Hist} = \left(\frac{24\sigma^3\sqrt{\pi}}{n}\right)^{1/3} \approx 3.5\sigma n^{-1/3}.\qquad(8.12)$$

Scott [1979, 1992] proposed the sample standard deviation as an estimate of $\sigma$ in Equation 8.12 to get the following bin width rule.

*SCOTT'S RULE*

$$\hat{h}^*_{Hist} \ = \ 3.5 \times s \times n^{-1/3}.$$

A robust rule was developed by Freedman and Diaconis [1981]. This uses the interquartile range (IQR) instead of the sample standard deviation.

*FREEDMAN-DIACONIS RULE*

$$\hat{h}^{*}_{Hist} = 2 \times IQR \times n^{-1/3}.$$

It turns out that when the data are skewed or heavy-tailed, the bin widths are too large using the Normal Reference Rule. Scott [1979, 1992] derived the following correction factor for skewed data:

$$\text{skewness factor }_{Hist} = \frac{2^{1/3}\sigma}{e^{5\sigma^2/4}(\sigma^2 + 2)^{1/3}(e^{\sigma^2} - 1)^{1/2}}. \qquad (8.13)$$

The bin width obtained from Equation 8.12 should be multiplied by this factor when there is evidence that the data come from a skewed distribution. A factor for heavy-tailed distributions can be found in Scott [1992]. If one suspects the data come from a skewed or heavy-tailed distribution, as indicated by calculating the corresponding sample statistics (Chapter 3) or by graphical exploratory data analysis (Chapter 5), then the Normal Reference Rule bin widths should be multiplied by these factors. Scott [1992] shows that the modification to the bin widths is greater for skewness and is not so critical for kurtosis.

## Example 8.2

Data representing the waiting times (in minutes) between eruptions of the Old Faithful geyser at Yellowstone National Park were collected [Hand, et al, 1994]. These data are contained in the file **geyser**. In this example, we use an alternative MATLAB function (available in the standard MATLAB package) for finding a histogram, called **histc**. This takes the *bin edges* as one of the arguments. This is in contrast to the **hist** function that takes the *bin centers* as an optional argument. The following MATLAB code will construct a histogram density estimate for the Old Faithful geyser data.

```
load geyser
n = length(geyser);
% Use Normal Reference Rule for bin width.
h = 3.5*std(geyser)*n^(-1/3);
% Get the bin mesh.
t0 = min(geyser)-1;
tm = max(geyser)+1;
rng = tm - t0;
nbin = ceil(rng/h);
bins = t0:h:(nbin*h + t0);
% Get the bin counts vk.
vk = histc(geyser,bins);
% Normalize to make it a bona fide density.
```

```
% We do not need the last count in fhat.
fhat(end) = [];
fhat = vk/(n*h);
```

We have to use the following to create a plot of our histogram density. The MATLAB **bar** function takes the bin centers as the argument, so we convert our mesh to bin centers before plotting. The plot is shown in Figure 8.2, and the existence of two modes is apparent.

```
% To plot this, use bar with the bin centers.
tm = max(bins);
bc = (t0+h/2):h:(tm-h/2);
bar(bc,fhat,1,'w')
```

❑

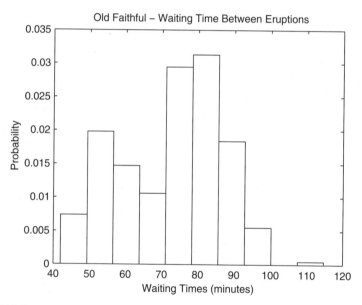

**FIGURE 8.2**
Histogram of Old Faithful **geyser** data. Here we are using Scott's Rule for the bin widths.

## Multivariate Histograms

Given a data set that contains $d$-dimensional observations $\mathbf{X}_i$, we would like to estimate the probability density $\hat{f}(\mathbf{x})$. We can extend the univariate histogram to $d$ dimensions in a straightforward way. We first partition the $d$-dimensional space into hyper-rectangles of size $h_1 \times h_2 \times \ldots \times h_d$. We denote the $k$-th bin by $B_k$ and the number of observations falling into that bin by $v_k$, with $\sum v_k = n$. The multivariate histogram is then defined as

$$\hat{f}_{Hist}(\mathbf{x}) = \frac{v_k}{nh_1h_2...h_d}; \qquad \mathbf{x} \text{ in } B_k. \tag{8.14}$$

If we need an estimate of the probability density at $\mathbf{x}$, we first determine the bin that the observation falls into. The estimate of the probability density would be given by the number of observations falling into that same bin divided by the sample size and the bin widths of the partitions. The MATLAB code to create a bivariate histogram was given in Chapter 5. This could be easily extended to the general multivariate case.

For a density function that is sufficiently smooth [Scott, 1992], we can write the asymptotic MISE for a multivariate histogram as

$$\text{AMISE}_{Hist}(\mathbf{h}) = \frac{1}{nh_1h_2...h_d} + \frac{1}{12}\sum_{j=1}^{d} h_j^2 R(f_j), \tag{8.15}$$

where $\mathbf{h} = (h_1, ..., h_d)$. As before, the first term indicates the asymptotic integrated variance and the second term provides the asymptotic integrated squared bias. This has the same general form as the 1-D histogram and shows the same bias-variance trade-off. Minimizing Equation 8.15 with respect to $h_i$ provides the following equation for optimal bin widths in the multivariate case

$$h_{i_{Hist}}^{*} = R(f_i)^{-1/2}\left(6\prod_{j=1}^{d} R(f_j)^{1/2}\right)^{\frac{1}{2+d}} n^{\frac{-1}{2+d}}, \tag{8.16}$$

where

$$R(f_i) = \int_{\mathfrak{R}^d} \left(\frac{\partial}{\partial x_i}f(\mathbf{x})\right)^2 d\mathbf{x}.$$

We can get a multivariate Normal Reference Rule by looking at the special case where the data are distributed as multivariate normal with the covariance equal to a diagonal matrix with $\sigma_1^2, ..., \sigma_d^2$ along the diagonal. The Normal Reference Rule in the multivariate case is given below [Scott, 1992].

*NORMAL REFERENCE RULE - MULTIVARIATE HISTOGRAMS*

$$h^*_{i_{Hist}} \approx 3.5\sigma_i n^{\frac{-1}{2+d}}; \qquad i = 1, ..., d.$$

Notice that this reduces to the same univariate Normal Reference Rule when $d = 1$. As before, we can use a suitable estimate for $\sigma_i$.

## Frequency Polygons

Another method for estimating probability density functions is to use a frequency polygon. A univariate frequency polygon approximates the density by linearly interpolating between the bin midpoints of a histogram with equal bin widths. Because of this, the frequency polygon extends beyond the histogram to empty bins at both ends.

The univariate probability density estimate using the frequency polygon is obtained from the following,

$$\hat{f}_{FP}(x) = \left(\frac{1}{2} - \frac{x}{h}\right)\hat{f}_k + \left(\frac{1}{2} + \frac{x}{h}\right)\hat{f}_{k+1}; \qquad \bar{B}_k \le x \le \bar{B}_{k+1}, \tag{8.17}$$

where $\hat{f}_k$ and $\hat{f}_{k+1}$ are adjacent univariate histogram values and $\bar{B}_k$ is the center of bin $B_k$. An example of a *section* of a frequency polygon is shown in Figure 8.3.

As is the case with the univariate histogram, under certain assumptions, we can write the asymptotic MISE as [Scott, 1992, 1985],

$$\text{AMISE}_{FP}(h) = \frac{2}{3nh} + \frac{49}{2880}h^4 R(f''), \tag{8.18}$$

where $f''$ is the second derivative of $f(x)$. The optimal bin width that minimizes the AMISE for the frequency polygon is given by

$$h^*_{FP} = 2\left(\frac{15}{49nR(f'')}\right)^{1/5}. \tag{8.19}$$

If $f(x)$ is the probability density function for the standard normal, then $R(f'') = 3/(8\sqrt{\pi}\sigma^5)$. Substituting this in Equation 8.19, we obtain the following Normal Reference Rule for a frequency polygon.

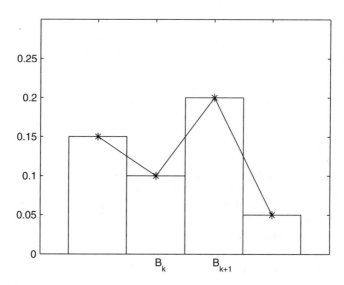

**FIGURE 8.3**
The frequency polygon is obtained by connecting the center of adjacent bins using straight lines. This figure illustrates a *section* of the frequency polygon.

*NORMAL REFERENCE RULE - FREQUENCY POLYGON*

$$h_{FP}^{*} = 2.15 \sigma n^{-1/5}.$$

We can use the sample standard deviation in this rule as an estimate of $\sigma$ or choose a robust estimate based on the interquartile range. If we choose the $IQR$ and use $\hat{\sigma} = IQR/1.348$, then we obtain a bin width of

$$\hat{h}_{FP}^{*} = 1.59 \times IQR \times n^{-1/5}.$$

As for the case of histograms, Scott [1992] provides a skewness factor for frequency polygons, given by

$$\text{skewness factor}_{FP} = \frac{12^{1/5} \sigma}{e^{7\sigma^{2}/4}(e^{\sigma^{2}} - 1)^{1/2}(9\sigma^{4} + 20\sigma^{2} + 12)^{1/5}}. \qquad (8.20)$$

If there is evidence that the data come from a skewed distribution, then the bin width should be multiplied by this factor. The kurtosis factor for frequency polygons can be found in Scott [1992].

## Example 8.3

Here we show how to create a frequency polygon using the Old Faithful **geyser** data. We must first create the histogram from the data, where we use the frequency polygon Normal Reference Rule to choose the smoothing parameter.

```
load geyser
n = length(geyser);
% Use Normal Reference Rule for bin width
% of frequency polygon.
h = 2.15*sqrt(var(geyser))*n^(-1/5);
t0 = min(geyser)-1;
tm = max(geyser)+1;
bins = t0:h:tm;
vk = histc(geyser,bins);
vk(end) = [];
fhat = vk/(n*h);
```

We then use the MATLAB function called **interp1** to interpolate between the bin centers. This function takes three arguments (and an optional fourth argument). The first two arguments to **interp1** are the **xdata** and **ydata** vectors that contain the observed data. In our case, these are the bin centers and the bin heights from the density histogram. The third argument is a vector of **xinterp** values for which we would like to obtain interpolated **yinterp** values. There is an optional fourth argument that allows the user to select the type of interpolation (**linear, cubic, nearest** and **spline**). The default is **linear**, which is what we need for the frequency polygon. The following code constructs the frequency polygon for the **geyser** data.

```
% For frequency polygon, get the bin centers,
% with empty bin center on each end.
bc2 = (t0-h/2):h:(tm+h/2);
binh = [0 fhat 0];
% Use linear interpolation between bin centers
% Get the interpolated values at x.
xinterp = linspace(min(bc2),max(bc2));
fp = interp1(bc2, binh, xinterp);
```

To see how this looks, we can plot the frequency polygon and underlying histogram, which is shown in Figure 8.4.

```
% To plot this, use bar with the bin centers
tm = max(bins);
bc = (t0+h/2):h:(tm-h/2);
bar(bc,fhat,1,'w')
hold on
plot(xinterp,fp)
hold off
```

```
axis([30 120 0 0.035])
xlabel('Waiting Time (minutes)')
ylabel('Probability Density Function')
title('Old Faithful-Waiting Times Between Eruptions')
```

To ensure that we have a valid probability density function, we can verify that the area under the curve is approximately one by using the **trapz** function.

```
area = trapz(xinterp,fp);
```

We get an approximate area under the curve of 0.9998, indicating that the frequency polygon is indeed a *bona fide* density estimate.
☐

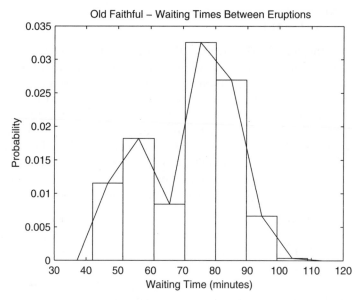

**FIGURE 8.4**
Frequency polygon for the Old Faithful data.

The frequency polygon can be extended to the multivariate case. The interested reader is referred to Scott [1985, 1992] for more details on the multivariate frequency polygon. He proposes an approximate Normal Reference Rule for the multivariate frequency polygon given by the following formula.

NORMAL REFERENCE RULE - FREQUENCY POLYGON (MULTIVARIATE)

$$h_i^* = 2\sigma_i n^{-1/(4+d)},$$

where a suitable estimate for $\sigma_i$ can be used. This is derived using the assumption that the true probability density function is multivariate normal with covariance equal to the identity matrix. The following example illustrates the procedure for obtaining a bivariate frequency polygon in MATLAB.

## Example 8.4

We first generate some random variables that are bivariate standard normal and then calculate the surface heights corresponding to the linear interpolation between the histogram density bin heights.

```
% First get the constants.
bin0 = [-4 -4];
n = 1000;
% Normal Reference Rule with sigma = 1.
h = 3*n^(-1/4)*ones(1,2);
% Generate bivariate standard normal variables.
x = randn(n,2);
% Find the number of bins.
nb1 = ceil((max(x(:,1))-bin0(1))/h(1));
nb2 = ceil((max(x(:,2))-bin0(2))/h(2));
% Find the mesh or bin edges.
t1 = bin0(1):h(1):(nb1*h(1)+bin0(1));
t2 = bin0(2):h(2):(nb2*h(2)+bin0(2));
[X,Y] = meshgrid(t1,t2);
```

Now that we have the random variables and the bin edges, the next step is to find the number of observations that fall into each bin. This is easily done with the MATLAB function **inpolygon**. This function can be used with any polygon (e.g., triangle or hexagon), and it returns the indices to the points that fall into that polygon.

```
% Find bin frequencies.
[nr,nc] = size(X);
vu = zeros(nr-1,nc-1);
for i = 1:(nr-1)
  for j = 1:(nc-1)
    xv = [X(i,j) X(i,j+1) X(i+1,j+1) X(i+1,j)];
    yv = [Y(i,j) Y(i,j+1) Y(i+1,j+1) Y(i+1,j)];
    in = inpolygon(x(:,1),x(:,2),xv,yv);
    vu(i,j) = sum(in(:));
  end
end
```

```
fhat = vu/(n*h(1)*h(2));
```

Now that we have the histogram density, we can use the MATLAB function interp2 to linearly interpolate at points between the bin centers.

```
% Now get the bin centers for the frequency polygon.
% We add bins at the edges with zero height.
t1 = (bin0(1)-h(1)/2):h(1):(max(t1)+h(1)/2);
t2 = (bin0(2)-h(2)/2):h(2):(max(t2)+h(2)/2);
[bcx,bcy] = meshgrid(t1,t2);
[nr,nc] = size(fhat);
binh = zeros(nr+2,nc+2);    % add zero bin heights
binh(2:(1+nr),2:(1+nc))=fhat;
% Get points where we want to interpolate to get
% the frequency polygon.
[xint,yint]=meshgrid(linspace(min(t1),max(t1),30),...
    linspace(min(t2),max(t2),30));
fp = interp2(bcx,bcy,binh,xint,yint,'linear');
```

We can verify that this is a valid density by estimating the area under the curve.

```
df1 = xint(1,2)-xint(1,1);
df2 = yint(2,1)-yint(1,1);
area = sum(sum(fp))*df1*df2;
```

This yields an area of 0.9976. A surface plot of the frequency polygon is shown in Figure 8.5.
□


### Averaged Shifted Histograms

When we create a histogram or a frequency polygon, we need to specify a complete mesh determined by the bin width $h$ and the starting point $t_0$. The reader should have noticed that the parameter $t_0$ did not appear in any of the asymptotic integrated squared bias or integrated variance expressions for the histograms or frequency polygons. The MISE is affected more by the choice of bin width than the choice of starting point $t_0$. The averaged shifted histogram (ASH) was developed to account for different choices of $t_0$, with the added benefit that it provides a *'smoother'* estimate of the probability density function.

The idea is to create many histograms with different bin origins $t_0$ (but with the same $h$) and average the histograms together. The histogram is a piecewise constant function, and the average of piecewise constant functions will also be the same type of function. Therefore, the ASH is also in the form of a histogram, and the following discussion treats it as such. The ASH is often implemented in conjunction with the frequency polygon, where the latter is used to linearly interpolate between the smaller bin widths of the ASH.

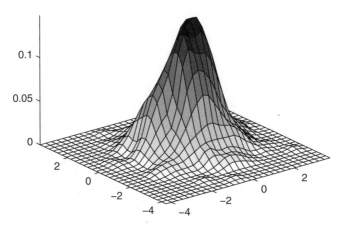

**FIGURE 8.5.**
Frequency polygon of bivariate standard normal data.

To construct an ASH, we have a set of $m$ histograms, $\hat{f}_1, ..., \hat{f}_m$ with constant bin width $h$. The origins are given by the sequence

$$t'_0 = t_0 + 0, t_0 + \frac{h}{m}, t_0 + \frac{2h}{m}, ..., t_0 + \frac{(m-1)h}{m}.$$

In the univariate case, the unweighted or naive ASH is given by

$$\hat{f}_{ASH}(x) = \frac{1}{m} \sum_{i=1}^{m} \hat{f}_i(x), \tag{8.21}$$

which is just the average of the histogram estimates at each point $x$. It should be clear that the $\hat{f}_{ASH}$ is a piecewise function over smaller bins, whose width is given by $\delta = h/m$. This is shown in Figure 8.6 where we have a single histogram $\hat{f}_i$ and the ASH estimate.

In what follows, we consider the ASH as a histogram over the narrower intervals given by $B'_k = [k\delta, (k+1)\delta)$, with $\delta = h/m$. As before we denote the bin counts for these bins by $v_k$. An alternative expression for the naive ASH can be written as

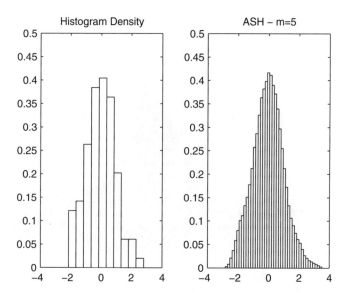

**FIGURE 8.6**
On the left is a histogram density based on 100 standard normal random variables, where we used the MATLAB default of 10 bins. On the right is an ASH estimate for the same data set, with $m = 5$.

$$\hat{f}_{ASH}(x) = \frac{1}{nh} \sum_{i = 1 - m}^{m - 1} \left(1 - \frac{|i|}{m}\right)v_{k + i}; \qquad x \text{ in } B'_k. \qquad (8.22)$$

To make this a little clearer, let's look at a simple example of the naive ASH, with $m = 3$. In this case, our estimate at a point $x$ is

$$\hat{f}_{ASH}(x) = \frac{1}{nh}\left[\left(1 - \frac{2}{3}\right)v_{k - 2} + \left(1 - \frac{1}{3}\right)v_{k - 1} + \left(1 - \frac{0}{3}\right)v_{k - 0} + \right.$$
$$\left.\left(1 - \frac{1}{3}\right)v_{k + 1} + \left(1 - \frac{2}{3}\right)v_{k + 2}\right]; \qquad x \text{ in } B'_k.$$

We can think of the factor $(1 - |i|/m)$ in Equation 8.22 as weights on the bin counts. We can use arbitrary weights instead, to obtain the general ASH.

*GENERAL AVERAGED SHIFTED HISTOGRAM*

$$\hat{f}_{ASH} = \frac{1}{nh} \sum_{|i| < m} w_m(i)v_{k + i}; \qquad x \text{ in } B'_k \ . \qquad (8.23)$$

A general formula for the weights is given by

$$w_m(i) = m \times \frac{K(i/m)}{\sum\limits_{j=1-m}^{m-1} K(j/m)}; \qquad i = 1-m, \ldots, m-1, \qquad (8.24)$$

with $K$ a continuous function over the interval $[-1, 1]$. This function $K$ is sometimes chosen to be a probability density function. In Example 8.5, we use the biweight function:

$$K(t) = \frac{15}{16}(1-t^2)^2 I_{[-1,1]}(t) \qquad (8.25)$$

for our weights. Here $I_{[-1,1]}$ is the indicator function over the interval $[-1, 1]$.

The algorithm for the general univariate ASH [Scott, 1992] is given here and is also illustrated in MATLAB in Example 8.5. This algorithm requires at least $m-1$ empty bins on either end.

*UNIVARIATE ASH - ALGORITHM:*

1. Generate a mesh over the range $(t_0, nbin \times \delta + t_0)$ with bin widths of size $\delta$, $\delta \ll h$ and $h = m\delta$. The quantity *nbin* is the number of bins - see the comments below for more information on this number. Include at least $m$ - 1 empty bins on either end of the range.
2. Compute the bin counts $v_k$.
3. Compute the weight vector $w_m(i)$ given in Equation 8.24.
4. Set all $\hat{f}_k = 0$.
5. Loop over $k = 1$ to *nbin*

   Loop over $i = max\{1, k-m+1\}$ to $min\{nbin, k+m-1\}$

$$\text{Calculate: } \hat{f}_i = \hat{f}_i + v_k w_m(i-k).$$

6. Divide all $\hat{f}_k$ by *nh*, these are the ASH heights.
7. Calculate the bin centers using $\bar{B}_k = t_0 + (k-0.5)\delta$.

In practice, one usually chooses the *m* and *h* by setting the number of narrow (size $\delta$) bins between 50 and 500 over the range of the sample. This is then extended to put some empty bins on either end of the range.

## Example 8.5

In this example, we construct an ASH probability density estimate of the Buffalo `snowfall` data [Scott, 1992]. These data represent the annual snowfall in inches in Buffalo, New York over the years 1910-1972. First load the data and get the appropriate parameters.

```
load snowfall
n = length(snowfall);
m = 30;
h = 14.6;
delta = h/m;
```

The next step is to construct a mesh using the smaller bin widths of size $\delta$ over the desired range. Here we start the density estimate at zero.

```
% Get the mesh.
t0 = 0;
tf = max(snowfall)+20;
nbin = ceil((tf-t0)/delta);
binedge = t0:delta:(t0+delta*nbin);
```

We need to obtain the bin counts for these smaller bins, and we use the `histc` function since we want to use the bin edges rather than the bin centers.

```
% Get the bin counts for the smaller binwidth delta.
vk = histc(snowfall,binedge);
% Put into a vector with m-1 zero bins on either end.
fhat = [zeros(1,m-1),vk,zeros(1,m-1)];
```

Next, we construct our weight vector according to Equation 8.24, where we use the biweight kernel given in Equation 8.25. Instead of writing the kernel as a separate function, we will use the MATLAB `inline` function to create a function object. We can then call that `inline` function just as we would an M-file function.

```
% Get the weight vector.
% Create an inline function for the kernel.
kern = inline('(15/16)*(1-x.^2).^2');
ind = (1-m):(m-1);
% Get the denominator.
den = sum(kern(ind/m));
% Create the weight vector.
wm = m*(kern(ind/m))/den;
```

The following section of code essentially implements steps 5 - 7 of the ASH algorithm.

```
% Get the bin heights over smaller bins.
fhatk = zeros(1,nbin);
for k = 1:nbin
```

```
    ind = k:(2*m+k-2);
    fhatk(k) = sum(wm.*fhat(ind));
end
fhatk = fhatk/(n*h);
bc = t0+((1:k)-0.5)*delta;
```

We use the following steps to obtain Figure 8.7, where we use a different type of MATLAB plot to show the ASH estimate. We use the bin edges with the `stairs` plot, so we must append an extra bin height at the end to ensure that the last bin is drawn and to make it dimensionally correct for plotting.

```
% To use the stairs plot, we need to use the
% bin edges.
stairs(binedge,[fhatk fhatk(end)])
axis square
title('ASH - Buffalo Snowfall Data')
xlabel('Snowfall (inches)')
```

❑

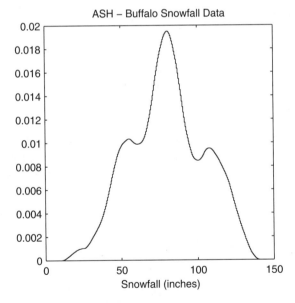

**FIGURE 8.7**
ASH estimate for the Buffalo snowfall data. The parameters used to obtain this were $h = 14.6$ inches and $m = 30$. Notice that the ASH estimate reveals evidence of three modes.

The multivariate ASH is obtained by averaging shifted multivariate histograms. Each histogram has the same bin dimension $h_1 \times \ldots \times h_d$, and each is constructed using shifts along the coordinates given by multiples of $\delta_i/m_i$, $i = 1, ..., d$. Scott [1992] provides a detailed algorithm for the bivariate ASH.

## 8.3 Kernel Density Estimation

Scott [1992] shows that as the number of histograms $m$ approaches infinity, the ASH becomes a kernel estimate of the probability density function. The first published paper describing nonparametric probability density estimation was by Rosenblatt [1956], where he described the general kernel estimator. Many papers that expanded the theory followed soon after. A partial list includes Parzen [1962], Cencov [1962] and Cacoullos [1966]. Several references providing surveys and summaries of nonparametric density estimation are provided in Section 8.7. The following treatment of kernel density estimation follows that of Silverman [1986] and Scott [1992].

### Univariate Kernel Estimators

The kernel estimator is given by

$$\hat{f}_{Ker}(x) = \frac{1}{nh} \sum_{i=1}^{n} K\left(\frac{x - X_i}{h}\right), \tag{8.26}$$

where the function $K(t)$ is called a **kernel**. This must satisfy the condition that $\int K(t)dt = 1$ to ensure that our estimate in Equation 8.26 is a *bona fide* density estimate. If we define $K_h(t) = K(t/h)/h$, then we can also write the kernel estimate as

$$\hat{f}_{Ker}(x) = \frac{1}{n} \sum_{i=1}^{n} K_h(x - X_i). \tag{8.27}$$

Usually, the kernel is a symmetric probability density function, and often a standard normal density is used. However, this does not have to be the case, and we will present other choices later in this chapter. From the definition of a kernel density estimate, we see that our estimate $\hat{f}_{Ker}(x)$ inherits all of the properties of the kernel function, such as continuity and differentiability..

From Equation 8.26, the estimated probability density function is obtained by placing a weighted kernel function, centered at each data point and then taking the average of them. See Figure 8.8 for an illustration of this procedure.

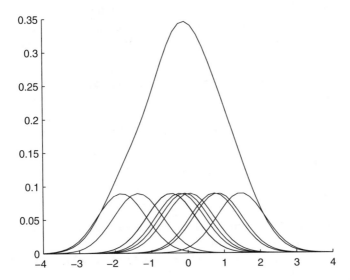

**FIGURE 8.8.**
We obtain the above kernel density estimate for $n = 10$ random variables. A weighted kernel is centered at each data point, and the curves are averaged together to obtain the estimate. Note that there are two 'bumps' where there is a higher concentration of smaller densities.

Notice that the places where there are more curves or kernels yield *'bumps'* in the final estimate. An alternative implementation is discussed in the exercises.

*PROCEDURE - UNIVARIATE KERNEL*

   1. Choose a kernel, a smoothing parameter $h$, and the domain (the set of $x$ values) over which to evaluate $\hat{f}(x)$.

   2. For each $X_i$, evaluate the following kernel at all $x$ in the domain:

$$K_i = K\left(\frac{x - X_i}{h}\right); \qquad i = 1, \ldots, n.$$

   The result from this is a set of $n$ curves, one for each data point $X_i$.

   3. Weight each curve by $1/h$.

   4. For each $x$, take the average of the weighted curves.

## Example 8.6

In this example, we show how to obtain the kernel density estimate for a data set, using the standard normal density as our kernel. We use the procedure outlined above. The resulting probability density estimate is shown in Figure 8.8.

```
% Generate standard normal random variables.
n = 10;
data = randn(1,n);
% We will get the density estimate at these x values.
x = linspace(-4,4,50);
fhat = zeros(size(x));
h = 1.06*n^(-1/5);
hold on
for i=1:n
    % get each kernel function evaluated at x
    % centered at data
    f = exp(-(1/(2*h^2))*(x-data(i)).^2)/sqrt(2*pi)/h;
    plot(x,f/(n*h));
    fhat = fhat+f/(n);
end
plot(x,fhat);
hold off
```

❑

As in the histogram, the parameter $h$ determines the amount of smoothing we have in the estimate $\hat{f}_{Ker}(x)$. In kernel density estimation, the $h$ is usually called the *window width*. A small value of $h$ yields a rough curve, while a large value of $h$ yields a smoother curve. This is illustrated in Figure 8.9, where we show kernel density estimates $\hat{f}_{Ker}(x)$ at various window widths. Notice that when the window width is small, we get a lot of noise or spurious structure in the estimate. When the window width is larger we get a smoother estimate, but there is the possibility that we might obscure bumps or other interesting structure in the estimate. In practice, it is recommended that the analyst examine kernel density estimates for different window widths to explore the data and to search for structures such as modes or bumps.

As with the other univariate probability density estimators, we are interested in determining appropriate values for the parameter $h$. These can be obtained by choosing values for $h$ that minimize the asymptotic MISE. Scott [1992] shows that, under certain conditions, the AMISE for a nonnegative univariate kernel density estimator is

$$\text{AMISE}_{Ker}(h) = \frac{R(K)}{nh} + \frac{1}{4}\sigma_k^4 h^4 R(f''), \quad (8.28)$$

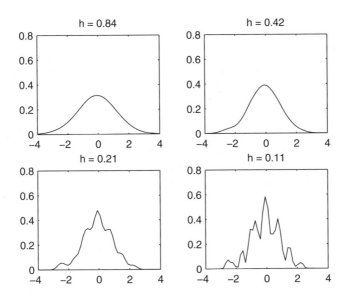

**FIGURE 8.9**
Four kernel density estimates using $n = 100$ standard normal random variables. Four different window widths are used. Note that as $h$ gets smaller, the estimate gets rougher.

where the kernel $K$ is a continuous probability density function with $\mu_K = 0$ and $0 < \sigma_K^2 < \infty$. The window width that minimizes this is given by

$$h_{Ker}^* = \left( \frac{R(K)}{n\sigma_k^4 R(f'')} \right)^{1/5} . \tag{8.29}$$

Parzen [1962] and Scott [1992] describe the conditions under which this holds. Notice in Equation 8.28 that we have the same bias-variance trade-off with $h$ that we had in previous density estimates.

For a kernel that is equal to the normal density $R(f'') = 3/(8\sqrt{\pi}\sigma^5)$, we have the following Normal Reference Rule for the window width $h$.

*NORMAL REFERENCE RULE - KERNELS*

$$h_{Ker}^* = \left( \frac{4}{3} \right)^{1/5} \sigma n^{-1/5} \approx 1.06 \sigma n^{-1/5} .$$

We can use some suitable estimate for $\sigma$, such as the standard deviation, or $\hat{\sigma} = IQR/1.348$. The latter yields a window width of

$$\hat{h}_{Ker}^{*} = 0.786 \times IQR \times n^{-1/5}.$$

Silverman [1986] recommends that one use whichever is smaller, the sample standard deviation or $IQR/1.348$ as an estimate for $\sigma$.

We now turn our attention to the problem of what kernel to use in our estimate. It is known [Scott, 1992] that the choice of smoothing parameter $h$ is more important than choosing the kernel. This arises from the fact that the effects from the choice of kernel (e.g., kernel tail behavior) are reduced by the averaging process. We discuss the efficiency of the kernels below, but what really drives the choice of a kernel are computational considerations or the amount of differentiability required in the estimate.

In terms of efficiency, the optimal kernel was shown to be [Epanechnikov, 1969]

$$K(t) = \begin{cases} \dfrac{3}{4}(1 - t^2); & -1 \le t \le 1 \\ 0; & \text{otherwise.} \end{cases}$$

It is illustrated in Figure 8.10 along with some other kernels.

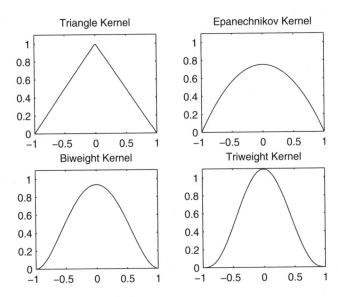

**FIGURE 8.10**
These illustrate four kernels that can be used in probability density estimation.

Several choices for kernels are given in Table 8.1. Silverman [1986] and Scott [1992] show that these kernels have efficiencies close to that of the Epanechnikov kernel, the least efficient being the normal kernel. Thus, it seems that efficiency should not be the major consideration in deciding what kernel to use. It is recommended that one choose the kernel based on other considerations as stated above.

**TABLE 8.1**

Examples of Kernels for Density Estimation

| Kernel Name | Equation |
|---|---|
| Triangle | $K(t) = (1 - |t|) \qquad -1 \le t \le 1$ |
| Epanechnikov | $K(t) = \frac{3}{4}(1 - t^2) \qquad -1 \le t \le 1$ |
| Biweight | $K(t) = \frac{15}{16}(1 - t^2)^2 \qquad -1 \le t \le 1$ |
| Triweight | $K(t) = \frac{35}{32}(1 - t^2)^3 \qquad -1 \le t \le 1$ |
| Normal | $K(t) = \frac{1}{\sqrt{2\pi}} \exp\left\{\frac{-t^2}{2}\right\} \qquad -\infty < t < \infty$ |

## Multivariate Kernel Estimators

Here we assume that we have a sample of size $n$, where each observation is a $d$-dimensional vector, $\mathbf{X}_i, i = 1, ..., n$. The simplest case for the multivariate kernel estimator is the product kernel. Descriptions of the general kernel density estimate can be found in Scott [1992] and in Silverman [1986]. The *product kernel* is

$$\hat{f}_{Ker}(\mathbf{x}) = \frac{1}{nh_1...h_d} \sum_{i=1}^{n} \left\{ \prod_{j=1}^{d} K\left(\frac{x_j - X_{ij}}{h_j}\right) \right\}, \tag{8.30}$$

where $X_{ij}$ is the $j$-th component of the $i$-th observation. Note that this is the product of the same univariate kernel, with a (possibly) different window width in each dimension. Since the product kernel estimate is comprised of univariate kernels, we can use any of the kernels that were discussed previously.

Scott [1992] gives expressions for the asymptotic integrated squared bias and asymptotic integrated variance for the multivariate product kernel. If the normal kernel is used, then minimizing these yields a normal reference rule for the multivariate case, which is given below.

*NORMAL REFERENCE RULE - KERNEL (MULTIVARIATE)*

$$h^*_{j_{Ker}} = \left(\frac{4}{n(d+2)}\right)^{\frac{1}{d+4}} \sigma_j; \qquad j = 1, \dots, d,$$

where a suitable estimate for $\sigma_j$ can be used. If there is any skewness or kurtosis evident in the data, then the window widths should be narrower, as discussed previously. The skewness factor for the frequency polygon (Equation 8.20) can be used here.

## Example 8.7

In this example, we construct the product kernel estimator for the **iris** data. To make it easier to visualize, we use only the first two variables (sepal length and sepal width) for each species. So, we first create a data matrix comprised of the first two columns for each species.

```
load iris
% Create bivariate data matrix with all three species.
data = [setosa(:,1:2)];
data(51:100,:) = versicolor(:,1:2);
data(101:150,:) = virginica(:,1:2);
```

Next we obtain the smoothing parameter using the Normal Reference Rule.

```
% Get the window width using the Normal Ref Rule.
[n,p] = size(data);
s = sqrt(var(data));
hx = s(1)*n^(-1/6);
hy = s(2)*n^(-1/6);
```

The next step is to create a grid over which we will construct the estimate.

```
% Get the ranges for x and y & construct grid.
num_pts = 30;
minx = min(data(:,1));
maxx = max(data(:,1));
miny = min(data(:,2));
maxy = max(data(:,2));
```

```
gridx = ((maxx+2*hx)-(minx-2*hx))/num_pts
gridy = ((maxy+2*hy)-(miny-2*hy))/num_pts
[X,Y]=meshgrid((minx-2*hx):gridx:(maxx+2*hx),...
    (miny-2*hy):gridy:(maxy+2*hy));
x = X(:);    %put into col vectors
y = Y(:);
```

We are now ready to get the estimates. Note that in this example, we are changing the form of the loop. Instead of evaluating each weighted curve and then averaging, we will be looping over each point in the domain.

```
z = zeros(size(x));
for i=1:length(x)
    xloc = x(i)*ones(n,1);
    yloc = y(i)*ones(n,1);
    argx = ((xloc-data(:,1))/hx).^2;
    argy = ((yloc-data(:,2))/hy).^2;
    z(i) = (sum(exp(-.5*(argx+argy))))/(n*hx*hy*2*pi);
end
[mm,nn] = size(X);
Z = reshape(z,mm,nn);
```

We show the surface plot for this estimate in Figure 8.11. As before, we can verify that our estimate is a *bona fide* by estimating the area under the curve. In this example, we get an area of 0.9994.

```
area = sum(sum(Z))*gridx*gridy;
```

❑

Before leaving this section, we present a summary of univariate probability density estimators and their corresponding Normal Reference Rule for the smoothing parameter $h$. These are given in Table 8.2.

## 8.4 Finite Mixtures

So far, we have been discussing nonparametric density estimation methods that require a choice of smoothing parameter $h$. In the previous section, we showed that we can get different estimates of our probability density depending on our choice for $h$. It would be helpful if we could avoid choosing a smoothing parameter. In this section, we present a method called finite mixtures that does not require a smoothing parameter. However, as is often the case, when we eliminate one parameter we end up replacing it with another. In finite mixtures, we do not have to worry about the smoothing parameter. Instead, we have to determine the number of terms in the mixture.

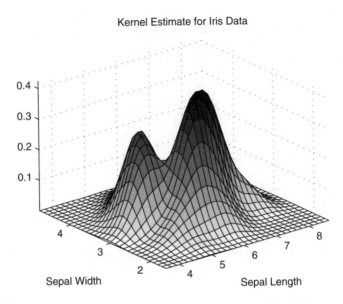

**FIGURE 8.11**
This is the product kernel density estimate for the sepal length and sepal width of the `iris` data. These data contain all three species. The presence of peaks in the data indicate that two of the species might be distinguishable based on these two variables.

**TABLE 8.2**

Summary of Univariate Probability Density Estimators and the Normal Reference Rule for the Smoothing Parameter

| Method | Estimator | Normal Reference Rule |
|---|---|---|
| Histogram | $\hat{f}_{Hist}(x) = \dfrac{v_k}{nh}$ <br><br> $x$ in $B_k$ | $h^*_{Hist} = 3.5 \sigma n^{-1/3}$ |
| Frequency Polygon | $\hat{f}_{FP}(x) = \left(\dfrac{1}{2} - \dfrac{x}{h}\right)\hat{f}_k + \left(\dfrac{1}{2} + \dfrac{x}{h}\right)\hat{f}_{k+1}$ <br><br> $\bar{B}_K \le x \le \bar{B}_{k+1}$ | $h^*_{FP} = 2.15 \sigma n^{-1/5}$ |
| Kernel | $\hat{f}_{Ker}(x) = \dfrac{1}{nh} \sum_{i=1}^{n} K\left(\dfrac{x - X_i}{h}\right)$ | $h^*_{Ker} = 1.06 \sigma n^{-1/5}$; <br><br> $K$ is the normal kernel. |

Finite mixtures offer advantages in the area of the computational load put on the system. Two issues to consider with many probability density estimation methods are the computational burden in terms of the amount of information we have to store and the computational effort needed to obtain the probability density estimate at a point. We can illustrate these ideas using the kernel density estimation method. To evaluate the estimate at a point $x$ (in the univariate case) we have to retain all of the data points, because the estimate is a weighted sum of $n$ kernels centered at each sample point. In addition, we must calculate the value of the kernel $n$ times. The situation for histograms and frequency polygons is a little better. The amount of information we must store to provide an estimate of the probability density is essentially driven by the number of bins. Of course, the situation becomes worse when we move to multivariate kernel estimates, histograms, and frequency polygons. With the massive, high-dimensional data sets we often work with, the computational effort and the amount of information that must be stored to use the density estimates is an important consideration. Finite mixtures is a technique for estimating probability density functions that can require relatively little computer storage space or computations to evaluate the density estimates.

## Univariate Finite Mixtures

The finite mixture method assumes the density $f(x)$ can be modeled as the sum of $c$ weighted densities, with $c \ll n$. The most general case for the univariate finite mixture is

$$f(x) = \sum_{i=1}^{c} p_i g(x; \theta_i),$$   (8.31)

where $p_i$ represents the *weight* or *mixing coefficient* for the $i$-th term, and $g(x; \theta_i)$ denotes a probability density, with parameters represented by the vector $\theta_i$. To make sure that this is a *bona fide* density, we must impose the condition that $p_1 + \ldots + p_c = 1$ and $p_i > 0$. To evaluate $f(x)$, we take our point $x$, find the value of the component densities $g(x; \theta_i)$ at that point, and take the weighted sum of these values.

## Example 8.8

The following example shows how to evaluate a finite mixture model at a given $x$. We construct the curve for a three term finite mixture model, where the component densities are taken to be normal. The model is given by

$$f(x) = 0.3 \times \phi(x; -3, 1) + 0.3 \times \phi(x; 0, 1) + 0.4 \times \phi(x; 2, 0.5),$$

where $\phi(x;\mu, \sigma^2)$ represents the normal probability density function at $x$. We see from the model that we have three terms or component densities, centered at -3, 0, and 2. The mixing coefficient or weight for the first two terms are 0.3 leaving a weight of 0.4 for the last term. The following MATLAB code produces the curve for this model and is shown in Figure 8.12.

```
% Create a domain x for the mixture.
x = linspace(-6,5);
% Create the model - normal components used.
mix = [0.3 0.3 0.4];          % mixing coefficients
mus = [-3 0 2];               % term means
vars = [1 1 0.5];
nterm = 3;
% Use Statistics Toolbox function to evaluate
% normal pdf.
fhat = zeros(size(x));
for i = 1:nterm
    fhat = fhat+mix(i)*normpdf(x,mus(i),vars(i));
end
plot(x,fhat)
title('3 Term Finite Mixture')
```

❑

Hopefully, the reader can see the connection between finite mixtures and kernel density estimation. Recall that in the case of univariate kernel density estimators, we obtain these by evaluating a weighted kernel centered at each sample point, and adding these $n$ terms. So, a kernel estimate can be considered a special case of a finite mixture where $c = n$.

The component densities of the finite mixture can be any probability density function, continuous or discrete. In this book, we confine our attention to the continuous case and use the normal density for the component function. Therefore, the estimate of a finite mixture would be written as

$$\hat{f}_{FM}(x) = \sum_{i=1}^{c} \hat{p}_i \phi(x;\hat{\mu}_i, \hat{\sigma}_i^2), \tag{8.32}$$

where $\phi(x;\hat{\mu}_i, \hat{\sigma}_i^2)$ denotes the normal probability density function with mean $\hat{\mu}_i$ and variance $\hat{\sigma}_i^2$. In this case, we have to estimate $c-1$ independent mixing coefficients, as well as the $c$ means and $c$ variances using the data. Note that to evaluate the density estimate at a point $x$, we only need to retain these $3c - 1$ parameters. Since $c \ll n$, this can be a significant computational savings over evaluating density estimates using the kernel method. With finite mixtures much of the computational burden is shifted to the estimation part of the problem.

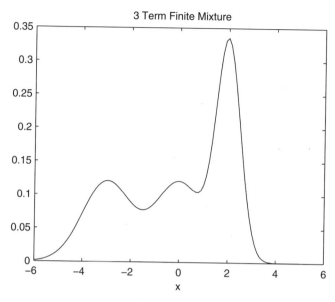

**FIGURE 8.12**
This shows the probability density function corresponding to the three-term finite mixture model from Example 8.8.

## Visualizing Finite Mixtures

The methodology used to estimate the parameters for finite mixture models will be presented later on in this section ( page 296 ). We first show a method for visualizing the underlying structure of finite mixtures with normal component densities [Priebe, et al. 1994], because it is used to help visualize and explain another approach to density estimation (adaptive mixtures). Here, structure refers to the number of terms in the mixture, along with the component means and variances. In essence, we are trying to visualize the high-dimensional parameter space (recall there are $3c$-1 parameters for the univariate mixture of normals) in a 2-D representation. This is called a $dF$ plot, where each component is represented by a circle. The circles are centered at the mean $\mu_i$ and the mixing coefficient $p_i$. The size of the radius of the circle indicates the standard deviation. An example of a $dF$ plot is given in Figure 8.13 and is discussed in the following example.

### Example 8.9

We construct a $dF$ plot for the finite mixture model discussed in the previous example. Recall that the model is given by

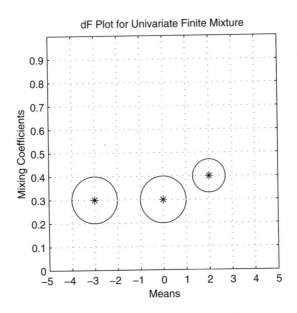

**FIGURE 8.13**
This shows the *dF* plot for the three term finite mixture model of Figure 8.12.

$$f(x) = 0.3 \times \phi(x;-3, 1) + 0.3 \times \phi(x;0, 1) + 0.4 \times \phi(x;2, 0.5).$$

Our first step is to set up the model consisting of the number of terms, the component parameters and the mixing coefficients.

```
% Recall the model - normal components used.
mix = [0.3 0.3 0.4];      % mixing coefficients
mus = [-3 0 2];           % term means
vars = [1 1 0.5];
nterm = 3;
```

Next we set up the figure for plotting. Note that we re-scale the mixing coefficients for easier plotting on the vertical axis and then map the labels to the corresponding value.

```
t = 0:.05:2*pi+eps;   % values to create circle
% To get some scales right.
minx = -5;
maxx = 5;
scale = maxx-minx;
lim = [minx maxx minx maxx];
% Set up the axis limits.
```

```
figure
axis equal
axis(lim)
grid on
% Create and plot a circle for each term.
hold on
for i=1:nterm
    % rescale for plotting purposes
    ycord = scale*mix(i)+minx;
    xc = mus(i)+sqrt(vars(i))*cos(t);
    yc = ycord+sqrt(vars(i))*sin(t);
    plot(xc,yc,mus(i),ycord,'*')
end
hold off
% Relabel the axis to show the right coefficient.
tick = (maxx-minx)/10;
set(gca,'Ytick',minx:tick:maxx)
set(gca,'XTick',minx:tick:maxx)
set(gca,'YTickLabel',...
   '0|0.1|0.2|0.3|0.4|0.5|0.6|0.7|0.8|0.9|1')
xlabel('Means'),ylabel('Mixing Coefficients')
title('dF Plot for Univariate Finite Mixture')
```

The first circle on the left corresponds to the component with $p_i = 0.3$ and $\mu_i = -3$. Similarly, the middle circle of Figure 8.13 represents the second term of the model. Note that this representation of the mixture makes it easier to see which terms carry more weight and where they are located in the domain.
□

## Multivariate Finite Mixtures

Finite mixtures is easily extended to the multivariate case. Here we define the multivariate finite mixture model as the weighted sum of multivariate component densities,

$$f(\mathbf{x}) = \sum_{i=1}^{c} p_i g(\mathbf{x}; \theta_i).$$

As before, the mixing coefficients or weights must be nonnegative and sum to one, and the component density parameters are represented by $\theta_i$. When we are estimating the function, we often use the multivariate normal as the component density. This gives the following equation for an estimate of a multivariate finite mixture

$$\hat{f}_{FM}(\mathbf{x}) = \sum_{i=1}^{c} \hat{p}_i \phi(\mathbf{x}; \hat{\mu}_i, \hat{\Sigma}_i), \qquad (8.33)$$

where $\mathbf{x}$ is a $d$-dimensional vector, $\hat{\mu}_i$ is a $d$-dimensional vector of means, and $\hat{\Sigma}_i$ is a $d \times d$ covariance matrix. There are still $c$-1 mixing coefficients to estimate. However, there are now $c \times d$ values that have to be estimated for the means and $(cd(c+1))/2$ values for the component covariance matrices.

The $dF$ representation has been extended [Solka, Poston, Wegman, 1995] to show the structure of a multivariate finite mixture, when the data are 2-D or 3-D. In the 2-D case, we represent each term by an ellipse centered at the mean of the component density $\hat{\mu}_i$, with the eccentricity of the ellipse showing the covariance structure of the term. For example, a term with a covariance that is close to the identity matrix will be shown as a circle. We label the center of each ellipse with text identifying the mixing coefficient. An example is illustrated in Figure 8.14.

A $dF$ plot for a trivariate finite mixture can be fashioned by using color to represent the values of the mixing coefficients. In this case, we use the three dimensions in our plot to represent the means for each term. Instead of ellipses, we move to ellipsoids, with eccentricity determined by the covariance as before. See Figure 8.15 for an example of a trivariate $dF$ plot. The $dF$ plots are particularly useful when working with the adaptive mixtures density estimation method that will be discussed shortly. We provide a function called **csdfplot** that will implement the $dF$ plots for univariate, bivariate and trivariate data.

**Example 8.10**

In this example, we show how to implement the function called **csdfplot** and illustrate its use with bivariate and trivariate models. The bivariate case is the following three component model:

$$p_1 = 0.5 \qquad p_2 = 0.3 \qquad p_3 = 0.2,$$

$$\mu_1 = \begin{bmatrix} -1 \\ -1 \end{bmatrix} \qquad \mu_2 = \begin{bmatrix} 1 \\ 1 \end{bmatrix} \qquad \mu_3 = \begin{bmatrix} 5 \\ 6 \end{bmatrix},$$

$$\Sigma_1 = \begin{bmatrix} 1 & 0 \\ 0 & 1 \end{bmatrix} \qquad \Sigma_2 = \begin{bmatrix} 0.5 & 0 \\ 0 & 0.5 \end{bmatrix} \qquad \Sigma_3 = \begin{bmatrix} 1 & 0.5 \\ 0.5 & 1 \end{bmatrix}.$$

```
% First create the model.
% The function expects a vector of weights;
% a matrix of means, where each column of the matrix
```

```
% corresponds to a d-D mean; a 3-D array of
% covariances, where each page of the array is a
% covariance matrix.
pies = [0.5 0.3 0.2]; % mixing coefficients
mus = [-1 1 5; -1 1 6];
% Delete any previous variances in the workspace.
clear vars
vars(:,:,1) = eye(2);
vars(:,:,2) = eye(2)*.5
vars(:,:,3) = [1 0.5; 0.5 1];
figure
csdfplot(mus,vars,pies)
```

The resulting plot is shown in Figure 8.14. Note that the covariance of two of the component densities are represented by circles, with one larger than the other. These correspond to the first two terms of the model. The third component density has an elliptical covariance structure indicating non-zero off-diagonal elements in the covariance matrix. We now do the same thing for the trivariate case, where the model is

$$\mu_1 = \begin{bmatrix} -1 \\ -1 \\ -1 \end{bmatrix} \qquad \mu_2 = \begin{bmatrix} 1 \\ 1 \\ 1 \end{bmatrix} \qquad \mu_3 = \begin{bmatrix} 5 \\ 6 \\ 2 \end{bmatrix},$$

$$\Sigma_1 = \begin{bmatrix} 1 & 0 & 0 \\ 0 & 1 & 0 \\ 0 & 0 & 1 \end{bmatrix} \qquad \Sigma_2 = \begin{bmatrix} 0.5 & 0 & 0 \\ 0 & 0.5 & 0 \\ 0 & 0 & 0.5 \end{bmatrix} \qquad \Sigma_3 = \begin{bmatrix} 1 & 0.7 & 0.2 \\ 0.7 & 1 & 0.5 \\ 0.2 & 0.5 & 1 \end{bmatrix}.$$

The mixing coefficients are the same as before. We need only to adjust the means and the covariance accordingly.

```
mus(3,:) = [-1 1 2];
% Delete previous vars array or you will get an error.
clear vars
vars(:,:,1) = eye(3);
vars(:,:,2) = eye(3)*.5;
vars(:,:,3)=[1 0.7 0.2;
             0.7 1 0.5;
             0.2 0.5 1];
figure
csdfplot(mus,vars,pies)
% get a different viewpoint
view([-34,9])
```

The trivariate *dF* plot for this model is shown in Figure 8.15. Two terms (the first two) are shown as spheres and one as an ellipsoid.
❏

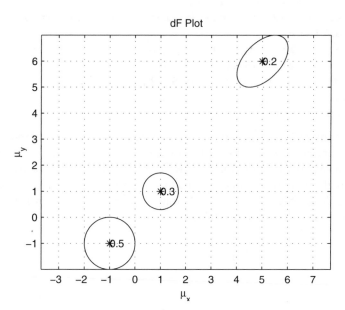

**FIGURE 8.14**
Bivariate *dF* plot for the three term mixture model of Example 8.10.

### EM Algorithm for Estimating the Parameters

The problem of estimating the parameters in a finite mixture has been studied extensively in the literature. The book by Everitt and Hand [1981] provides an excellent overview of this topic and offers several methods for parameter estimation. The technique we present here is called the Expectation-Maximization (EM) method. This is a general method for optimizing likelihood functions and is useful in situations where data might be missing or simpler optimization methods fail. The seminal paper on this topic is by Dempster, Laird and Rubin [1977], where they formalize the EM algorithm and establish its properties. Redner and Walker [1984] apply it to mixture densities. The EM methodology is now a standard tool for statisticians and is used in many applications.

In this section, we discuss the EM algorithm as it can be applied to estimating the parameters of a finite mixture of normal densities. To use the EM algo-

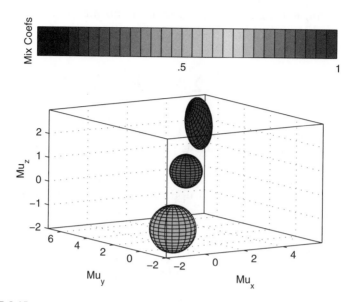

**FIGURE 8.15**
Trivariate *dF* plot for the three term mixture model of Example 8.10.

rithm, we must have a value for the number of terms $c$ in the mixture. This is usually obtained using prior knowledge of the application (the analyst expects a certain number of groups), using graphical exploratory data analysis (looking for clusters or other group structure) or using some other method of estimating the number of terms. The approach called adaptive mixtures [Priebe, 1994] offers a way to address the problem of determining the number of component densities to use in the finite mixture model. This approach is discussed later.

Besides the number of terms, we must also have an initial guess for the value of the component parameters. Once we have an initial estimate, we update the parameter estimates using the data and the equations given below. These are called the iterative EM update equations, and we provide the multivariate case as the most general one. The univariate case follows easily.

The first step is to determine the posterior probabilities given by

$$\hat{\tau}_{ij} = \frac{\hat{p}_i \phi(\mathbf{x}_j; \hat{\mu}_i, \hat{\Sigma}_i)}{\hat{f}(\mathbf{x}_j)}; \qquad i = 1, \ldots, c\,;\, j = 1, \ldots, n\,. \qquad (8.34)$$

where $\hat{\tau}_{ij}$ represents the estimated posterior probability that point $\mathbf{x}_j$ belongs to the $i$-th term, $\phi(\mathbf{x}_j;\hat{\mu}_i, \hat{\Sigma}_i)$ is the multivariate normal density for the $i$-th term evaluated at $\mathbf{x}_j$, and

$$\hat{f}(\mathbf{x}_j) = \sum_{k=1}^{c} \hat{p}_k \phi(\mathbf{x}_j;\hat{\mu}_k, \hat{\Sigma}_k) \tag{8.35}$$

is the finite mixture estimate at point $\mathbf{x}_j$.

The posterior probability tells us the likelihood that a point belongs to each of the separate component densities. We can use this estimated posterior probability to obtain a weighted update of the parameters for each component. This yields the iterative EM update equations for the mixing coefficients, the means and the covariance matrices. These are

$$\hat{p}_i = \frac{1}{n}\sum_{j=1}^{n} \hat{\tau}_{ij} \tag{8.36}$$

$$\hat{\mu}_i = \frac{1}{n}\sum_{j=1}^{n} \frac{\hat{\tau}_{ij}\mathbf{x}_j}{\hat{p}_i} \tag{8.37}$$

$$\hat{\Sigma}_i = \frac{1}{n}\sum_{j=1}^{n} \frac{\hat{\tau}_{ij}(\mathbf{x}_j - \hat{\mu}_i)(\mathbf{x}_j - \hat{\mu}_i)^T}{\hat{p}_i}. \tag{8.38}$$

Note that if $d = 1$, then the update equation for the variance is

$$\hat{\sigma}_i^2 = \frac{1}{n}\sum_{j=1}^{n} \frac{\hat{\tau}_{ij}(x_j - \hat{\mu}_i)^2}{\hat{p}_i}. \tag{8.39}$$

The steps for the EM algorithm to estimate the parameters for a finite mixture with multivariate normal components are given here and are illustrated in Example 8.11.

*FINITE MIXTURES - EM PROCEDURE*

1. Determine the number of terms or component densities $c$ in the mixture.

2. Determine an initial guess at the component parameters. These are the mixing coefficients, means and covariance matrices for each normal density.
3. For each data point $\mathbf{x}_j$, calculate the posterior probability using Equation 8.34.
4. Update the mixing coefficients, the means and the covariance matrices for the individual components using Equations 8.36 through 8.38.
5. Repeat steps 3 through 4 until the estimates converge.

Typically, step 5 is implemented by continuing the iteration until the changes in the estimates at each iteration are less than some pre-set tolerance. Note that with the iterative EM algorithm, we need to use the entire data set to simultaneously update the parameter estimates. This imposes a high computational load when dealing with massive data sets.

## Example 8.11

In this example, we provide the MATLAB code that implements the multivariate EM algorithm for estimating the parameters of a finite mixture probability density model. To illustrate this, we will generate a data set that is a mixture of two terms with equal mixing coefficients. One term is centered at the point $(-2, 2)$ and the other is centered at $(2, 0)$. The covariance of each component density is given by the identity matrix. Our first step is to generate 200 data points from this distribution.

```
% Create some artificial two-term mixture data.
n = 200;
data = zeros(n,2);
% Now generate 200 random variables. First find
% the number that come from each component.
r = rand(1,n);
% Find the number generated from component 1.
ind = length(find(r <= 0.5));
% Create some mixture data. Note that the
% component densities are multivariate normals.
% Generate the first term.
data(1:ind,1) = randn(ind,1) - 2;
data(1:ind,2) = randn(ind,1) + 2;
% Generate the second term.
data(ind+1:n,1) = randn(n-ind,1) + 2;
data(ind+1:n,2) = randn(n-ind,1);
```

We must then specify various parameters for the EM algorithm, such as the number of terms.

```
c = 2;    % number of terms
```

```
[n,d] = size(data);  % n=# pts, d=# dims
tol = 0.00001;   % set up criterion for stopping EM
max_it = 100;
totprob = zeros(n,1);
```

We also need an initial guess at the component density parameters.

```
% Get the initial parameters for the model to start EM
mu(:,1) = [-1 -1]';   % each column represents a mean
mu(:,2) = [1 1]';
mix_cof = [0.3 0.7];
var_mat(:,:,1) = eye(d);
var_mat(:,:,2) = eye(d);
varup = zeros(size(var_mat));
muup = zeros(size(mu));
% Just to get started.
num_it = 1;
deltol = tol+1;% to get started
```

The following steps implement the EM update formulas found in Equations 8.34 through 8.38.

```
while num_it <= max_it & deltol > tol
    % get the posterior probabilities
    totprob = zeros(n,1);
    for i=1:c
      posterior(:,i) = mix_cof(i)*...
          csevalnorm(data,mu(:,i)',var_mat(:,:,i));
        totprob = totprob+posterior(:,i);
    end
    den = totprob*ones(1,c);
    posterior = posterior./den;
    % Update the mixing coefficients.
    mix_cofup = sum(posterior)/n;
    % Update the means.
    mut = data'*posterior;
    MIX = ones(d,1)*mix_cof;
    muup = mut./(MIX*n);
    % Update the means and the variances.
    for i=1:c
       cen_data = data-ones(n,1)*mu(:,i)';
       mat = cen_data'*...
          diag(posterior(:,i))*cen_data;
       varup(:,:,i)=mat./(mix_cof(i)*n);
    end
    % Get the tolerances.
    delvar = max(max(max(abs(varup-var_mat))));
    delmu = max(max(abs(muup-mu)));
```

```
     delpi = max(abs(mix_cof-mix_cofup));
     deltol = max([delvar,delmu,delpi]);
     % Reset parameters.
     num_it = num_it+1;
     mix_cof = mix_cofup;
     mu = muup;
     var_mat = varup;
   end  % while loop
```

For our data set, it took 37 iterations to converge to an answer. The convergence of the EM algorithm to a solution and the number of iterations depends on the tolerance, the initial parameters, the data set, etc. The estimated model returned by the EM algorithm is

$$\hat{p}_1 = 0.498 \qquad \hat{p}_2 = 0.502,$$

$$\hat{\mu}_1 = \begin{bmatrix} -2.08 \\ 2.03 \end{bmatrix} \qquad \hat{\mu}_2 = \begin{bmatrix} 1.83 \\ -0.03 \end{bmatrix}.$$

For brevity, we omit the estimated covariances, but we can see from these results that the model does match the data that we generated.
❑

## Adaptive Mixtures

The adaptive mixtures [Priebe, 1994] method for density estimation uses a data-driven approach for estimating the number of component densities in a mixture model. This technique uses the recursive EM update equations that are provided below. The basic idea behind adaptive mixtures is to take one point at a time and determine the distance from the observation to each component density in the model. If the distance to each component is larger than some threshold, then a new term is created. If the distance is less than the threshold for all terms, then the parameter estimates are updated based on the recursive EM equations.

We start our explanation of the adaptive mixtures approach with a description of the recursive EM algorithm for mixtures of multivariate normal densities. This method recursively updates the parameter estimates based on a new observation. As before, the first step is to determine the posterior probability that the new observation belongs to each term:

$$\hat{\tau}_i^{(n+1)} = \frac{\hat{p}_i^{(n)} \phi(\mathbf{x}^{(n+1)}; \hat{\mu}_i^{(n)}, \hat{\Sigma}_i^{(n)})}{\hat{f}^{(n)}(\mathbf{x}^{(n+1)})}; \qquad i = 1, ..., c, \qquad (8.40)$$

where $\hat{\tau}_i^{(n+1)}$ represents the estimated posterior probability that the new observation $\mathbf{x}^{(n+1)}$ belongs to the $i$-th term, and the superscript $(n)$ denotes the estimated parameter values based on the previous $n$ observations. The denominator is the finite mixture density estimate

$$\hat{f}^{(n)}(\mathbf{x}^{(n+1)}) = \sum_{i=1}^{c} \hat{p}_i \phi(\mathbf{x}^{(n+1)}; \hat{\mu}_i^{(n)}, \hat{\Sigma}_i^{(n)})$$

for the new observation using the mixture from the previous $n$ points.

The remainder of the recursive EM update equations are given by Equations 8.41 through 8.43. Note that recursive equations are typically in the form of the old value for an estimate plus an update term using the new observation. The recursive update equations for mixtures of multivariate normals are:

$$\hat{p}_i^{(n+1)} = \hat{p}_i^{(n)} + \frac{1}{n}(\hat{\tau}_i^{(n+1)} - \hat{p}_i^{(n)}) \tag{8.41}$$

$$\hat{\mu}_i^{(n+1)} = \hat{\mu}_i^{(n)} + \frac{\hat{\tau}_i^{(n+1)}}{n\hat{p}_i^{(n)}}(\mathbf{x}^{(n+1)} - \hat{\mu}_i^{(n)}) \tag{8.42}$$

$$\hat{\Sigma}_i^{(n+1)} = \hat{\Sigma}_i^{(n)} + \frac{\hat{\tau}_i^{(n+1)}}{n\hat{p}_i^{(n)}}\left[(\mathbf{x}^{(n+1)} - \hat{\mu}_i^{(n)})(\mathbf{x}^{(n+1)} - \hat{\mu}_i^{(n)})^T - \hat{\Sigma}_i^{(n)}\right] . \tag{8.43}$$

This reduces to the 1-D case in a straightforward manner, as was the case with the iterative EM update equations.

The adaptive mixtures approach updates our probability density estimate $\hat{f}(\mathbf{x})$ and also provides the opportunity to expand the parameter space (i.e., the model) if the data indicate that should be done. To accomplish this, we need a way to determine when a new component density should be added. This could be done in several ways, but the one we present here is based on the Mahalanobis distance. If this distance is too large for all of the terms (or alternatively if the minimum distance is larger than some threshold), then we can consider the new point too far away from the existing terms to update the current model. Therefore, we create a new term.

The squared Mahalanobis distance between the new observation $\mathbf{x}^{(n+1)}$ and the $i$-th term is given by

$$MD_i^2(\mathbf{x}^{(n+1)}) = (\mathbf{x}^{(n+1)} - \hat{\mu}_i^{(n)})^T \left(\hat{\Sigma}_i^{(n)}\right)^{-1}(\mathbf{x}^{(n+1)} - \hat{\mu}_i^{(n)}) . \tag{8.44}$$

We create a new term if

$$\min_i\{MD_i^2(\mathbf{x}^{(n+1)})\} > t_C, \tag{8.45}$$

where $t_C$ is a threshold to create a new term. The rule in Equation 8.45 states that if the smallest squared Mahalanobis distance is greater than the threshold, then we create a new term. In the univariate case, if $t_C = 1$ is used, then a new term is created if a new observation is more than one standard deviation away from the mean of each term. For $t_C = 4$, a new term would be created for an observation that is at least two standard deviations away from the existing terms. For multivariate data, we would like to keep the same term creation rate as in the 1-D case. Solka [1995] provides thresholds $t_C$ based on the squared Mahalanobis distance for the univariate, bivariate, and trivariate cases. These are shown in Table 8.3.

TABLE 8.3

Recommended Thresholds for Adaptive Mixtures

| Dimensionality | Create Threshold |
|:---:|:---:|
| 1 | 1 |
| 2 | 2.34 |
| 3 | 3.54 |

When we create a new term, we initialize the parameters using Equations 8.46 through 8.48. We denote the current number of terms in the model by $N$.

$$\hat{\mu}_{N+1}^{(n+1)} = \mathbf{x}^{(n+1)}, \tag{8.46}$$

$$\hat{p}_{N+1}^{(n+1)} = \frac{1}{n+1}, \tag{8.47}$$

$$\hat{\Sigma}_{N+1}^{(n+1)} = \Im(\hat{\Sigma}_i), \tag{8.48}$$

where $\Im(\hat{\Sigma}_i)$ is a weighted average using the posterior probabilities. In practice, some other estimate or initial covariance can be used for the new term. To ensure that the mixing coefficients sum to one when a new term is added, the $\hat{p}_i^{(n+1)}$ must be rescaled using

$$\hat{p}_i^{(n+1)} = \frac{n\hat{p}_i^{(n)}}{n+1}; \qquad i = 1, ..., N.$$

We continue through the data set, one point at a time, adding new terms as necessary. Our density estimate is then given by

$$\hat{f}_{AM}(\mathbf{x}) = \sum_{i=1}^{N} \hat{p}_i \phi(\mathbf{x}; \hat{\mu}_i, \hat{\Sigma}_i). \tag{8.49}$$

This allows for a variable number of terms $N$, where usually $N \ll n$. The adaptive mixtures technique is captured in the procedure given here, and a function called **csadpmix** is provided with the Computational Statistics Toolbox. Its use in the univariate case is illustrated in Example 8.12.

*ADAPTIVE MIXTURES PROCEDURE:*

    1. Initialize the adaptive mixtures procedure using the first data point $\mathbf{x}^{(1)}$:

$$\hat{\mu}_1^{(1)} = \mathbf{x}^{(1)}, \ \hat{p}_1^{(1)} = 1, \text{ and } \hat{\Sigma}_1^{(1)} = \mathbf{I},$$

    where $\mathbf{I}$ denotes the identity matrix. In the univariate case, the variance of the initial term is one.

    2. For a new data point $\mathbf{x}^{(n+1)}$, calculate the squared Mahalanobis distance as in Equation 8.44.

    3. If the minimum squared distance is greater than $t_C$, then create a new term using Equations 8.46 through 8.48. Increase the number of terms $N$ by one.

    4. If the minimum squared distance is less than the create threshold $t_C$, then update the existing terms using Equations 8.41 through 8.43.

    5. Continue steps 2 through 4 using all data points.

  In practice, the adaptive mixtures method is used to get initial values for the parameters, as well as an estimate of the number of terms needed to model the density. One would then use these as a starting point and apply the iterative EM algorithm to refine the estimates.

## Example 8.12

In this example, we illustrate the MATLAB code that implements the univariate adaptive mixtures density estimation procedure. The source code for these functions are given in Appendix D. We generate random variables using the same three term mixture model that was discussed in Example 8.9. Recall that the model is given by

$$f(x) = 0.3 \times \phi(x;-3, 1) + 0.3 \times \phi(x;0, 1) + 0.4 \times \phi(x;2, 0.5).$$

```
% Get the true model to generate data.
pi_tru = [0.3 0.3 0.4];
n = 100;
x = zeros(n,1);
% Now generate 100 random variables. First find
% the number that fall in each one.
r = rand(1,100);
% Find the number generated from each component.
ind1 = length(find(r <= 0.3));
ind2 = length(find(r > 0.3 & r <= 0.6));
ind3 = length(find(r > 0.6));
% create some artificial  3 term mixture data
x(1:ind1) = randn(ind1,1) - 3;
x(ind1+1:ind2+ind1)=randn(ind2,1);
x(ind1+ind2+1:n) = randn(ind3,1)*sqrt(0.5)+2;
```

We now call the adaptive mixtures function **csadpmix** to estimate the model.

```
% Now call the adaptive mixtures function.
maxterms = 25;
[pihat,muhat,varhat] = csadpmix(x,maxterms);
```

The following MATLAB commands provide the plots shown in Figure 8.16.

```
% Get the plots.
csdfplot(muhat,varhat,pihat,min(x),max(x));
axis equal
nterms = length(pihat);
figure
csplotuni(pihat,muhat,varhat,...
    nterms,min(x)-5,max(x)+5,100)
```

We reorder the observations and repeat the process to get the plots in Figure 8.17.

```
% Now re-order the points and repeat
% the adaptive mixtures process.
ind = randperm(n);
x = x(ind);
[pihat,muhat,varhat] = csadpmix(x,maxterms);
```

❑

Our example above demonstrates some interesting things to consider with adaptive mixtures. First, the model complexity or the number of terms is sometimes greater than is needed. For example, in Figure 8.16, we show a *dF* plot for the three term mixture model in Example 8.12. Note that the adaptive mixture approach yields more than three terms. This is a problem with mixture models in general. Different models (i.e., number of terms and estimated component parameters) can produce essentially the same function estimate or curve for $\hat{f}(\mathbf{x})$. This is illustrated in Figures 8.16 and 8.17, where we see that similar curves are obtained from two different models for the same data set. These results are straight from the adaptive mixtures density estimation approach. In other words, we did not use this estimate as an initial starting point for the EM approach. If we had applied the iterative EM to these estimated models, then the curves should be the same.

The other issue that must be considered when using the adaptive mixtures approach is that the resulting model or estimated probability density function depends on the order in which the data are presented to the algorithm. This is also illustrated in Figures 8.16 and 8.17, where the second estimated model is obtained after re-ordering the data. These issues were addressed by Solka [1995].

---

## 8.5 Generating Random Variables

In the introduction, we discussed several uses of probability density estimates, and it is our hope that the reader will discover many more. One of the applications of density estimation is in the area of modeling and simulation. Recall that a key aspect of modeling and simulation is the collection of data generated according to some underlying random process and the desire to generate more random variables from the same process for simulation purposes. One option is to use one of the density estimation techniques discussed in this chapter and randomly sample from that distribution. In this section, we provide the methodology for generating random variables from finite or adaptive mixtures density estimates.

We have already seen an example of this procedure in Example 8.11 and Example 8.12. The procedure is to first choose the class membership of generated observations based on uniform (0,1) random variables. The number of random variables generated from each component density is given by the corresponding proportion of these uniform variables that are in the required range. The steps are outlined here.

*PROCEDURE - GENERATING RANDOM VARIABLES (FINITE MIXTURE)*

    1. We are given a finite mixture model $(p_i, \, g_i(\mathbf{x};\theta_i))$ with $c$ components, and we want to generate $n$ random variables from that distribution.

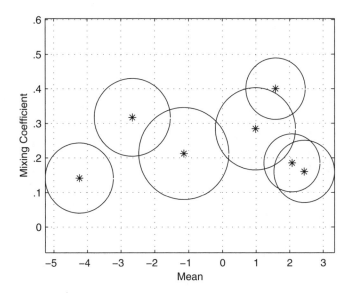

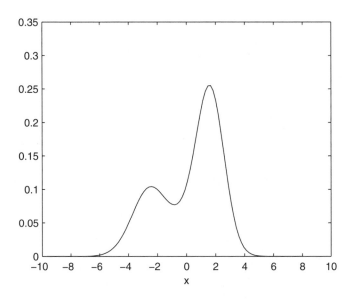

**FIGURE 8.16**
The upper plot shows the *dF* representation for Example 8.12. Compare this with Figure 8.17 for the same data. Note that the curves are essentially the same, but the number of terms and associated parameters are different. Thus, we can get different models for the same data.

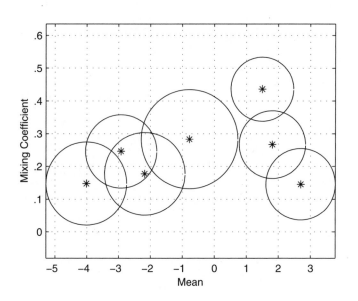

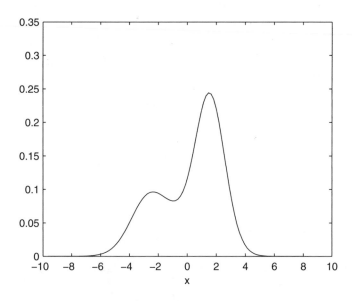

**FIGURE 8.17**
This is the second estimated model using adaptive mixtures for the data generated in Example 8.12. This second model was obtained by re-ordering the data set and then implementing the adaptive mixtures technique. This shows the dependence of the technique on the order in which the data are presented to the method.

2. First determine the component membership of each of the $n$ random variables. We do this by generating $n$ uniform $(0,1)$ random variables ($U_i$). Component membership is determined as follows

If $0 \le U_i < p_1$, then $X_i$ is from component density 1.

If $p_1 \le U_i < p_1 + p_2$, then $X_i$ is from component density 2.

$\cdots$

If $\sum_{j=1}^{c-1} p_j \le U_i \le 1$, then $X_i$ is from component density $c$.

3. Generate the $X_i$ from the corresponding $g_i(x;\theta_i)$ using the component membership found in step 2.

Note that with this procedure, one could generate random variables from a mixture of any component densities. For instance, the model could be a mixture of exponentials, betas, etc.

## Example 8.13

Generate a random sample of size $n$ from a finite mixture estimate of the Old Faithful Geyser data (**geyser**). First we have to load up the data and build a finite mixture model.

```
load geyser
% Expects rows to be observations.
data = geyser';
% Get the finite mixture.
% Use a two term model.
% Set initial model to means at 50 and 80.
muin = [50, 80];
% Set mixing coefficients equal.
piesin = [0.5, 0.5];
% Set initial variances to 1.
varin = [1, 1];
max_it = 100;
tol = 0.001;
% Call the finite mixtures.
[pies,mus,vars]=...
    csfinmix(data,muin,varin,piesin,max_it,tol);
```

Now generate some random variables according to this estimated model.

```
% Now generate some random variables from this model.
% Get the true model to generate data from this.
n = 300;
x = zeros(n,1);
```

```
% Now generate 300 random variables. First find
% the number that fall in each one.
r = rand(1,n);
% Find the number generated from component 1.
ind = length(find(r <= pies(1)));
% Create some mixture data. Note that the
% component densities are normals.
x(1:ind)    = randn(ind,1)*sqrt(vars(1))  + mus(1);
x(ind+1:n)  = randn(n-ind,1)*sqrt(vars(2)) + mus(2);
```

We can plot density histograms to compare the two data sets. These are shown in Figure 8.18. Not surprisingly, they look similar, but different. The user is asked to explore this further in the exercises.
❑

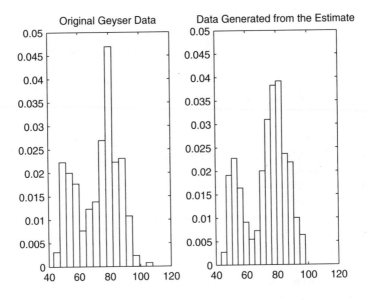

**FIGURE 8.18**
Histogram density estimates of the Old Faithful geyser data. The one on the right shows the estimate from the data that was sampled from the finite mixture density estimate of the original data.

## 8.6 MATLAB Code

The MATLAB Statistics Toolbox does not have any functions for nonparametric density estimation. The functions it has for estimating distribution parameters (e.g., **mle**, **normfit**, **expfit**, **betafit**, etc.) can be used for parametric density estimation. The standard MATLAB package has functions for frequency histograms, as explained in Chapter 5.

We provide several functions for nonparametric density estimation with the Computational Statistics Toolbox. These are listed in Table 8.4.

TABLE 8.4

List of Functions from Chapter 8 Included in the Computational Statistics Toolbox

| Purpose | MATLAB Function |
|---|---|
| These provide a bivariate histogram. | cshist2d<br>cshistden |
| This returns a frequency polygon density estimate. | csfreqpoly |
| This function returns the Averaged Shifted Histogram. | csash |
| These functions perform kernel density estimation. | cskernnd<br>cskern2d |
| Create plots | csdfplot<br>csplotuni |
| Functions for finite and adaptive mixtures | csfinmix<br>csadpmix |

## 8.7 Further Reading

The discussion of histograms, frequency polygons and averaged shifted histograms presented in this book follows that of Scott [1992]. Scott's book is an excellent resource for univariate and multivariate density estimation, and it describes many applications of the techniques. It includes a comprehensive treatment of the underlying theory on selecting smoothing parameters, analyzing the performance of density estimates in terms of the asymptotic mean integrated squared error, and also addresses high dimensional data.

The summary book by Silverman [1986] provides a relatively non-theoretical treatment of density estimation. He includes a discussion of histograms, kernel methods and others. This book is readily accessible to most statisticians, data analysts or engineers. It contains applications and computational details, making the subject easier to understand.

Other books on density estimation include Tapia and Thompson [1978], Devroye and Gyorfi [1985], Wand and Jones [1995], and Simonoff [1996]. The Tapia and Thompson book offers a theoretical foundation for density estimation and includes a discussion of Monte Carlo simulations. The Devroye and Gyorfi text describes the underlying theory of density estimation using the $L_1$ (absolute error) viewpoint instead of $L_2$ (squared error). The books by Wand and Jones and Simonoff look at using kernel methods for smoothing and exploratory data analysis.

A paper by Izenman [1991] provides a comprehensive review of many methods in univariate and multivariate density estimation and includes an extensive bibliography. Besides histograms and kernel methods, he discusses projection pursuit density estimation [Friedman, Stuetzle, and Schroeder, 1984], maximum penalized likelihood estimators, sieve estimators, and orthogonal estimators.

For the reader who would like more information on finite mixtures, we recommend Everitt and Hand [1981] for a general discussion of this topic. The book provides a summary of the techniques for obtaining mixture models (estimating the parameters) and illustrates them using applications. That text also discusses ways to handle the problem of determining the number of terms in the mixture and other methods for estimating the parameters. It is appropriate for someone with a general statistics or engineering background. For readers who would like more information on the theoretical details of finite mixtures, we refer them to McLachlan and Basford [1988] or Titterington, Smith and Makov [1985]. A recent book by McLachlan and Peel [2000] provides many examples of finite mixtures, linking them to machine learning, data mining, and pattern recognition.

The EM algorithm is described in the text by McLachlan and Krishnan [1997]. This offers a unified treatment of the subject, and provides numerous applications of the EM algorithm to regression, factor analysis, medical imaging, experimental design, finite mixtures, and others.

For a theoretical discussion of the adaptive mixtures approach, the reader is referred to Priebe [1993, 1994]. These examine the error in the adaptive mixtures density estimates and its convergence properties. A recent paper by Priebe and Marchette [2000] describes a data-driven method for obtaining parsimonious mixture model estimates. This methodology addresses some of the problems with the adaptive/finite mixtures approach: 1) that adaptive mixtures is not designed to yield a parsimonious model and 2) how many terms or component densities should be used in a finite mixture model.

Solka, Poston, and Wegman [1995] extend the static $dF$ plot to a dynamic one. References to MATLAB code are provided in this paper describing a dynamic view of the adaptive mixtures and finite mixtures estimation process in time (i.e., iterations of the EM algorithm).

## Exercises

8.1. Create a MATLAB function that will return the value of the histogram estimate for the probability density function. Do this for the 1-D case.

8.2. Generate a random sample of data from a standard normal. Construct a kernel density estimate of the probability density function and verify that the area under the curve is approximately 1 using **trapz**.

8.3. Generate 100 univariate normals and construct a histogram. Calculate the MSE at a point $x_0$ using Monte Carlo simulation. Do this for varying bin widths. What is the better bin width? Does the sample size make a difference? Does it matter whether $x_0$ is in the tails or closer to the mean? Repeat this experiment using the absolute error. Are your conclusions similar?

8.4. Generate univariate normal random variables. Using the Normal Reference Rules for $h$, construct a histogram, a frequency polygon and a kernel estimate of the data. Estimate the MSE at a point $x_0$ using Monte Carlo simulation.

8.5. Generate a random sample from the exponential distribution. Construct a histogram using the Normal Reference Rule. Using Monte Carlo simulation, estimate the MISE. Use the skewness factor to adjust $h$ and re-estimate the MISE. Which window width is better?

8.6. Use the **snowfall** data and create a MATLAB **movie** that shows how 1-D histograms change with bin width. See **help** on **movie** for information on how to do this. Also make a **movie** showing how changing the bin origin affects the histogram.

8.7. Repeat Example 8.2 for bin widths given by the Freedman-Diaconis Rule. Is there a difference in the results? What does the histogram look like if you use Sturge's Rule?

8.8. Write a MATLAB function that will return the value of a bivariate histogram at a point, given the bin counts, the sample size, and the window widths.

8.9. Write a MATLAB function that will evaluate the cumulative distribution function for a univariate frequency polygon. You can use the **trapz, quad,** or **quadl** functions.

8.10. Load the **iris** data. Create a $150 \times 2$ matrix by concatenating the first two columns of each species. Construct and plot a frequency polygon of these data. Do the same thing for all possible pairs of columns. You might also look at a **contour** plot of the frequency polygons. Is there evidence of groups in the plots?

8.11. In this chapter, we showed how you could construct a kernel density estimate by placing a weighted kernel at each data point, evaluating the kernels over the domain, and then averaging the $n$ curves. In that implementation, we are looping over all of the data points. An alternative implementation is to loop over all points in the domain where you want to get the value of the estimate, evaluate a weighted kernel at each point, and take the average. The following code shows you how to do this. Implement this using the Buffalo **snowfall** data. Verify that this is a valid density by estimating the area under the curve.

```
load snowfall
x = 0:140;
n = length(snowfall);
h = 1.06*sqrt(var(snowfall))*n^(-1/5);
fhat = zeros(size(x));
% Loop over all values of x in the domain
% to get the kernel evaluated at that point.
for i = 1:length(x)
 xloc = x(i)*ones(1,n);
 % Take each value of x and evaluate it at
 % n weighted kernels -
 % each one centered at a data point, then add them up.
 arg = ((xloc-snowfall)/h).^2;
 fhat(i) = (sum(exp(-.5*(arg)))/(n*h*sqrt(2*pi)));
end
```

8.12. Write a MATLAB function that will construct a kernel density estimate for the multivariate case.

8.13. Write a MATLAB function that will provide the finite mixture density estimate at a point in $d$ dimensions.

8.14. Implement the univariate adaptive mixtures density estimation procedure on the Buffalo **snowfall** data. Once you have your initial model, use the EM algorithm to refine the estimate.

8.15. In Example 8.13, we generate a random sample from the kernel estimate of the Old Faithful **geyser** data. Repeat this example to obtain a new random sample of **geyser** data from the estimated model and construct a new density estimate from the second sample. Find the integrated squared error between the two density estimates. Does the error between the curves indicate that the second random sample generates a similar density curve?

8.16. Say we have a kernel density estimate where the kernel used is a normal density. If we put this in the context of finite mixtures, then what are the values for the component parameters $(p_i, \mu_i, \sigma_i^2)$ in the corresponding finite mixture?

8.17. Repeat Example 8.12. Plot the curves from the estimated models. What is the ISE between the two estimates? Use the iterative EM algorithm on both models to refine the estimates. What is the ISE after you do this? What can you say about the two different models? Are your conclusions different if you use the IAE?

8.18. Write a MATLAB function that will generate random variables (univariate or multivariate) from a finite mixture of normals.

8.19. Using the method for generating random variables from a finite mixture that was discussed in this chapter, develop and implement an algorithm for generating random variables based on a kernel density estimate.

8.20. Write a function that will estimate the MISE between two functions. Convert it to also estimate the MIAE between two functions.

8.21. Apply some of the univariate density estimation techniques from this chapter to the **forearm** data.

8.22. The **elderly** data set contains the height measurements (in centimeters) of 351 elderly females [Hand, et al., 1994]. Use some of the univariate density estimation techniques from this chapter to explore the data. Is there evidence of bumps and modes?

8.23. Apply the multivariate techniques of this chapter to the **nfl** data [Csorgo and Welsh, 1989; Hand, et al., 1994]. These data contain bivariate measurements of the game time to the first points scored by kicking the ball between the end posts ( $X_1$ ), and the game time to the first points scored by moving the ball into the end zone ( $X_2$ ). The times are in minutes and seconds. Plot your results.

# Chapter 9

## Statistical Pattern Recognition

### 9.1 Introduction

Statistical pattern recognition is an application in computational statistics that uses many of the concepts we have covered so far, such as probability density estimation and cross-validation. Examples where statistical pattern recognition techniques can be used are numerous and arise in disciplines such as medicine, computer vision, robotics, military systems, manufacturing, finance and many others. Some of these include the following:

- A doctor diagnoses a patient's illness based on the symptoms and test results.
- A radiologist locates areas where there is non-healthy tissue in x-rays.
- A military analyst classifies regions of an image as natural or man-made for use in targeting systems.
- A geologist determines whether a seismic signal represents an impending earthquake.
- A loan manager at a bank must decide whether a customer is a good credit risk based on their income, past credit history and other variables.
- A manufacturer must classify the quality of materials before using them in their products.

In all of these applications, the human is often assisted by statistical pattern recognition techniques.

Statistical methods for pattern recognition are covered in this chapter. In this section, we first provide a brief introduction to the goals of pattern recognition and a broad overview of the main steps of building classifiers. In Section 9.2 we present a discussion of Bayes classifiers and pattern recognition in an hypothesis testing framework. Section 9.3 contains techniques for evaluating the classifier. In Section 9.4, we illustrate how to construct classification trees. Section 9.5 contains methods for unsupervised classification or clustering, including agglomerative methods and $k$-means clustering.

We first describe the process of statistical pattern recognition in a **_supervised learning_** setting. With supervised learning, we have cases or observations where we know which class each case belongs to. Figure 9.1 illustrates the major steps of statistical pattern recognition.

The first step in pattern recognition is to select **_features_** that will be used to distinguish between the classes. As the reader might suspect, the choice of features is perhaps the most important part of the process. Building accurate classifiers is much easier with features that allow one to readily distinguish between classes.

Once features are selected, we obtain a sample of these features for the different classes. This means that we find objects that belong to the classes of interest and then measure the features. Each observed set of feature measurements (sometimes also called a **_case_** or **_pattern_**) has a class label attached to it. Now that we have data that are known to belong to the different classes, we can use this information to create the methodology that will take as input a set of feature measurements and output the class that it belongs to. How these classifiers are created will be the topic of this chapter.

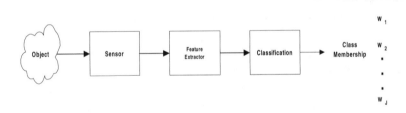

**FIGURE 9.1**
This shows a schematic diagram of the major steps for statistical pattern recognition.

One of the main examples we use to illustrate these ideas is one that we encountered in Chapter 5. In the `iris` data set, we have three species of iris: _Iris setosa, Iris versicolor_ and _Iris virginica_. The data were used by Fisher [1936] to develop a classifier that would take measurements from a new iris and determine its species based on the features [Hand, et al., 1994]. The four features that are used to distinguish the species of iris are sepal length, sepal width, petal length and petal width. The next step in the pattern recognition process is to find many flowers from each species and measure the corresponding sepal length, sepal width, petal length, and petal width. For each set of measured features, we attach a class label that indicates which species it belongs to. We build a classifier using these data and (possibly) one of the techniques that are described in this chapter. To use the classifier, we measure the four features for an iris of unknown species and use the classifier to assign the species membership.

Sometimes we are in a situation where we do not know the class membership for our observations. Perhaps we are unable or unwilling to assume how many groups are represented by the data. In this case, we are in the *unsupervised learning* mode. To illustrate this, say we have data that comprise measurements of a type of insect called *Chaetocnema* [Lindsey, Herzberg, and Watts, 1987; Hand, et al., 1994]. These variables measure the width of the first joint of the first tarsus, the width of the first joint of the second tarsus, and the maximal width of the aedegus. All measurements are in microns. We suspect that there are three species represented by these data. To explore this hypothesis further, we could use one of the unsupervised learning or clustering techniques that will be covered in Section 9.5.

## 9.2 Bayes Decision Theory

The Bayes approach to pattern classification is a fundamental technique, and we recommend it as the starting point for most pattern recognition applications. If this method is not adequate, then more complicated techniques may be used (e.g., neural networks, classification trees). Bayes decision theory poses the classification problem in terms of probabilities; therefore, all of the probabilities must be known or estimated from the data. We will see that this is an excellent application of the probability density estimation methods from Chapter 8.

We have already seen an application of Bayes decision theory in Chapter 2. There we wanted to know the probability that a piston ring came from a particular manufacturer given that it failed. It makes sense to make the decision that the part came from the manufacturer that has the highest posterior probability. To put this in the pattern recognition context, we could think of the part failing as the feature. The resulting classification would be the manufacturer ($M_A$ or $M_B$) that sold us the part. In the following, we will see that Bayes decision theory is an application of Bayes' Theorem, where we will classify observations using the posterior probabilities.

We start off by fixing some notation. Let the class membership be represented by $\omega_j$, $j = 1, ..., J$ for a total of $J$ classes. For example, with the **iris** data, we have $J = 3$ classes:

$\omega_1$ = *Iris setosa*

$\omega_2$ = *Iris versicolor*

$\omega_3$ = *Iris virginica.*

The features we are using for classification are denoted by the $d$-dimensional vector $\mathbf{x}$, $d = 1, 2, \ldots$ . With the `iris` data, we have four measurements, so $d = 4$. In the supervised learning situation, each of the observed feature vectors will also have a class label attached to it.

Our goal is to use the data to create a decision rule or classifier that will take a feature vector $\mathbf{x}$ whose class membership is unknown and return the class it most likely belongs to. A logical way to achieve this is to assign the class label to this feature vector using the class corresponding to the highest *posterior probability*. This probability is given by

$$P(\omega_j|\mathbf{x}); \qquad j = 1, \ldots, J. \tag{9.1}$$

Equation 9.1 represents the probability that the case belongs to the $j$-th class given the observed feature vector $\mathbf{x}$. To use this rule, we would evaluate all of the $J$ posterior probabilities, and the one with the highest probability would be the class we choose. We can find the posterior probabilities using Bayes' Theorem:

$$P(\omega_j|\mathbf{x}) = \frac{P(\omega_j)P(\mathbf{x}|\omega_j)}{P(\mathbf{x})}, \tag{9.2}$$

where

$$P(\mathbf{x}) = \sum_{j=1}^{J} P(\omega_j)P(\mathbf{x}|\omega_j). \tag{9.3}$$

We see from Equation 9.2 that we must know the *prior probability* that it would be in class $j$ given by

$$P(\omega_j); \qquad j = 1, \ldots, J, \tag{9.4}$$

and the *class-conditional probability* (sometimes called the *state-conditional probability*)

$$P(\mathbf{x}|\omega_j); \qquad j = 1, \ldots, J. \tag{9.5}$$

The class-conditional probability in Equation 9.5 represents the probability distribution of the features for each class. The prior probability in Equation 9.4 represents our initial degree of belief that an observed set of features is a case from the $j$-th class. The process of estimating these probabilities is how we build the classifier.

We start our explanation with the prior probabilities. These can either be inferred from prior knowledge of the application, estimated from the data or assumed to be equal. In the piston ring example, we know how many parts we buy from each manufacturer. So, the prior probability that the part came from a certain manufacturer would be based on the percentage of parts obtained from that manufacturer. In other applications, we might know the prevalence of some class in our population. This might be the case in medical diagnosis, where we have some idea of the percentage of the population who are likely to have a certain disease or medical condition. In the case of the `iris` data, we could estimate the prior probabilities using the proportion of each class in our sample. We had 150 observed feature vectors, with 50 coming from each class. Therefore, our estimated prior probabilities would be

$$\hat{P}(\omega_j) = \frac{n_j}{N} = \frac{50}{150} = 0.33; \qquad j = 1, 2, 3 .$$

Finally, we might use equal priors when we believe each class is equally likely.

Now that we have our prior probabilities, $\hat{P}(\omega_j)$, we turn our attention to the class-conditional probabilities $P(\mathbf{x}|\omega_j)$. We can use the density estimation techniques covered in Chapter 8 to obtain these probabilities. In essence, we take all of the observed feature vectors that are known to come from class $\omega_j$ and estimate the density using only those cases. We will cover two approaches: parametric and nonparametric.

### Estimating Class-Conditional Probabilities: Parametric Method

In parametric density estimation, we assume a distribution for the class-conditional probability densities and estimate them by estimating the corresponding distribution parameters. For example, we might assume the features come from a multivariate normal distribution. To estimate the density, we have to estimate $\hat{\mu}_j$ and $\hat{\Sigma}_j$ for each class. This procedure is illustrated in Example 9.1 for the `iris` data.

### Example 9.1

In this example, we estimate our class-conditional probabilities using the `iris` data. We assume that the required probabilities are multivariate normal for each class. The following MATLAB code shows how to get the class-conditional probabilities for each species of iris.

```
load iris
% This loads up three matrices:
% setosa, virginica and versicolor
% We will assume each class is multivariate normal.
% To get the class-conditional probabilities, we
% get estimates for the parameters for each class.
muset = mean(setosa);
```

```
covset = cov(setosa);
muvir = mean(virginica);
covvir = cov(virginica);
muver = mean(versicolor);
covver = cov(versicolor);
```

❏

### Estimating Class-Conditional Probabilities: Nonparametric

If it is not appropriate to assume the features for a class follow a known distribution, then we can use the nonparametric density estimation techniques from Chapter 8. These include the averaged shifted histogram, the frequency polygon, kernel densities, finite mixtures and adaptive mixtures. To obtain the class-conditional probabilities, we take the set of measured features from each class and estimate the density using one of these methods. This is illustrated in Example 9.2, where we use the product kernel to estimate the probability densities for the **iris** data.

### Example 9.2

We estimate the class-conditional probability densities for the **iris** data using the product kernel, where the univariate normal kernel is used for each dimension. We illustrate the use of two functions for estimating the product kernel. One is called **cskern2d** that can only be used for bivariate data. The output arguments from this function are matrices for use in the MATLAB plotting functions **surf** and **mesh**. The **cskern2d** function should be used when the analyst wants to plot the resulting probability density. We use it on the first two dimensions of the **iris** data and plot the surface for *Iris virginica* in Figure 9.2.

```
load iris
% This loads up three matrices:
% setosa, virginica and versicolor
% We will use the product kernel to estimate densities.
% To try this, get the kernel estimate for the first
% two features and plot.
% The arguments of 0.1 indicate the grid size in
% each dimension. This creates the domain over
% which we will estimate the density.
[xset,yset,pset]=cskern2d(setosa(:,1:2),0.1,0.1);
[xvir,yvir,pvir]=cskern2d(virginica(:,1:2),0.1,0.1);
[xver,yver,pver]=cskern2d(versicolor(:,1:2),0.1,0.1);
mesh(xvir,yvir,pvir)
colormap(gray(256))
```

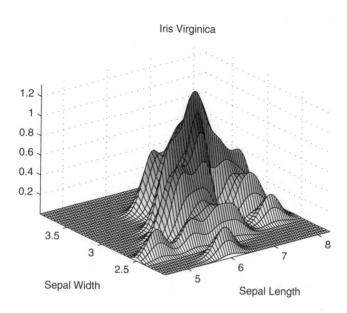

**FIGURE 9.2**
Using only the first two features of the data for *Iris virginica*, we construct an estimate of the corresponding class-conditional probability density using the product kernel. This is the output from the function **cskern2d**.

A more useful function for statistical pattern recognition is **cskernmd**, which returns the value of the probability density $\hat{f}(\mathbf{x})$ for a given $d$-dimensional vector **x**.

```
% If one needs the value of the probability curve,
% then use this.
ps = cskernmd(setosa(1,1:2),setosa(:,1:2));
pver = cskernmd(setosa(1,1:2),versicolor(:,1:2));
pvir = cskernmd(setosa(1,1:2),virginica(:,1:2));
```
❏

## Bayes Decision Rule

Now that we know how to get the prior probabilities and the class-conditional probabilities, we can use Bayes' Theorem to obtain the posterior probabilities. Bayes Decision Rule is based on these posterior probabilities.

*BAYES DECISION RULE:*
*Given a feature vector* **x**, *assign it to class* $\omega_j$ *if*

$$P(\omega_j|\mathbf{x}) > P(\omega_i|\mathbf{x}); \qquad i = 1, ..., J; \; i \neq j. \tag{9.6}$$

This states that we will classify an observation **x** as belonging to the class that has the highest posterior probability. It is known [Duda and Hart, 1973] that the decision rule given by Equation 9.6 yields a classifier with the minimum probability of error.

We can use an equivalent rule by recognizing that the denominator of the posterior probability (see Equation 9.2) is simply a normalization factor and is the same for all classes. So, we can use the following alternative decision rule:

$$P(\mathbf{x}|\omega_j)P(\omega_j) > P(\mathbf{x}|\omega_i)P(\omega_i); \qquad i = 1, ..., J; \; i \neq j. \tag{9.7}$$

Equation 9.7 is Bayes Decision Rule in terms of the class-conditional and prior probabilities. If we have equal priors for each class, then our decision is based only on the class-conditional probabilities. In this case, the decision rule partitions the feature space into $J$ decision regions $\Omega_1, \Omega_2, ..., \Omega_J$. If **x** is in region $\Omega_j$, then we will say it belongs to class $\omega_j$.

We now turn our attention to the error we have in our classifier when we use Bayes Decision Rule. An error is made when we classify an observation as class $\omega_i$ when it is really in the $j$-th class. We denote the complement of region $\Omega_i$ as $\Omega_i^c$, which represents every region except $\Omega_i$. To get the probability of error, we calculate the following integral over all values of **x** [Duda and Hart, 1973; Webb, 1999]

$$P(\text{error}) = \sum_{i=1}^{J} \int_{\Omega_i^c} P(\mathbf{x}|\omega_i)P(\omega_i)d\mathbf{x}. \tag{9.8}$$

Thus, to find the probability of making an error (i.e., assigning the wrong class to an observation), we find the probability of error for each class and add the probabilities together. In the following example, we make this clearer by looking at a two class case and calculating the probability of error.

### Example 9.3
We will look at a univariate classification problem with equal priors and two classes. The class-conditionals are given by the normal distributions as follows:

$$P(x|\omega_1) = \phi(\mathbf{x}; -1, 1)$$
$$P(x|\omega_2) = \phi(\mathbf{x}; 1, 1).$$

The priors are

$$P(\omega_1) = 0.6$$
$$P(\omega_2) = 0.4.$$

The following MATLAB code creates the required curves for the decision rule of Equation 9.7.

```
% This illustrates the 1-D case for two classes.
% We will shade in the area where there can be
% misclassified observations.
% Get the domain for the densities.
dom = -6:.1:8;
dom = dom';
% Note: could use csnormp or normpdf.
pxg1 = csevalnorm(dom,-1,1);
pxg2 = csevalnorm(dom,1,1);
plot(dom,pxg1,dom,pxg2)
% Find decision regions - multiply by priors
ppxg1 = pxg1*0.6;
ppxg2 = pxg2*0.4;
plot(dom,ppxg1,'k',dom,ppxg2,'k')
xlabel('x')
```

The resulting plot is given in Figure 9.3, where we see that the decision regions given by Equation 9.7 are obtained by finding where the two curves intersect. If we observe a value of a feature given by $x = -2$, then we would classify that object as belonging to class $\omega_1$. If we observe $x = 4$, then we would classify that object as belonging to class $\omega_2$. Let's see what happens when $x = -0.75$. We can find the probabilities using

```
x = -0.75;
% Evaluate each un-normalizd posterior.
po1 = csevalnorm(x,-1,1)*0.6;
po2 = csevalnorm(x,1,1)*0.4;
```

$$P(-0.75|\omega_1)P(\omega_1) = 0.23$$
$$P(-0.75|\omega_2)P(\omega_2) = 0.04.$$

These are shown in Figure 9.4. Note that there is non-zero probability that the case corresponding to $x = -0.75$ could belong to class 2. We now turn our attention to how we can estimate this error.

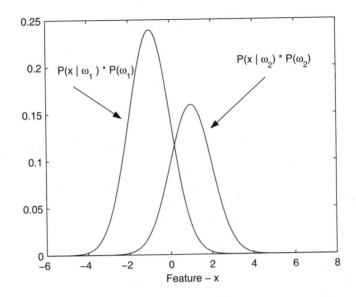

**FIGURE 9.3**
Here we show the univariate, two class case from Example 9.3. Note that each curve represents the probabilities in Equation 9.7. The point where the two curves intersect partitions the domain into one where we would classify observations as class 1 $(\omega_1)$ and another where we would classify observations as class 2 $(\omega_2)$.

```
% To get estimates of the error, we can
% estimate the integral as follows
% Note that 0.1 is the step size and we
% are approximating the integral using a sum.
% The decision boundary is where the two curves meet.
ind1 = find(ppxg1 >= ppxg2);
% Now find the other part.
ind2 = find(ppxg1<ppxg2);
pmis1 = sum(ppxg1(ind2))*.1;
pmis2 = sum(ppxg2(ind1))*.1;
errorhat = pmis1 + pmis2;
```

From this, we estimate the probability of error as 0.15. To get this probability, we find the shaded area under the curves as shown in Figure 9.5.
❑

We would like to note several points regarding Bayes Decision Rule and the classification error. First, as we already saw in Example 9.3, the boundaries

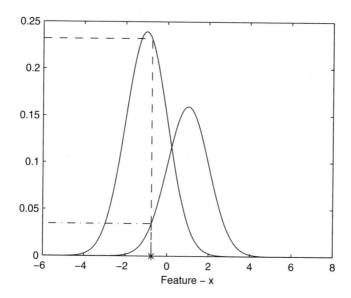

**FIGURE 9.4**
The vertical dotted line represents $x = -0.75$. The probabilities needed for the decision rule of Equation 9.7 are represented by the horizontal dotted lines. We would classify this case as belonging to class 1 ($\omega_1$), but there is a possibility that it could belong to class 2 ($\omega_2$).

for the decision regions are found as the **x** such that the following equation is satisfied:

$$P(\mathbf{x}|\omega_j)P(\omega_j) = P(\mathbf{x}|\omega_i)P(\omega_i); \qquad i \neq j.$$

Secondly, we can change this decision region as we will see shortly when we discuss the likelihood ratio approach to classification. If we change the decision boundary, then the error will be greater, illustrating that Bayes Decision Rule is one that minimizes the probability of misclassification [Duda and Hart, 1973].

## Example 9.4

We continue Example 9.3, where we show what happens when we change the decision boundary to $x = -0.5$. This means that if a feature has a value of $x < -0.5$, then we classify it as belonging to class 1. Otherwise, we say it belongs to class 2. The areas under the curves that we need to calculate are shown in Figure 9.6. As we see from the following MATLAB code, where we estimate the error, that the probability of error increases.

```
% Change the decision boundary.
```

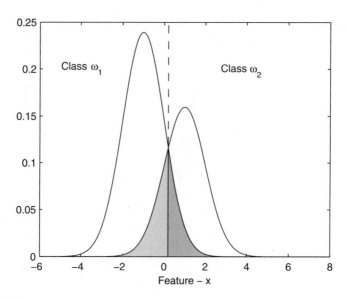

**FIGURE 9.5**
The shaded regions show the probability of misclassifying an object. The lighter region shows the probability of classifying as class 1 when it is really class 2. The darker region shows the probability of classifying as class 2, when it belongs to class 1.

```
bound = -0.5;
ind1 = find(dom <= bound);
ind2 = find(dom > bound);
pmis1 = sum(ppxg1(ind2))*.1;
pmis2 = sum(ppxg2(ind1))*.1;
errorhat = pmis1 + pmis2;
```

This yields an estimated error of 0.20.
❏

Bayes decision theory can address more general situations where there might be a variable cost or risk associated with classifying something incorrectly or allowing actions in addition to classifying the observation. For example, we might want to penalize the error of classifying some section of tissue in an image as cancerous when it is not, or we might want to include the action of not making a classification if our uncertainty is too great. We will provide references at the end of the chapter for those readers who require the more general treatment of statistical pattern recognition.

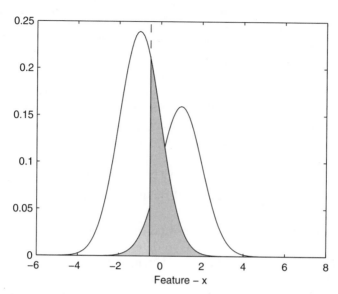

**FIGURE 9.6**
If we move the decision boundary to $x = -0.5$, then the probability of error is given by the shaded areas. Not surprisingly, the error increases when we change from the boundary given by Bayes Decision Rule.

## Likelihood Ratio Approach

The likelihood ratio technique addresses the issue of variable misclassification costs in a hypothesis testing framework. This methodology does not assign an explicit cost to making an error as in the Bayes approach, but it enables us to set the amount of error we will tolerate for misclassifying one of the classes.

Recall from Chapter 6 that in hypothesis testing we have two types of errors. One type of error is when we wrongly reject the null hypothesis when it is really true. This is the Type I error. The other way we can make a wrong decision is to not reject the null hypothesis when we should. Typically, we try to control the probability of Type I error by setting a desired significance level $\alpha$, and we use this level to determine our decision boundary. We can fit our pattern recognition process into the same framework.

In the rest of this section, we consider only two classes, $\omega_1$ and $\omega_2$. First, we have to determine what class corresponds to the null hypothesis and call this the non-target class. The other class is denoted as the target class. In this book, we use $\omega_1$ to represent the target class and $\omega_2$ to represent the non-target class. The following examples should clarify these concepts.

- We are building a classifier for a military command and control system that will take features from images of objects and classify them as targets or non-targets. If an object is classified as a target, then we will destroy it. Target objects might be tanks or military trucks. Non-target objects are such things as school buses or automobiles. We would want to make sure that when we build a classifier we do not classify an object as a tank when it is really a school bus. So, we will control the amount of acceptable error in wrongly saying it (a school bus or automobile) is in the target class. This is the same as our Type I error, if we write our hypotheses as

$H_0$       Object is a school bus, automobile, etc.

$H_1$       Object is a tank, military vehicle, etc.

- Another example, where this situation arises is in medical diagnosis. Say that the doctor needs to determine whether a patient has cancer by looking at radiographic images. The doctor does not want to classify a region in the image as cancer when it is not. So, we might want to control the probability of wrongly deciding that there is cancer when there is none. However, failing to identify a cancer when it is really there is more important to control. Therefore, in this situation, the hypotheses are

$H_0$       X-ray shows cancerous tissue

$H_1$       X-ray shows only healthy tissue

The terminology that is sometimes used *for the Type I error in pattern recognition is **false alarms** or **false positives**. A false alarm is wrongly classifying something as a target $(\omega_1)$, when it should be classified as non-target $(\omega_2)$. The probability of making a false alarm (or the probability of making a Type I error) is denoted as

$$P(FA) = \alpha.$$

This probability is represented as the shaded area in Figure 9.7.

Recall that Bayes Decision Rule gives a rule that yields the minimum probability of incorrectly classifying observed patterns. We can change this boundary to obtain the desired probability of false alarm $\alpha$. Of course, if we do this, then we must accept a higher probability of misclassification as shown in Example 9.4.

In the two class case, we can put our Bayes Decision Rule in a different form. Starting from Equation 9.7, we have our decision as

$$P(\mathbf{x}|\omega_1)P(\omega_1) > P(\mathbf{x}|\omega_2)P(\omega_2) \Rightarrow \mathbf{x} \text{ is in } \omega_1, \qquad (9.9)$$

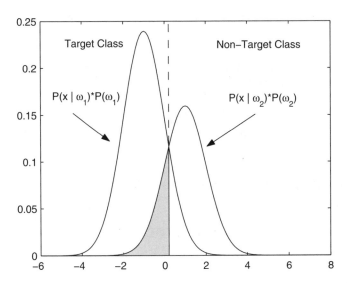

**FIGURE 9.7**
The shaded region shows the probability of false alarm or the probability of wrongly classifying as target (class $\omega_1$) when it really belongs to class $\omega_2$.

or else we classify **x** as belonging to $\omega_2$. Rearranging this inequality yields the following decision rule

$$L_R(\mathbf{x}) = \frac{P(\mathbf{x}|\omega_1)}{P(\mathbf{x}|\omega_2)} > \frac{P(\omega_2)}{P(\omega_1)} = \tau_C \Rightarrow \mathbf{x} \text{ is in } \omega_1. \tag{9.10}$$

The ratio on the left of Equation 9.10 is called the likelihood ratio, and the quantity on the right is the threshold. If $L_R > \tau_C$, then we decide that the case belongs to class $\omega_1$. If $L_R < \tau_C$, then we group the observation with class $\omega_2$.

If we have equal priors, then the threshold is one ($\tau_C = 1$). Thus, when $L_R > 1$, we assign the observation or pattern to $\omega_1$, and if $L_R < 1$, then we classify the observation as belonging to $\omega_2$. We can also adjust this threshold to obtain a desired probability of false alarm, as we show in Example 9.5.

**Example 9.5**
We use the class-conditional and prior probabilities of Example 9.3 to show how we can adjust the decision boundary to achieve the desired probability of false alarm. Looking at Figure 9.7, we see that

$$P(FA) = \int_{-\infty}^{C} P(x|\omega_2)P(\omega_2)dx,$$

where $C$ represents the value of $x$ that corresponds to the decision boundary. We can factor out the prior, so

$$P(FA) = P(\omega_2)\int_{-\infty}^{C} P(x|\omega_2)dx.$$

We then have to find the value for C such that

$$\int_{-\infty}^{C} P(x|\omega_2)dx = \frac{P(FA)}{P(\omega_2)}.$$

From Chapter 3, we recognize that $C$ is a quantile. Using the probabilities in Example 9.3, we know that $P(\omega_2) = 0.4$ and $P(x|\omega_2)$ is normal with mean 1 and variance of 1. If our desired $P(FA) = 0.05$, then

$$\int_{-\infty}^{C} P(x|\omega_2)dx = \frac{0.05}{0.40} = 0.125.$$

We can find the value for C using the inverse cumulative distribution function for the normal distribution. In MATLAB, this is

```
c = norminv(0.05/0.4,1,1);
```

This yields a decision boundary of $x = -0.15$.
❑

## 9.3 Evaluating the Classifier

Once we have our classifier, we need to evaluate its usefulness by measuring the percentage of observations that we correctly classify. This yields an estimate of the probability of correctly classifying cases. It is also important to report the probability of false alarms, when the application requires it (e.g., when there is a target class). We will discuss two methods for estimating the probability of correctly classifying cases and the probability of false alarm: the use of an independent test sample and cross-validation.

### Independent Test Sample

If our sample is large, we can divide it into a training set and a testing set. We use the training set to build our classifier and then we classify observations in the test set using our classification rule. The proportion of correctly classified observations is the ***estimated classification rate***. Note that the classifier has not seen the patterns in the test set, so the classification rate estimated in this way is not biased. Of course, we could collect more data to be used as the independent test set, but that is often impossible or impractical.

By biased we mean that the estimated probability of correctly classifying a pattern is not overly optimistic. A common mistake that some researchers make is to build a classifier using their sample and then use the same sample to determine the proportion of observations that are correctly classified. That procedure typically yields much higher classification success rates, because the classifier has already *seen* the patterns. It does not provide an accurate idea of how the classifier recognizes patterns it has not seen before. However, for a thorough discussion on these issues, see Ripley [1996]. The steps for evaluating the classifier using an independent test set are outlined below.

*PROBABILITY OF CORRECT CLASSIFICATION- INDEPENDENT TEST SAMPLE*

1. Randomly separate the sample into two sets of size $n_{TEST}$ and $n_{TRAIN}$, where $n_{TRAIN} + n_{TEST} = n$. One is for building the classifier (the training set), and one is used for testing the classifier (the testing set).

2. Build the classifier (e.g., Bayes Decision Rule, classification tree, etc.) using the training set.

3. Present each pattern from the test set to the classifier and obtain a class label for it. Since we know the correct class for these observations, we can count the number we have successfully classified. Denote this quantity as $N_{CC}$.

4. The rate at which we correctly classified observations is

$$P(CC) = \frac{N_{CC}}{n_{TEST}}.$$

The higher this proportion, the better the classifier. We illustrate this procedure in Example 9.6.

### Example 9.6

We first load the data and then divide the data into two sets, one for building the classifier and one for testing it. We use the two species of **iris** that are hard to separate: *Iris versicolor* and *Iris virginica*.

```
load iris
% This loads up three matrices:
% setosa, versicolor and virginica.
% We will use the versicolor and virginica.
% To make it interesting, we will use only the
% first two features.
% Get the data for the training and testing set. We
% will just pick every other one for the testing set.
indtrain = 1:2:50;
indtest = 2:2:50;
versitest = versicolor(indtest,1:2);
versitrain = versicolor(indtrain,1:2);
virgitest = virginica(indtest,1:2);
virgitrain = virginica(indtrain,1:2);
```

We now build the classifier by estimating the class-conditional probabilities. We use the parametric approach, making the assumption that the class-conditional densities are multivariate normal. In this case, the estimated priors are equal.

```
% Get the classifier. We will assume a multivariate
% normal model for these data.
muver = mean(versitrain);
covver = cov(versitrain);
muvir = mean(virgitrain);
covvir = cov(virgitrain);
```

Note that the classifier is obtained using the training set only. We use the testing set to estimate the probability of correctly classifying observations.

```
% Present each test case to the classifier. Note that
% we are using equal priors, so the decision is based
% only on the class-conditional probabilities.
% Put all of the test data into one matrix.
X = [versitest;virgitest];
% These are the probability of x given versicolor.
pxgver = csevalnorm(X,muver,covver);
% These are the probability of x given virginica.
pxgvir = csevalnorm(X,muvir,covvir);
% Check which are correctly classified.
% In the first 25, pxgver > pxgvir are correct.
ind = find(pxgver(1:25)>pxgvir(1:25));
ncc = length(ind);
% In the last 25, pxgvir > pxgver are correct.
ind = find(pxgvir(26:50) > pxgver(26:50));
ncc = ncc + length(ind);
pcc = ncc/50;
```

Using this type of classifier and this partition of the learning sample, we estimate the probability of correct classification to be 0.74.
◻

## Cross-Validation

The cross-validation procedure is discussed in detail in Chapter 7. Recall that with cross-validation, we systematically partition the data into testing sets of size $k$. The $n - k$ observations are used to build the classifier, and the remaining $k$ patterns are used to test it. We continue in this way through the entire data set. When the sample is too small to partition it into a single testing and training set, then cross-validation is the recommended approach. The following is the procedure for calculating the probability of correct classification using cross-validation with $k = 1$.

*PROBABILITY OF CORRECT CLASSIFICATION - CROSS-VALIDATION*

1. Set the number of correctly classified patterns to 0, $N_{CC} = 0$.
2. Keep out one observation, call it $x_i$.
3. Build the classifier using the remaining $n - 1$ observations.
4. Present the observation $x_i$ to the classifier and obtain a class label using the classifier from the previous step.
5. If the class label is correct, then increment the number correctly classified using

$$N_{CC} = N_{CC} + 1.$$

6. Repeat steps 2 through 5 for each pattern in the sample.
7. The probability of correctly classifying an observation is given by

$$P(CC) = \frac{N_{CC}}{n}.$$

## Example 9.7
We return to the **iris** data of Example 9.6, and we estimate the probability of correct classification using cross-validation with $k = 1$. We first set up some preliminary variables and load the data.

```
load iris
% This loads up three matrices:
% setosa, versicolor and virginica.
% We will use the versicolor and virginica.
% Note that the priors are equal, so the decision is
```

```
% based on the class-conditional probabilities.
ncc = 0;
% We will use only the first two features of
% the iris data for our classification.
% This should make it more difficult to
% separate the classes.
% Delete 3rd and 4th features.
virginica(:,3:4) = [];
versicolor(:,3:4) = [];
[nver,d] = size(versicolor);
[nvir,d] = size(virginica);
n = nvir + nver;
```

First, we will loop through all of the **versicolor** observations. We build a classifier, leaving out one pattern at a time for testing purposes. Throughout this loop, the class-conditional probability for **virginica** remains the same, so we find that first.

```
% Loop first through all of the patterns corresponding
% to versicolor. Here correct classification
% is obtained if pxgver > pxgvir;
muvir = mean(virginica);
covvir = cov(virginica);
% These will be the same for this part.
for i = 1:nver
    % Get the test point and the training set
    versitrain = versicolor;
    % This is the testing point.
    x = versitrain(i,:);
    % Delete from training set.
    % The result is the training set.
    versitrain(i,:)=[];
    muver = mean(versitrain);
    covver = cov(versitrain);
    pxgver = csevalnorm(x,muver,covver);
    pxgvir = csevalnorm(x,muvir,covvir);
    if pxgver > pxgvir
    % then we correctly classified it
        ncc = ncc+1;
    end
end
```

We repeat the same procedure leaving out each **virginica** observation as the test pattern.

```
% Loop through all of the patterns of virginica notes.
% Here correct classification is obtained when
% pxgvir > pxxgver
```

```
muver = mean(versicolor);
covver = cov(versicolor);
% Those remain the same for the following.
for i = 1:nvir
    % Get the test point and training set.
    virtrain = virginica;
    x = virtrain(i,:);
    virtrain(i,:)=[];
    muvir = mean(virtrain);
    covvir = cov(virtrain);
    pxgver = csevalnorm(x,muver,covver);
    pxgvir = csevalnorm(x,muvir,covvir);
    if pxgvir > pxgver
        % then we correctly classified it
        ncc = ncc+1;
    end
end
```

Finally, the probability of correct classification is estimated using

```
pcc = ncc/n;
```

The estimated probability of correct classification for the **iris** data using cross-validation is 0.68.

❑

## Receiver Operating Characteristic (ROC) Curve

We now turn our attention to how we can use cross-validation to evaluate a classifier that uses the likelihood approach with varying decision thresholds $\tau_C$. It would be useful to understand how the classifier performs for various thresholds (corresponding to the probability of false alarm) of the likelihood ratio. This will tell us what performance degradation we have (in terms of correctly classifying the target class) if we limit the probability of false alarm to some level.

We start by dividing the sample into two sets: one with all of the target observations and one with the non-target patterns. Denote the observations as follows

$$x_i^{(1)} \Rightarrow \text{Target pattern } (\omega_1)$$

$$x_i^{(2)} \Rightarrow \text{Non-target pattern } (\omega_2).$$

Let $n_1$ represent the number of target observations (class $\omega_1$) and $n_2$ denote the number of non-target (class $\omega_2$) patterns. We work first with the non-target observations to determine the threshold we need to get a desired probability of false alarm. Once we have the threshold, we can determine the probability of correctly classifying the observations belonging to the target class.

Before we go on to describe the receiver operating characteristic (ROC) curve, we first describe some terminology. For any boundary we might set for the decision regions, we are likely to make mistakes in classifying cases. There will be some target patterns that we correctly classify as targets and some we misclassify as non-targets. Similarly, there will be non-target patterns that are correctly classified as non-targets and some that are misclassified as targets. This is summarized as follows:

- <u>True Positives - TP</u>: This is the fraction of patterns correctly classified as target cases.
- <u>False Positives - FP</u>: This is the fraction of non-target patterns incorrectly classified as target cases.
- <u>True Negatives - TN</u>: This is the fraction of non-target cases correctly classified as non-target.
- <u>False Negatives - FN</u>: This is the fraction of target cases incorrectly classified as non-target.

In our previous terminology, the false positives (FP) correspond to the false alarms. Figure 9.8 shows these areas for a given decision boundary.

A **ROC** *curve* is a plot of the true positive rate against the false positive rate. ROC curves are used primarily in signal detection and medical diagnosis [Egan, 1975; Lusted, 1971; McNeil, et. al., 1975; Hanley and McNeil, 1983; Hanley and Hajian-Tilaki, 1997]. In their terminology, the true positive rate is also called the sensitivity. *Sensitivity* is the probability that a classifier will classify a pattern as a target when it really is a target. *Specificity* is the probability that a classifier will correctly classify the true non-target cases. Therefore, we see that a ROC curve is also a plot of sensitivity against 1 minus specificity.

One of the purposes of a ROC curve is to measure the discriminating power of the classifier. It is used in the medical community to evaluate the diagnostic power of tests for diseases. By looking at a ROC curve, we can understand the following about a classifier:

- It shows the trade-off between the probability of correctly classifying the target class (sensitivity) and the false alarm rate (1 – specificity).
- The area under the ROC curve can be used to compare the performance of classifiers.

We now show in more detail how to construct a ROC curve. Recall that the likelihood ratio is given by

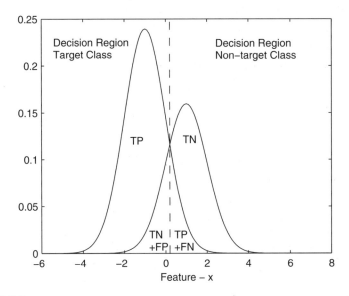

**FIGURE 9.8**
In this figure, we see the decision regions for deciding whether a feature corresponds to the target class or the non-target class.

$$L_R(\mathbf{x}) = \frac{P(\mathbf{x}|\omega_1)}{P(\mathbf{x}|\omega_2)}.$$

We start off by forming the likelihood ratios using the non-target ($\omega_2$) observations and cross-validation to get the distribution of the likelihood ratios when the class membership is truly $\omega_2$. We use these likelihood ratios to set the threshold that will give us a specific probability of false alarm.

Once we have the thresholds, the next step is to determine the rate at which we correctly classify the target cases. We first form the likelihood ratio for each target observation using cross-validation, yielding a distribution of likelihood ratios for the target class. For each given threshold, we can determine the number of target observations that would be correctly classified by counting the number of $L_R$ that are greater than that threshold. These steps are described in detail in the following procedure.

*CROSS-VALIDATION FOR SPECIFIED FALSE ALARM RATE*

1. Given observations with class labels $\omega_1$ (target) and $\omega_2$ (non-target), set desired probabilities of false alarm and a value for $k$.

2. Leave $k$ points out of the non-target class to form a set of test cases denoted by *TEST*. We denote cases belonging to class $\omega_2$ as $\mathbf{x}_i^{(2)}$.

3. Estimate the class-conditional probabilities using the remaining $n_2 - k$ non-target cases and the $n_1$ target cases.

4. For each of those $k$ observations, form the likelihood ratios

$$L_R(\mathbf{x}_i^{(2)}) = \frac{P(\mathbf{x}_i^{(2)}|\omega_1)}{P(\mathbf{x}_i^{(2)}|\omega_2)}; \qquad \mathbf{x}_i^{(2)} \text{ in } TEST.$$

5. Repeat steps 2 through 4 using all of the non-target cases.

6. Order the likelihood ratios for the non-target class.

7. For each probability of false alarm, find the threshold that yields that value. For example, if the $P(FA) = 0.1$, then the threshold is given by the quantile $\hat{q}_{0.9}$ of the likelihood ratios. Note that higher values of the likelihood ratios indicate the target class. We now have an array of thresholds corresponding to each probability of false alarm.

8. Leave $k$ points out of the target class to form a set of test cases denoted by *TEST*. We denote cases belonging to $\omega_1$ by $\mathbf{x}_i^{(1)}$.

9. Estimate the class-conditional probabilities using the remaining $n_1 - k$ target cases and the $n_2$ non-target cases.

10. For each of those $k$ observations, form the likelihood ratios

$$L_R(\mathbf{x}_i^{(1)}) = \frac{P(\mathbf{x}_i^{(1)}|\omega_1)}{P(\mathbf{x}_i^{(1)}|\omega_2)}; \qquad \mathbf{x}_i^1 \text{ in } TEST.$$

11. Repeat steps 8 through 10 using all of the target cases.

12. Order the likelihood ratios for the target class.

13. For each threshold and probability of false alarm, find the proportion of target cases that are correctly classified to obtain the $P(CC_{Target})$. If the likelihood ratios $L_R(\mathbf{x}_i^{(1)})$ are sorted, then this would be the number of cases that are greater than the threshold.

This procedure yields the rate at which the target class is correctly classified for a given probability of false alarm. We show in Example 9. 8 how to implement this procedure in MATLAB and plot the results in a ROC curve.

### Example 9.8

In this example, we illustrate the cross-validation procedure and ROC curve using the univariate model of Example 9.3. We first use MATLAB to generate some data.

```
% Generate some data, use the model in Example 9.3.
% p(x|w1) ~ N(-1,1), p(w1) = 0.6
% p(x|w2) ~ N(1,1),p(w2) = 0.4;
% Generate the random variables.
n = 1000;
u = rand(1,n);% find out what class they are from
n1 = length(find(u <= 0.6));% # in target class
n2 = n-n1;
x1 = randn(1,n1) - 1;
x2 = randn(1,n2) + 1;
```

We set up some arrays to store the likelihood ratios and estimated probabilities. We also specify the values for the $P(FA)$. For each $P(FA)$, we will be estimating the probability of correctly classifying objects from the target class.

```
% Set up some arrays to store things.
lr1 = zeros(1,n1);
lr2 = zeros(1,n2);
pfa = 0.01:.01:0.99;
pcc = zeros(size(pfa));
```

We now implement steps 2 through 7 of the cross-validation procedure. This is the part where we find the thresholds that provide the desired probability of false alarm.

```
% First find the threshold corresponding
% to each false alarm rate.
% Build classifier using target data.
mu1 = mean(x1);
var1 = cov(x1);
% Do cross-validation on non-target class.
for i = 1:n2
    train = x2;
    test = x2(i);
    train(i) = [];
    mu2 = mean(train);
    var2 = cov(train);
    lr2(i) = csevalnorm(test,mu1,var1)./...
        csevalnorm(test,mu2,var2);
end
% sort the likelihood ratios for the non-target class
lr2 = sort(lr2);
% Get the thresholds.
thresh = zeros(size(pfa));
for i = 1:length(pfa)
    thresh(i) = csquantiles(lr2,1-pfa(i));
end
```

For the given thresholds, we now find the probability of correctly classifying the target cases. This corresponds to steps 8 through 13.

```
% Now find the probability of correctly
% classifying targets.
mu2 = mean(x2);
var2 = cov(x2);
% Do cross-validation on target class.
for i = 1:n1
    train = x1;
    test = x1(i);
    train(i) = [];
    mu1 = mean(train);
    var1 = cov(train);
    lr1(i) = csevalnorm(test,mu1,var1)./...
        csevalnorm(test,mu2,var2);
end
% Find the actual pcc.
for i = 1:length(pfa)
    pcc(i) = length(find(lr1 >= thresh(i)));
end
pcc = pcc/n1;
```

The ROC curve is given in Figure 9.9. We estimate the area under the curve as 0.91, using

```
area = sum(pcc)*.01;
```

❑

## 9.4 Classification Trees

In this section, we present another technique for pattern recognition called classification trees. Our treatment of classification trees follows that in the book called *Classification and Regression Trees* by Breiman, Friedman, Olshen and Stone [1984]. For ease of exposition, we do not include the MATLAB code for the classification tree in the main body of the text, but we do include it in Appendix D. There are several main functions that we provide to work with trees, and these are summarized in Table 9.1. We will be using these functions in the text when we discuss the classification tree methodology.

While Bayes decision theory yields a classification rule that is intuitively appealing, it does not provide insights about the structure or the nature of the classification rule or help us determine what features are important. Classification trees can yield complex decision boundaries, and they are appropriate for ordered data, categorical data or a mixture of the two types. In this book,

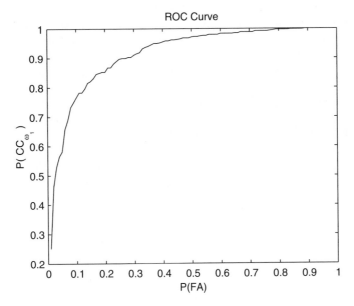

**FIGURE 9.9**
This shows the ROC curve for Example 9.8.

**TABLE 9.1**

Matlab Functions for Working with Classification Trees

| Purpose | Matlab Function |
| --- | --- |
| Grows the initial large tree | csgrowc |
| Gets a sequence of minimal complexity trees | csprunec |
| Returns the class for a set of features, using the decision tree | cstreec |
| Plots a tree | csplotreec |
| Given a sequence of subtrees and an index for the best tree, extract the tree (also cleans out the tree) | cspicktreec |

we will be concerned only with the case where all features are continuous random variables. The interested reader is referred to Breiman, et al. [1984], Webb [1999], and Duda, Hart and Stork [2001] for more information on the other cases.

A decision or classification tree represents a multi-stage decision process, where a binary decision is made at each stage. The tree is made up of *nodes* and *branches*, with nodes being designated as an internal or a terminal node. *Internal nodes* are ones that split into two children, while *terminal nodes* do not have any children. A terminal node has a class label associated with it, such that observations that fall into the particular terminal node are assigned to that class.

To use a classification tree, a feature vector is presented to the tree. If the value for a feature is less than some number, then the decision is to move to the left child. If the answer to that question is no, then we move to the right child. We continue in that manner until we reach one of the terminal nodes, and the class label that corresponds to the terminal node is the one that is assigned to the pattern. We illustrate this with a simple example.

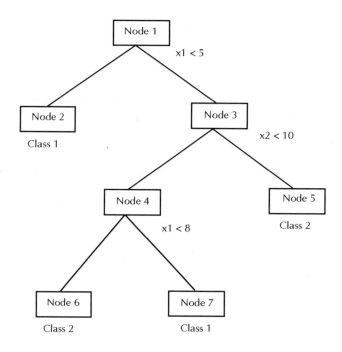

**FIGURE 9.10**
This simple classification tree for two classes is used in Example 9.9. Here we make decisions based on two features, $x_1$ and $x_2$.

## Example 9.9

We show a simple classification tree in Figure 9.10, where we are concerned with only two features. Note that all internal nodes have two children and a splitting rule. The split can occur on either variable, with observations that are less than that value being assigned to the left child and the rest going to the right child. Thus, at node 1, any observation where the first feature is less than 5 would go to the left child. When an observation stops at one of the terminal nodes, it is assigned to the corresponding class for that node. We illustrate these concepts with several cases. Say that we have a feature vector given by $\mathbf{x} = (4, 6)$, then passing this down the tree, we get

$$\text{node } 1 \rightarrow \text{node } 2 \Rightarrow \omega_1 .$$

If our feature vector is $\mathbf{x} = (6, 6)$, then we travel the tree as follows:

$$\text{node } 1 \rightarrow \text{node } 3 \rightarrow \text{node } 4 \rightarrow \text{node } 6 \Rightarrow \omega_2 .$$

For a feature vector given by $\mathbf{x} = (10, 12)$, we have

$$\text{node } 1 \rightarrow \text{node } 3 \rightarrow \text{node } 5 \Rightarrow \omega_2 .$$

□

We give a brief overview of the steps needed to create a tree classifier and then explain each one in detail. To start the process, we must grow an overly large tree using a criterion that will give us optimal splits for the tree. It turns out that these large trees fit the training data set very well. However, they do not generalize, so the rate at which we correctly classify new patterns is low. The proposed solution [Breiman, et al., 1984] to this problem is to continually prune the large tree using a minimal cost complexity criterion to get a sequence of sub-trees. The final step is to choose a tree that is the 'right size' using cross-validation or an independent test sample. These three main procedures are described in the remainder of this section. However, to make things easier for the reader, we first provide the notation that will be used to describe classification trees.

*CLASSIFICATION TREES - NOTATION*

**L** denotes a learning set made up of observed feature vectors and their class label.

*J* denotes the number of classes.

*T* is a classification tree.

*t* represents a node in the tree.

$t_L$ and $t_R$ are the left and right child nodes.

$\{t_1\}$ is the tree containing only the root node.

$T_t$ is a branch of tree $T$ starting at node $t$.

$\widehat{T}$ is the set of terminal nodes in the tree.

$|\widehat{T}|$ is the number of terminal nodes in tree $T$.

$t_k^*$ is the node that is the weakest link in tree $T_k$.

$n$ is the total number of observations in the learning set.

$n_j$ is the number of observations in the learning set that belong to the $j$-th class $\omega_j$, $j = 1, ..., J$.

$n(t)$ is the number of observations that fall into node $t$.

$n_j(t)$ is the number of observations at node $t$ that belong to class $\omega_j$.

$\pi_j$ is the prior probability that an observation belongs to class $\omega_j$. This can be estimated from the data as

$$\hat{\pi}_j = \frac{n_j}{n}. \tag{9.11}$$

$p(\omega_j, t)$ represents the joint probability that an observation will be in node $t$ and it will belong to class $\omega_j$. It is calculated using

$$p(\omega_j, t) = \frac{\pi_j n_j(t)}{n_j}. \tag{9.12}$$

$p(t)$ is the probability that an observation falls into node $t$ and is given by

$$p(t) = \sum_{j=1}^{J} p(\omega_j, t). \tag{9.13}$$

$p(\omega_j|t)$ denotes the probability that an observation is in class $\omega_j$ given it is in node $t$. This is calculated from

$$p(\omega_j|t) = \frac{p(\omega_j, t)}{p(t)}. \tag{9.14}$$

$r(t)$ represents the resubstitution estimate of the probability of misclassification for node $t$ and a given classification into class $\omega_j$. This is found by subtracting the maximum conditional probability $p(\omega_j|t)$ for the node from 1:

$$r(t) = 1 - \max_{j} \{p(\omega_j|t)\}. \tag{9.15}$$

$R(t)$ is the resubstitution estimate of risk for node $t$. This is

$$R(t) = r(t)p(t). \tag{9.16}$$

$R(T)$ denotes a resubstitution estimate of the overall misclassification rate for a tree $T$. This can be calculated using every terminal node in the tree as follows

$$R(T) = \sum_{t \in \widehat{T}} r(t)p(t) = \sum_{t \in \widehat{T}} R(t). \tag{9.17}$$

$\alpha$ is the complexity parameter.

$i(t)$ denotes a measure of impurity at node $t$.

$\Delta i(s, t)$ represents the decrease in impurity and indicates the goodness of the split $s$ at node $t$. This is given by

$$\Delta i(s, t) = i(t) - p_R i(t_R) - p_L i(t_L). \tag{9.18}$$

$p_L$ and $p_R$ are the proportion of data that are sent to the left and right child nodes by the split $s$.

### Growing the Tree

The idea behind binary classification trees is to split the $d$-dimensional space into smaller and smaller partitions, such that the partitions become purer in terms of the class membership. In other words, we are seeking partitions where the majority of the members belong to one class. To illustrate these ideas, we use a simple example where we have patterns from two classes, each one containing two features, $x_1$ and $x_2$. How we obtain these data are discussed in the following example.

### Example 9.10

We use synthetic data to illustrate the concepts of classification trees. There are two classes, and we generate 50 points from each class. From Figure 9.11, we see that each class is a two term mixture of bivariate uniform random variables.

```
% This shows how to generate the data that will be used
% to illustrate classification trees.
deln = 25;
data(1:deln,:) = rand(deln,2)+.5;
so=deln+1; sf = 2*deln;
data(so:sf,:) = rand(deln,2)-.5;
so=sf+1; sf = 3*deln;
data(so:sf,1) = rand(deln,1)-.5;
data(so:sf,2) = rand(deln,1)+.5;
so=sf+1; sf = 4*deln;
data(so:sf,1) = rand(deln,1)+.5;
data(so:sf,2) = rand(deln,1)-.5;
```

A scatterplot of these data is given in Figure 9.11. One class is depicted by the '*' and the other is represented by the 'o'. These data are available in the file called **cartdata**, so the user can load them and reproduce the next several examples.

❑

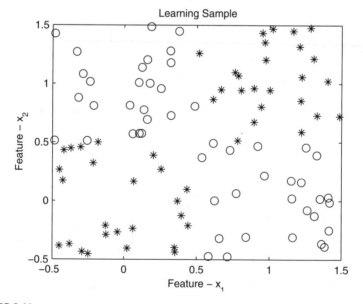

**FIGURE 9.11**
This shows a scatterplot of the data that will be used in our classification tree examples. Data that belong to class 1 are shown by the '*', and those that belong to class 2 are denoted by an 'o'.

To grow a tree, we need to have some criterion to help us decide how to split the nodes. We also need a rule that will tell us when to stop splitting the nodes, at which point we are finished growing the tree. The stopping rule can be quite simple, since we first grow an overly large tree. One possible choice is to continue splitting terminal nodes until each one contains observations from the same class, in which case some nodes might have only one observation in the node. Another option is to continue splitting nodes until there is some maximum number of observations left in a node or the terminal node is pure (all observations belong to one class). Recommended values for the maximum number of observations left in a terminal node are between 1 and 5.

We now discuss the splitting rule in more detail. When we split a node, our goal is to find a split that reduces the impurity in some manner. So, we need a measure of impurity $i(t)$ for a node $t$. Breiman, et al. [1984] discuss several possibilities, one of which is called the **Gini diversity index**. This is the one we will use in our implementation of classification trees. The Gini index is given by

$$i(t) = \sum_{i \neq j} p(\omega_i | t) p(\omega_j | t), \tag{9.19}$$

which can also be written as

$$i(t) = 1 - \sum_{j=1}^{J} p^2(\omega_j | t). \tag{9.20}$$

Equation 9.20 is the one we code in the MATLAB function **csgrowc** for growing classification trees.

Before continuing with our description of the splitting process, we first note that our use of the term *'best'* does not necessarily mean that the split we find is the optimal one out of all the infinite possible splits. To grow a tree at a given node, we search for the best split (in terms of decreasing the node impurity) by first searching through each variable or feature. We have $d$ possible best splits for a node (one for each feature), and we choose the best one out of these $d$ splits. The problem now is to search through the infinite number of possible splits. We can limit our search by using the following convention. For all feature vectors in our learning sample, we search for the best split at the $k$-th feature by proposing splits that are halfway between consecutive values for that feature. For each proposed split, we evaluate the impurity criterion and choose the split that yields the largest decrease in impurity.

Once we have finished growing our tree, we must assign class labels to the terminal nodes and determine the corresponding misclassification rate. It makes sense to assign the class label to a node according to the likelihood that it is in class $\omega_j$ given that it fell into node $t$. This is the posterior probability

$p(\omega_j|t)$ given by Equation 9.14. So, using Bayes decision theory, we would classify an observation at node $t$ with the class $\omega_j$ that has the highest posterior probability. The error in our classification is then given by Equation 9.15. We summarize the steps for growing a classification tree in the following procedure. In the learning set, each observation will be a row in the matrix $\mathbf{X}$, so this matrix has dimensionality $n \times (d + 1)$, representing $d$ features and a class label. The measured value of the $k$-th feature for the $i$-th observation is denoted by $x_{ik}$.

### PROCEDURE - GROWING A TREE

1. Determine the maximum number of observations $n_{max}$ that will be allowed in a terminal node.
2. Determine the prior probabilities of class membership $\pi_j$. These can be estimated from the data (Equation 9.11), or they can be based on prior knowledge of the application.
3. If a terminal node in the current tree contains more than the maximum allowed observations and contains observations from several classes, then search for the best split. For each feature $k$,

   a. Put the $x_{ik}$ in ascending order to give the ordered values $x_{(i)k}$.

   b. Determine all splits $s_{(i)k}$ in the $k$-th feature using

   $$s_{(i)k} = x_{(i)k} + (x_{(i)k} - x_{(i+1)k})/2$$

   c. For each proposed split, evaluate the impurity function $i(t)$ and the goodness of the split using Equations 9.20 and 9.18.

   d. Pick the best, which is the one that yields the largest decrease in impurity.

4. Out of the $k$ best splits in step 3, split the node on the variable that yields the best overall split.
5. For that split found in step 4, determine the observations that go to the left child and those that go to the right child.
6. Repeat steps 3 through 5 until each terminal node satisfies the stopping rule (has observations from only one class or has the maximum allowed cases in the node).

### Example 9.11

In this example, we grow the initial large tree on the data set given in the previous example. We stop growing the tree when each terminal node has a maximum of 5 observations or the node is pure. We first load the data that we generated in the previous example. This file contains the data matrix, the inputs to the function **csgrowc**, and the resulting tree.

```
load cartdata
% Loads up data.
% Inputs to function - csgrowc.
maxn = 5;     % maximum number in terminal nodes
clas = [1 2];  % class labels
pies = [0.5 0.5];  % optional prior probabilities
Nk = [50, 50];   % number in each class
```

The following MATLAB commands grow the initial tree and plot the results in Figure 9.12.

```
tree = csgrowc(X,maxn,clas,Nk,pies);
csplotreec(tree)
```

We see from Figure 9.12, that the tree has partitioned the feature space into eight decision regions or eight terminal nodes.
❑

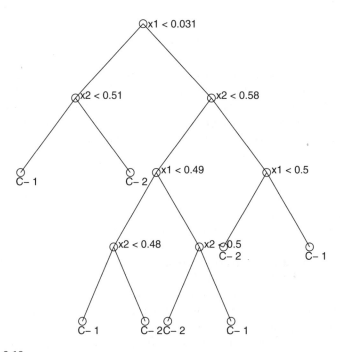

**FIGURE 9.12**
This is the classification tree for the data shown in Figure 9.11. This tree partitions the feature space into 8 decision regions.

## Pruning the Tree

Recall that the classification error for a node is given by Equation 9.15. If we grow a tree until each terminal node contains observations from only one class, then the error rate will be zero. Therefore, if we use the classification error as a stopping criterion or as a measure of when we have a good tree, then we would grow the tree until there are pure nodes. However, as we mentioned before, this procedure over fits the data and the classification tree will not generalize well to new patterns. The suggestion made in Breiman, et al. [1984] is to grow an overly large tree, denoted by $T_{max}$, and then to find a nested sequence of subtrees by successively pruning branches of the tree. The best tree from this sequence is chosen based on the misclassification rate estimated by cross-validation or an independent test sample. We describe the two approaches after we discuss how to prune the tree.

The pruning procedure uses the misclassification rates along with a cost for the complexity of the tree. The complexity of the tree is based on the number of terminal nodes in a subtree or branch. The cost complexity measure is defined as

$$R_\alpha(T) = R(T) + \alpha |\widehat{T}|; \qquad \alpha \geq 0. \qquad (9.21)$$

We look for a tree that minimizes the cost complexity given by Equation 9.21. The $\alpha$ is a parameter that represents the complexity cost per terminal node. If we have a large tree where every terminal node contains observations from only one class, then $R(T)$ will be zero. However, there will be a penalty paid because of the complexity, and the cost complexity measure becomes

$$R_\alpha(T) = \alpha |\widehat{T}|.$$

If $\alpha$ is small, then the penalty for having a complex tree is small, and the resulting tree is large. The tree that minimizes $R_\alpha(T)$ will tend to have few nodes and large $\alpha$.

Before we go further with our explanation of the pruning procedure, we need to define what we mean by the branches of a tree. A branch $T_t$ of a tree $T$ consists of the node $t$ and all its descendent nodes. When we prune or delete this branch, then we remove all descendent nodes of $t$, leaving the branch root node $t$. For example, using the tree in Figure 9.10, the branch corresponding to node 3 contains nodes 3, 4, 5, 6, and 7, as shown in Figure 9.13. If we delete that branch, then the remaining nodes are 1, 2, and 3.

Minimal complexity pruning searches for the branches that have the weakest link, which we then delete from the tree. The pruning process produces a sequence of subtrees with fewer terminal nodes and decreasing complexity.

We start with our overly large tree and denote this tree as $T_{max}$. We are searching for a finite sequence of subtrees such that

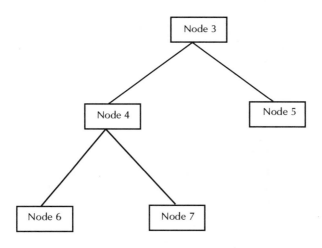

**FIGURE 9.13**
These are the nodes that comprise the branch corresponding to node 3.

$$T_{max} > T_1 > T_2 > \ldots > T_K = \{t_1\}.$$

Note that the starting point for this sequence is the tree $T_1$. Tree $T_1$ is found in a way that is different from the other subtrees in the sequence. We start off with $T_{max}$, and we look at the misclassification rate for the terminal node pairs (both sibling nodes are terminal nodes) in the tree. It is shown in Breiman, et al. [1984] that

$$R(t) \geq R(t_L) + R(t_R). \tag{9.22}$$

Equation 9.22 indicates that the misclassification error in the parent node is greater than or equal to the sum of the error in the children. We search through the terminal node pairs in $T_{max}$ looking for nodes that satisfy

$$R(t) = R(t_L) + R(t_R), \tag{9.23}$$

and we prune off those nodes. These splits are ones that do not improve the overall misclassification rate for the descendants of node $t$. Once we have completed this step, the resulting tree is $T_1$.

There is a continuum of values for the complexity parameter $\alpha$, but if a tree $T(\alpha)$ is a tree that minimizes $R_\alpha(T)$ for a given $\alpha$, then it will continue to minimize it until a jump point for $\alpha$ is reached. Thus, we will be looking for a sequence of complexity values $\alpha$ and the trees that minimize the cost complexity measure for each level. Once we have our tree $T_1$, we start pruning off the branches that have the weakest link. To find the weakest link, we first define a function on a tree as follows

$$g_k(t) = \frac{R(t) - R(T_{kt})}{|\widehat{T}_{kt}| - 1} \qquad t \text{ is an internal node,} \qquad (9.24)$$

where $T_{kt}$ is the branch $T_t$ corresponding to the internal node $t$ of subtree $T_k$. From Equation 9.24, for every internal node in tree $T_k$, we determine the value for $g_k(t)$. We define the weakest link $t_k^*$ in tree $T_k$ as the internal node $t$ that minimizes Equation 9.24,

$$g_k(t_k^*) = min_t\{g_k(t)\}. \qquad (9.25)$$

Once we have the weakest link, we prune the branch defined by that node. The new tree in the sequence is obtained by

$$T_{k+1} = T_k - T_{t_k^*}, \qquad (9.26)$$

where the subtraction in Equation 9.26 indicates the pruning process. We set the value of the complexity parameter to

$$\alpha_{k+1} = g_k(t_k^*). \qquad (9.27)$$

The result of this pruning process will be a decreasing sequence of trees,

$$T_{max} > T_1 > T_2 > \dots > T_K = \{t_1\},$$

along with an increasing sequence of values for the complexity parameter

$$0 = \alpha_1 < \dots < \alpha_k < \alpha_{k+1} < \dots < \alpha_K.$$

We need the following key fact when we describe the procedure for choosing the best tree from the sequence of subtrees:

For $k \geq 1$, the tree $T_k$ is the minimal cost complexity tree for the interval $\alpha_k \leq \alpha < \alpha_{k+1}$, and

$$T(\alpha) = T(\alpha_k) = T_k.$$

*PROCEDURE - PRUNING THE TREE*

1. Start with a large tree $T_{max}$.
2. Find the first tree in the sequence $T_1$ by searching through all terminal node pairs. For each of these pairs, if $R(t) = R(t_L) + R(t_R)$, then delete nodes $t_L$ and $t_R$.
3. For all internal nodes in the current tree, calculate $g_k(t)$ as given in Equation 9.24.
4. The weakest link is the node that has the smallest value for $g_k(t)$.
5. Prune off the branch that has the weakest link.
6. Repeat steps 3 through 5 until only the root node is left.

## Example 9.12

We continue with the same data set from the previous examples. We apply the pruning procedure to the large tree obtained in Example 9.11. The pruning function for classification trees is called **csprunec**. The input argument is a tree, and the output argument is a cell array of subtrees, where the first tree corresponds to tree $T_1$ and the last tree corresponds to the root node.

```
treeseq = csprunec(tree);
K = length(treeseq);
alpha = zeros(1,K);
% Find the sequence of alphas.
% Note that the root node corresponds to K,
% the last one in the sequence.
for i = 1:K
    alpha(i) = treeseq{i}.alpha;
end
```

The resulting sequence for $\alpha$ is

```
alpha = 0, 0.01, 0.03, 0.07, 0.08, 0.10.
```

We see that as $k$ increases (or, equivalently, the complexity of the tree decreases), the complexity parameter increases. We plot two of the subtrees in Figures 9.14 and 9.15. Note that tree $T_5$ with $\alpha = 0.08$ has fewer terminal nodes than tree $T_3$ with $\alpha = 0.03$.
□

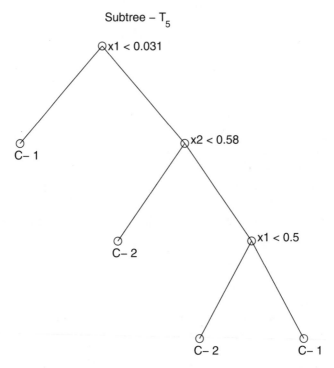

**FIGURE 9.14**
This is the subtree corresponding to $k = 5$ from Example 9.12. For this tree, $\alpha = 0.08$.

### Choosing the Best Tree

In the previous section, we discussed the importance of using independent test data to evaluate the performance of our classifier. We now use the same procedures to help us choose the right size tree. It makes sense to choose a tree that yields the smallest true misclassification cost, but we need a way to estimate this.

The values for misclassification rates that we get when constructing a tree are really estimates using the learning sample. We would like to get less biased estimates of the true misclassification costs, so we can use these values to choose the tree that has the smallest estimated misclassification rate. We can get these estimates using either an independent test sample or cross-validation. In this text, we cover the situation where there is a unit cost for misclassification and the priors are estimated from the data. For a general treatment of the procedure, the reader is referred to Breiman, et al. [1984].

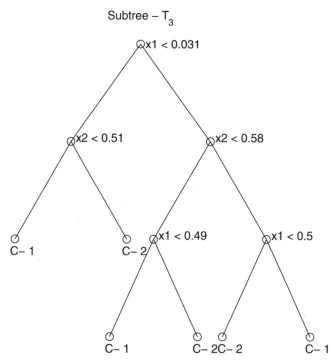

**FIGURE 9.15**
Here is the subtree corresponding to $k = 3$ from Example 9.12. For this tree, $\alpha = 0.03$.

## Selecting the Best Tree Using an Independent Test Sample

We first describe the independent test sample case, because it is easier to understand. The notation that we use is summarized below.

*NOTATION - INDEPENDENT TEST SAMPLE METHOD*

$\mathbf{L}_1$ is the subset of the learning sample $\mathbf{L}$ that will be used for building the tree.

$\mathbf{L}_2$ is the subset of the learning sample $\mathbf{L}$ that will be used for testing the tree and choosing the best subtree.

$n^{(2)}$ is the number of cases in $\mathbf{L}_2$.

$n_j^{(2)}$ is the number of observations in $\mathbf{L}_2$ that belong to class $\omega_j$.

$n_{ij}^{(2)}$ is the number of observations in $\mathbf{L}_2$ that belong to class $\omega_j$ that were classified as belonging to class $\omega_i$.

$\hat{Q}^{TS}(\omega_i | \omega_j)$ represents the estimate of the probability that a case belonging to class $\omega_j$ is classified as belonging to class $\omega_i$, using the independent test sample method.

$\hat{R}^{TS}(\omega_j)$ is an estimate of the expected cost of misclassifying patterns in class $\omega_j$, using the independent test sample.

$\hat{R}^{TS}(T_k)$ is the estimate of the expected misclassification cost for the tree represented by $T_k$ using the independent test sample method.

If our learning sample is large enough, we can divide it into two sets, one for building the tree and one for estimating the misclassification costs. We use the set $\mathbf{L}_1$ to build the tree $T_{max}$ and to obtain the sequence of pruned subtrees. This means that the trees have never seen any of the cases in the second sample $\mathbf{L}_2$. So, we present all observations in $\mathbf{L}_2$ to each of the trees to obtain an honest estimate of the true misclassification rate of each tree.

Since we have unit cost and estimated priors given by Equation 9.11, we can write $\hat{Q}^{TS}(\omega_i | \omega_j)$ as

$$\hat{Q}^{TS}(\omega_i | \omega_j) = \frac{n_{ij}^{(2)}}{n_j^{(2)}}. \tag{9.28}$$

Note that if it happens that the number of cases belonging to class $\omega_j$ is zero (i.e., $n_j^{(2)} = 0$), then we set $\hat{Q}^{TS}(\omega_i | \omega_j) = 0$. We can see from Equation 9.28 that this estimate is given by the proportion of cases that belong to class $\omega_j$ that are classified as belonging to class $\omega_i$.

The total proportion of observations belonging to class $\omega_j$ that are misclassified is given by

$$\hat{R}^{TS}(\omega_j) = \sum_i \hat{Q}^{TS}(\omega_i | \omega_j). \tag{9.29}$$

This is our estimate of the expected misclassification cost for class $\omega_j$. Finally, we use the total proportion of test cases misclassified by tree $T$ as our estimate of the misclassification cost for the tree classifier. This can be calculated using

$$\hat{R}^{TS}(T_k) = \frac{1}{n^{(2)}} \sum_{i,j} n_{ij}^{(2)}. \tag{9.30}$$

Equation 9.30 is easily calculated by simply counting the number of misclassified observations from $\mathbf{L}_2$ and dividing by the total number of cases in the test sample.

The rule for picking the best subtree requires one more quantity. This is the standard error of our estimate of the misclassification cost for the trees. In our case, the prior probabilities are estimated from the data, and we have unit cost for misclassification. Thus, the standard error is estimated by

$$\hat{SE}(\hat{R}^{TS}(T_k)) = \left\{ \hat{R}^{TS}(T_k)(1 - \hat{R}^{TS}(T_k))/n^{(2)} \right\}^{1/2}, \qquad (9.31)$$

where $n^{(2)}$ is the number of cases in the independent test sample.

To choose the right size subtree, Breiman, et al. [1984] recommend the following. First find the tree that gives the smallest value for the estimated misclassification error. Then we add the standard error given by Equation 9.31 to that misclassification error. Find the smallest tree (the tree with the largest subscript $k$) such that its misclassification cost is less than the minimum misclassification plus its standard error. In essence, we are choosing the least complex tree whose accuracy is comparable to the tree yielding the minimum misclassification rate.

PROCEDURE - CHOOSING THE BEST SUBTREE - TEST SAMPLE METHOD

1. Randomly partition the learning set into two parts, $\mathbf{L}_1$ and $\mathbf{L}_2$ or obtain an independent test set by randomly sampling from the population.

2. Using $\mathbf{L}_1$, grow a large tree $T_{max}$.

3. Prune $T_{max}$ to get the sequence of subtrees $T_k$.

4. For each tree in the sequence, take the cases in $\mathbf{L}_2$ and present them to the tree.

5. Count the number of cases that are misclassified.

6. Calculate the estimate for $\hat{R}^{TS}(T_k)$ using Equation 9.30.

7. Repeat steps 4 through 6 for each tree in the sequence.

8. Find the minimum error

$$\hat{R}^{TS}_{min} = \min_{k} \ \{\hat{R}^{TS}(T_k)\}.$$

9. Calculate the standard error in the estimate of $\hat{R}^{TS}_{min}$ using Equation 9.31.

10. Add the standard error to $\hat{R}^{TS}_{min}$ to get

$$\hat{R}^{TS}_{min} + \hat{SE}(\hat{R}^{TS}_{min}).$$

11. Find the tree with the fewest number of nodes (or equivalently, the largest $k$) such that its misclassification error is less than the amount found in step 10.

## Example 9.13

We implement this procedure using the sequence of trees found in Example 9.12. Since our sample was small, only 100 points, we will not divide this into a testing and training set. Instead, we will simply generate another set of random variables from the same distribution. The testing set we use in this example is contained in the file **cartdata**. First we generate the data that belong to class 1.

```
% Priors are 0.5 for both classes.
% Generate 200 data points for testing.
% Find the number in each class.
n = 200;
u = rand(1,n);
% Find the number in class 1.
n1 = length(find(u<=0.5));
n2 = n - n1;
% Generate the ones for class 1
% Half are upper right corner, half are lower left
data1 = zeros(n1,2);
u = rand(1,n1);
n11 = length(find(u<=0.5));
n12 = n1 - n11;
data1(1:n11,:) = rand(n11,2)+.5;
data1(n11+1:n1,:) = rand(n12,2)-.5;
```

Next we generate the data points for class 2.

```
% Generate the ones for class 2.
% Half are in lower right corner, half are upper left.
data2 = rand(n2,2);
u = rand(1,n2);
n21 = length(find(u<=0.5));
n22 = n2 - n21;
data2(1:n21,1) = rand(n21,1)-.5;
data2(1:n21,2) = rand(n21,1)+.5;
data2(n21+1:n2,1) = rand(n22,1)+.5;
data2(n21+1:n2,2) = rand(n22,1)-.5;
```

Now we determine the misclassification rate for each tree in the sequence using the independent test cases. The function **cstreec** returns the class label for a given feature vector.

```
% Now check the trees using independent test
% cases in data1 and data2.
```

```
% Keep track of the ones misclassified.
K = length(treeseq);
Rk = zeros(1,K-1); % we do not check the root
for k = 1:K-1
 nmis = 0;
 treek = treeseq{k};
 % loop through the cases from class 1
 for i = 1:n1
     [clas,pclass,node]=cstreec(data1(i,:),treek);
     if clas ~= 1
         nmis = nmis+1;  % misclassified
     end
 end
 % Loop through class 2 cases
 for i = 1:n2
   [clas,pclass,node] = cstreec(data2(i,:),treek);
   if clas ~= 2
     nmis = nmis+1; % misclassified
   end
 end
 Rk(k) = nmis/n;
end
```

The estimated misclassification errors are:

```
Rk = 0.01, 0.035, 0.050, 0.19, 0.32.
```

We see that the minimum estimated misclassification error is the tree $T_1$. We show below how to use Equation 9.31 to get the estimated standard error.

```
% Find the minimum Rk.
[mrk,ind] = min(Rk);
% The tree T_1 corresponds to the minimum Rk.
% Now find the se for that one.
semrk = sqrt(mrk*(1-mrk)/n);
% The SE is 0.0070. We add that to min(Rk).
Rk2 = mrk+semrk;
```

When we add the estimated standard error of 0.007 to the minimum estimated misclassification error, we get 0.017. None of the other trees in the sequence has an error less than this, so tree $T_1$ is the one we would select as the best tree.
☐

### Selecting the Best Tree Using Cross-Validation

We now turn our attention to the case where we use cross-validation to estimate our misclassification error for the trees. In cross-validation, we divide our learning sample into several training and testing sets. We use the training sets to build sequences of trees and then use the test sets to estimate the misclassification error.

In previous examples of cross-validation, our testing sets contained only one observation. In other words, the learning sample was sequentially partitioned into $n$ test sets. As we discuss shortly, it is recommended that far fewer than $n$ partitions be used when estimating the misclassification error for trees using cross-validation. We first provide the notation that will be used in describing the cross-validation method for choosing the right size tree.

*NOTATION - CROSS-VALIDATION METHOD*

$\mathbf{L}_v$ denotes a partition of the learning sample $\mathbf{L}$, such that

$$\mathbf{L}^{(v)} = \mathbf{L} - \mathbf{L}_v; \qquad v = 1, ..., V.$$

$T_k^{(v)}$ is a tree grown using the partition $\mathbf{L}^{(v)}$.

$\alpha_k^{(v)}$ denotes the complexity parameter for a tree grown using the partition $\mathbf{L}^{(v)}$.

$\hat{R}^{CV}(T)$ represents the estimate of the expected misclassification cost for the tree using cross-validation.

We start the procedure by dividing the learning sample $\mathbf{L}$ into $V$ partitions $\mathbf{L}_v$. Breiman, et al. [1984] recommend a value of $V = 10$ and show that cross-validation using finer partitions does not significantly improve the results. For better results, it is also recommended that systematic random sampling be used to ensure a fixed fraction of each class will be in $\mathbf{L}_v$ and $\mathbf{L}^{(v)}$. These partitions $\mathbf{L}_v$ are set aside and used to test our classification tree and to estimate the misclassification error. We use the remainder of the learning set $\mathbf{L}^{(v)}$ to get a sequence of trees

$$T_{max}^{(v)} > T_1^{(v)} > ... > T_k^{(v)} > T_{k+1}^{(v)} > ... > T_K^{(v)} = \{t_1\},$$

for each training partition. Keep in mind that we have our original sequence of trees that were created using the entire learning sample $\mathbf{L}$, and that we are going to use these sequences of trees $T_k^{(v)}$ to evaluate the classification performance of each tree in the original sequence $T_k$. Each one of these sequences will also have an associated sequence of complexity parameters

$$0 = \alpha_1^{(v)} < ... < \alpha_k^{(v)} < \alpha_{k+1}^{(v)} < ... < \alpha_K^{(v)}.$$

At this point, we have $V + 1$ sequences of subtrees and complexity parameters.

We use the test samples $\mathbf{L}_v$ along with the trees $T_k^{(v)}$ to determine the classification error of the subtrees $T_k$. To accomplish this, we have to find trees that have equivalent complexity to $T_k$ in the sequence of trees $T_k^{(v)}$.

Recall that a tree $T_k$ is the minimal cost complexity tree over the range $\alpha_k \leq \alpha < \alpha_{k+1}$. We define a representative complexity parameter for that interval using the geometric mean

$$\alpha'_k = \sqrt{\alpha_k \alpha_{k+1}}. \tag{9.32}$$

The complexity for a tree $T_k$ is given by this quantity. We then estimate the misclassification error using

$$\hat{R}^{CV}(T_k) = \hat{R}^{CV}(T(\alpha'_k)), \tag{9.33}$$

where the right hand side of Equation 9.33 is the proportion of test cases that are misclassified, using the trees $T_k^{(v)}$ that correspond to the complexity parameter $\alpha'_k$.

To choose the best subtree, we need an expression for the standard error of the misclassification error $\hat{R}^{CV}(T_k)$. When we present our test cases from the partition $\mathbf{L}_v$, we record a zero or a one, denoting a correct classification and an incorrect classification, respectively. We see then that the estimate in Equation 9.33 is the mean of the ones and zeros. We estimate the standard error of this from

$$\hat{SE}(\hat{R}^{CV}(T_k)) = \sqrt{\frac{s^2}{n}}, \tag{9.34}$$

where $s^2$ is $(n-1)/n$ times the sample variance of the ones and zeros.

The cross-validation procedure for estimating the misclassification error when we have unit cost and the priors are estimated from the data is outlined below.

*PROCEDURE - CHOOSING THE BEST SUBTREE (CROSS-VALIDATION)*

1. Obtain a sequence of subtrees $T_k$ that are grown using the learning sample $\mathbf{L}$.

2. Determine the cost complexity parameter $\alpha'_k$ for each $T_k$ using Equation 9.32.

3. Partition the learning sample into $V$ partitions, $\mathbf{L}_v$. These will be used to test the trees.

4. For each $\mathbf{L}_v$, build the sequence of subtrees using $\mathbf{L}^{(v)}$. We should now have $V+1$ sequences of trees.

5. Now find the estimated misclassification error $\hat{R}^{CV}(T_k)$. For $\alpha'_k$ corresponding to $T_k$, find all equivalent trees $T_k^{(v)}$, $v = 1, \ldots, V$. We do this by choosing the tree $T_k^{(v)}$ such that

$$\alpha'_k \in [\alpha_k^{(v)}, \alpha_{k+1}^{(v)}).$$

6. Take the test cases in each $\mathbf{L}_v$ and present them to the tree $T_k^{(v)}$ found in step 5. Record a one if the test case is misclassified and a zero if it is classified correctly. These are the classification costs.

7. Calculate $\hat{R}^{CV}(T_k)$ as the proportion of test cases that are misclassified (or the mean of the array of ones and zeros found in step 6).

8. Calculate the standard error as given by Equation 9.34.

9. Continue steps 5 through 8 to find the misclassification cost for each subtree $T_k$.

10. Find the minimum error

$$\hat{R}_{min}^{CV} = \min_k \ \{\hat{R}^{CV}(T_k)\}.$$

11. Add the estimated standard error to it to get

$$\hat{R}_{min}^{CV} + \hat{SE}(\hat{R}_{min}^{CV}).$$

12. Find the tree with the largest $k$ or fewest number of nodes such that its misclassification error is less than the amount found in step 11.

## Example 9.14

For this example, we return to the **iris** data, described at the beginning of this chapter. We implement the cross-validation approach using $V = 5$. We start by loading the data and setting up the indices that correspond to each partition. The fraction of cases belonging to each class is the same in all testing sets.

```
load iris
% Attach class labels to each group.
setosa(:,5)=1;
versicolor(:,5)=2;
virginica(:,5)=3;
X = [setosa;versicolor;virginica];
n = 150;% total number of data points
% These indices indicate the five partitions
% for cross-validation.
ind1 = 1:5:50;
```

```
ind2 = 2:5:50;
ind3 = 3:5:50;
ind4 = 4:5:50;
ind5 = 5:5:50;
```

Next we set up all of the testing and training sets. We use the MATLAB **eval** function to do this in a loop.

```
% Get the testing sets: test1, test2, ...
for i = 1:5
 eval(['test' int2str(i) '=[setosa(ind' int2str(i)
  ',:);versicolor(ind' int2str(i) ...
   ',:);virginica(ind' int2str(i) ',:)];'])
end
for i = 1:5
 tmp1 = setosa;
 tmp2 = versicolor;
 tmp3 = virginica;
 % Remove points that are in the test set.
 eval(['tmp1(ind' int2str(i) ',:) = [];'])
 eval(['tmp2(ind' int2str(i) ',:) = [];'])
 eval(['tmp3(ind' int2str(i) ',:) = [];'])
 eval(['train' int2str(i) '= [tmp1;tmp2;tmp3];'])
end
```

Now we grow the trees using all of the data and each training set.

```
% Grow all of the trees.
pies = ones(1,3)/3;
maxn = 2;% get large trees
clas = 1:3;
Nk = [50,50,50];
tree = csgrowc(X,maxn,clas,Nk,pies);
Nk1 = [40 40 40];
for i = 1:5
 eval(['tree' int2str(i) '= ,...
   csgrowc(train',...
   int2str(i) ',maxn,clas,Nk1,pies);'])
end
```

The following MATLAB code gets all of the sequences of pruned subtrees:

```
% Now prune each sequence.
treeseq = csprunec(tree);
for i = 1:5
 eval(['treeseq' int2str(i) '=,...
   csprunec(tree' int2str(i) ');'])
end
```

The complexity parameters must be extracted from each sequence of sub-trees. We show how to get this for the main tree and for the sequences of sub-trees grown on the first partition. This must be changed appropriately for each of the remaining sequences of subtrees.

```
K = length(treeseq);
alpha = zeros(1,K);
% Find the sequence of alphas.
for i = 1:K
 alpha(i) = treeseq{i}.alpha;
end
% For the other subtree sequences, change the
% 1 to 2, 3, 4, 5 and re-run.
K1 = length(treeseq1);
for i = 1:K1
 alpha1(i) = treeseq1{i}.alpha;
end
```

We need to obtain the equivalent complexity parameters for the main sequence of trees using Equation 9.32. We do this in MATLAB as follows:

```
% Get the akprime equivalent values for the main tree.
for i = 1:K-1
    akprime(i) = sqrt(alpha(i)*alpha(i+1));
end
```

We must now loop through all of the subtrees in the main sequence, find the equivalent subtrees in each partition and use those trees to classify the cases in the corresponding test set. We show a portion of the MATLAB code here to illustrate how we find the equivalent subtrees. The complete steps are contained in the M-file called **ex9_14.m** (downloadable with the Computational Statistics Toolbox). In addition, there is an alternative way to implement cross-validation using cell arrays (courtesy of Tom Lane, The MathWorks). The complete procedure can be found in **ex9_14alt.m**.

```
n = 150;
k = length(akprime);
misclass = zeros(1,n);
% For the first tree, find the
% equivalent tree from the first partition
ind = find(alpha1 <= akprime(1));
% Should be the last one.
% Get the tree that corresponds to that one.
tk = treeseq1{ind(end)};
% Get the misclassified points in the test set.
for j = 1:30  % loop through the points in test 1
 [c,pclass,node] = cstreec(test1(j,1:4),tk);
 if c ~= test1(j,5)
```

```
    misclass(j) = 1;
  end
end
```

We continue in this manner using all of the subtrees. The estimated misclassification error using cross-validation is

Rk = 0.047, 0.047, 0.047, 0.067, 0.21, 0.41,

and the estimated standard error for $\hat{R}_{min}^{CV}$ is 0.017. When we add this to the minimum of the estimated errors, we get 0.064. We see that the tree with the minimum complexity that has error less than this is tree $T_3$. All of the data and variables that are generated in this example can be loaded from irisexamp.mat.
❑

## 9.5 Clustering

Clustering methodology is used to explore a data set where the goal is to separate the sample into groups or to provide understanding about the underlying structure or nature of the data. The results from clustering methods can be used to prototype supervised classifiers or to generate hypotheses. Clustering is called unsupervised classification because we typically do not know what groups there are in the data or the group membership of an individual observation. In this section, we discuss two main methods for clustering. The first is hierarchical clustering, and the second method is called *k*-means clustering. First, however, we cover some preliminary concepts.

### Measures of Distance

The goal of clustering is to partition our data into groups such that the observations that are in one group are dissimilar to those in other groups. We need to have some way of measuring that dissimilarity, and there are several measures that fit our purpose.

The first measure of dissimilarity is the **Euclidean distance** given by

$$d_{rs} = \sqrt{(\mathbf{x}_r - \mathbf{x}_s)^T(\mathbf{x}_r - \mathbf{x}_s)}, \qquad (9.35)$$

where $\mathbf{x}_r$ is a column vector representing one observation. We could also use the **Mahalanobis distance** defined as

$$d_{rs} = \sqrt{(\mathbf{x}_r - \mathbf{x}_s)^T \Sigma^{-1}(\mathbf{x}_r - \mathbf{x}_s)}, \qquad (9.36)$$

where $\Sigma^{-1}$ denotes the inverse covariance matrix. The ***city block distance*** is found using absolute values rather than squared distances, and it is calculated using

$$d_{rs} = \sum_{j=1}^{d} |x_{rj} - x_{sj}|.$$ (9.37)

In Equation 9.37, we take the absolute value of the difference between the observations $\mathbf{x}_r$ and $\mathbf{x}_s$ componentwise and then add up the values. The final distance that we present covers the more general case of the Euclidean distance or the city block distance. This is called the ***Minkowski distance***, and it is found using

$$d_{rs} = \left\{ \sum_{j=1}^{d} |x_{rj} - x_{sj}|^p \right\}^{1/p}.$$ (9.38)

If $p = 1$, then we have the city block distance, and if $p = 2$ we have the Euclidean distance.

The researcher should be aware that distances might be affected by differing scales or magnitude among the variables. For example, suppose our data measured two variables: age and annual income in dollars. Because of its magnitude, the income variable could influence the distances between observations, and we would end up clustering mostly on the incomes. In some situations, we might want to standardize the observations. The MATLAB Statistics Toolbox contains a function called **zscore** that will perform this standardization.

The MATLAB Statistics Toolbox also has a function that calculates distances. It is called **pdist** and takes as its argument a matrix **X** that is dimension $n \times d$. Each row represents an observation in our data set. The **pdist** function returns a vector containing the distance information. The default distance is Euclidean, but the user can specify other distances as discussed above. We illustrate the use of this function in the following example.

**Example 9.15**
We use a small data set to illustrate the various distances available in the MATLAB Statistics Toolbox. We have only five data points. The following commands set up the matrix of values and plots the points in Figure 9.16.

```
% Let's make up a data set - 2-D.
x = [1 1; 1 2; 2 1; -1 -1; -1 -2];
plot(x(:,1),x(:,2),'kx') % plots the points.
axis([-3 3 -3 3])
text(x(:,1)+.1,x(:,2)+.1,'1|2|3|4|5');
```

We first find the Euclidean distance between the points using the **pdist** function. We also illustrate the use of the function **squareform** that puts the distances in a more familiar matrix form, where the *ij*-th element corresponds to the distance between the *i*-th and *j*-th observation.

```
% Find the Euclidean distance using pdist.
% Convert to matrix form for easier reading.
ye = pdist(x,'euclid');
ye_mat = squareform(ye);
```

The matrix we get from this is

```
ye_mat =
```

| 0 | 1.0000 | 1.0000 | 2.8284 | 3.6056 |
|---|--------|--------|--------|--------|
| 1.0000 | 0 | 1.4142 | 3.6056 | 4.4721 |
| 1.0000 | 1.4142 | 0 | 3.6056 | 4.2426 |
| 2.8284 | 3.6056 | 3.6056 | 0 | 1.0000 |
| 3.6056 | 4.4721 | 4.2426 | 1.0000 | 0 |

We contrast this with the city block distance.

```
% Contrast with city block metric.
ycb = pdist(x,'cityblock');
ycb_mat = squareform(ycb);
```

The result we get from this is

```
ycb_mat =
```

| 0 | 1 | 1 | 4 | 5 |
|---|---|---|---|---|
| 1 | 0 | 2 | 5 | 6 |
| 1 | 2 | 0 | 5 | 6 |
| 4 | 5 | 5 | 0 | 1 |
| 5 | 6 | 6 | 1 | 0 |

❑

## Hierarchical Clustering

There are two types of hierarchical clustering methods: agglomerative and divisive. *Divisive* methods start with one large group and successively split the groups until there are *n* groups with one observation per group. In general, methods for this type of hierarchical clustering are computationally inefficient [Webb, 1999], so we do not discuss them further. *Agglomerative* methods are just the opposite; we start with *n* groups (one observation per group) and successively merge the two most similar groups until we are left with only one group.

There are five commonly used methods for merging clusters in agglomerative clustering. These are single linkage, complete linkage, average linkage,

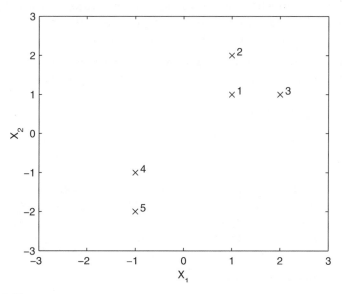

**FIGURE 9.16**
These are the observations used in Example 9.15. Two clusters are clearly seen.

centroid linkage and Ward's method. The MATLAB Statistics Toolbox provides a function called **linkage** that will perform agglomerative clustering using any of these methods. Its use is illustrated in the next example, but first we briefly describe each of the methods [Hair, et al., 1995].

The *single linkage* method uses minimum distance, where the distance between clusters is defined as the distance between the closest pair of observations. Pairs consisting of one case from each group are used in the calculation. The first cluster is formed by merging the two groups with the shortest distance. Then the next smallest distance is found between all of the clusters (keep in mind that an observation is also a cluster). The two clusters corresponding to the smallest distance are then merged. The process continues in this manner until there is one group. In some cases, single linkage can lead to chaining of the observations, where those on the ends of the chain might be very dissimilar.

The process for the *complete linkage* method is similar to single linkage, but the clustering criterion is different. The distance between groups is defined as the most distant pair of observations, with one coming from each group. The logic behind using this type of similarity criterion is that the maximum distance between observations in each cluster represents the smallest sphere that can enclose all of the objects in both clusters. Thus, the closest of these cluster pairs should be grouped together. The complete linkage method does not have the chaining problem that single linkage has.

The *average linkage* method for clustering starts out the same way as single and complete linkage. In this case, the cluster criterion is the average distance between all pairs, where one member of the pair comes from each cluster. Thus, we find all pairwise distances between observations in each cluster and take the average. This linkage method tends to combine clusters with small variances and to produce clusters with approximately equal variance.

*Centroid linkage* calculates the distance between two clusters as the distance between the centroids. The *centroid* of a cluster is defined as the $d$-dimensional sample mean for those observations that belong to the cluster. Whenever we merge clusters together or add an observation to a cluster, the centroid is recalculated.

The distance between two clusters using **Ward's linkage** method is defined as the incremental sum of the squares between two clusters. To merge clusters, the within-group sum-of-squares is minimized over all possible partitions obtained by combining two clusters. The within-group sum-of-squares is defined as the sum of the squared distances between all observations in a cluster and its centroid. This method tends to produce clusters with approximately the same number of observations in each one.

## Example 9.16

We illustrate the **linkage** function using the data and distances from the previous example. We look only at single linkage and complete linkage using the Euclidean distances. We show the results of the clustering in dendrograms given in Figures 9.17 and 9.18.

```
% Get the cluster output from the linkage function.
zsingle = linkage(ye,'single');
zcomplete = linkage(ye,'complete');
% Get the dendrogram.
dendrogram(zsingle)
title('Clustering - Single Linkage')
dendrogram(zcomplete)
title('Clustering - Complete Linkage')
```

A *dendrogram* shows the links between objects as inverted U-shaped lines, where the height of the U represents the distance between the objects. The cases are listed along the horizontal axis. Cutting the tree at various $y$ values of the dendrogram yields different clusters. For example, cutting the complete linkage tree at $y = 1.2$ would yield 3 clusters. As expected, if we choose to create two clusters, then the two linkage methods give the same cluster definitions.
❑

Now that we have our cases clustered, we would like to measure the validity of the clustering. One way to do this would be to compare the distances between all observations with the links in the dendrogram. If the clustering

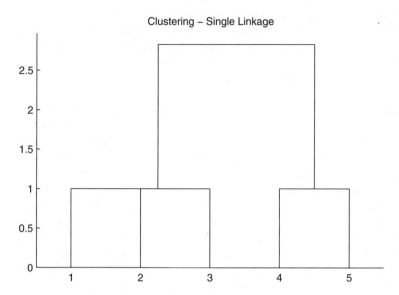

**FIGURE 9.17**
This is the dendrogram using Euclidean distances and single linkage.

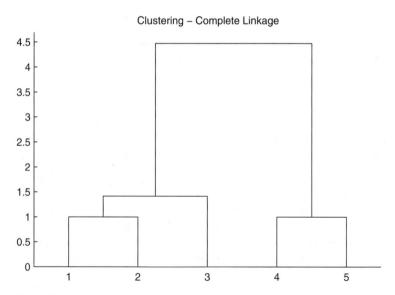

**FIGURE 9.18**
This is the dendrogram using Euclidean distances and complete linkage.

is a valid one, then there should be a strong correlation between them. We can measure this using the ***cophenetic correlation coefficient***. A cophenetic matrix is defined using the results of the linkage procedure. The *ij*-th entry of the cophenetic matrix is the fusion level at which the *i*-th and *j*-th objects appear together in the same cluster for the first time. The correlation coefficient between the distances and the corresponding cophenetic entries is the cophenetic correlation coefficient. Large values indicate that the linkage provides a reasonable clustering of the data. The MATLAB Statistics Toolbox provides a function that will calculate the cophenetic correlation coefficient. Its use is illustrated in the following example.

### Example 9.17

In this example, we show how to obtain the cophenetic correlation coefficient in MATLAB. We use the same small data set from before and calculate the cophenetic correlation coefficient when we have clusters based on different distances and linkages. First, we get the clusters using the following commands.

```
x = [1 1; 1 2; 2 1; -1 -1; -1 -2];
ye = pdist(x,'euclid');
ycb = pdist(x,'cityblock');
zsineu = linkage(ye,'single');
zcompeu = linkage(ye,'complete');
zsincb = linkage(ycb,'single');
zcomcb = linkage(ycb,'complete');
```

We now have four different cluster hierarchies. Their cophenetic correlation coefficients can be found from the following:

```
ccompeu = cophenet(zcompeu,ye);
csineu = cophenet(zsineu,ye);
csincb = cophenet(zsincb,ycb);
ccomcb = cophenet(zcomcb,ycb);
```

As expected, all of the resulting cophenetic correlation coefficients are large (approximately 0.95), with the largest corresponding to the complete linkage clustering based on the city block distance.
□


### *K*-Means Clustering

The goal of *k*-means clustering is to partition the data into *k* groups such that the within-group sum-of-squares is minimized. One way this technique differs from hierarchical clustering is that we must specify the number of groups or clusters that we are looking for. We briefly describe two algorithms for obtaining clusters via *k*-means.

One of the basic algorithms for $k$-means clustering is a two step procedure. First, we assign observations to its closest group, usually using the Euclidean distance between the observation and the cluster centroid. The second step of the procedure is to calculate the new cluster centroid using the assigned objects. These steps are alternated until there are no changes in cluster membership or until the centroids do not change. This algorithm is sometimes referred to as HMEANS [Spath, 1980] or the basic ISODATA method.

*PROCEDURE - HMEANS ALGORITHM*

1. Specify the number of clusters $k$.
2. Determine initial cluster centroids. These can be randomly chosen or the user can specify them.
3. Calculate the distance between each observation and each cluster centroid.
4. Assign every observation to the closest cluster.
5. Calculate the centroid (i.e., the $d$-dimensional mean) of every cluster using the observations that were just grouped there.
6. Repeat steps 3 through 5 until no more changes are made.

There are two problems with the HMEANS algorithm. The first one is that this method could lead to empty clusters, so users should be aware of this possibility. As the centroid is recalculated and observations are reassigned to groups, some clusters could become empty. The second issue concerns the optimality of the partitions. With $k$-means, we are searching for partitions where the within-group sum-of-squares is minimum. It can be shown [Webb, 1999] that in some cases, the final $k$-means cluster assignment is not optimal, in the sense that moving a single point from one cluster to another may reduce the sum of squared errors. The following procedure helps address the second problem.

*PROCEDURE - K-MEANS*

1. Obtain a partition of $k$ groups, possibly from the HMEANS algorithm.
2. Take each data point $x_i$ and calculate the Euclidean distance between it and every cluster centroid.
3. Here $x_i$ is in the $r$-th cluster, $n_r$ is the number of points in the $r$-th cluster, and $d_{ir}^2$ is the Euclidean distance between $x_i$ and the centroid of cluster $r$. If there is a group $s$ such that

$$\frac{n_r}{n_r - 1} d_{ir}^2 > \frac{n_s}{n_s + 1} d_{is}^2,$$

then move $\mathbf{x}_i$ to cluster $s$.

4. If there are several clusters that satisfy the above inequality, then move the $\mathbf{x}_i$ to the group that has the smallest value for

$$\frac{n_s}{n_s + 1} d_{is}^2 \, .$$

5. Repeat steps 2 through 4 until no more changes are made.

We note that there are many algorithms for $k$-means clustering described in the literature. We provide some references to these in the last section of this chapter.

## Example 9.18

We show how to implement HMEANS in MATLAB, using the **iris** data. Normally, clustering methods would be used on data where we do not know what groups are there, unlike the **iris** data. However, since we do know the true groups represented by the data, these will give us a way to verify that the clusters make sense. We first obtain the cluster centers by randomly picking observations from the data set. Note that initial cluster centers do not have to be actual observations.

```
load iris
k = 3;
% Put all of the data together.
x = [setosa;versicolor;virginica];
[n,d] = size(x);

% Pick some observations to be the cluster centers.
ind = randperm(n);
ind = ind(1:k);          k random observations
nc = x(ind,:);
% Set up storage.
% Integers 1,...,k indicating cluster membership
cid = zeros(1,n);
% Make this different to get the loop started.
oldcid = ones(1,n);
% The number in each cluster.
nr = zeros(1,k);
% Set up maximum number of iterations.
maxiter = 100;
iter = 1;

while ~isequal(cid,oldcid) & iter < maxiter
    oldcid = cid;
```

```
% Implement the hmeans algorithm.
% For each point, find the distance
% to all cluster centers.
for i = 1:n
    dist = sum((repmat(x(i,:),k,1)-nc).^2,2);
     % assign it to this cluster
    [m,ind] = min(dist);              ind ≤ k
    cid(i) = ind;
end
% Find the new cluster centers.
for i = 1:k
    % Find all points in this cluster.
    ind = find(cid==i);
    % Find the centroid.
    nc(i,:) = mean(x(ind,:));
    % Find the number in each cluster;
    nr(i) = length(ind);
end
iter = iter + 1
end
```

To check these results, we show a scatterplot of the *first two features* of the **iris** data in Figure 9.19, where the three classes are represented by different plotting symbols. The clusters we obtain from this implementation of *k*-means clustering (using the HMEANS procedure) are shown in Figure 9.20. The algorithm finds the one group, corresponding to *Iris setosa*, but has trouble separating the other two species. However, the results are certainly reasonable.

❑

## 9.6 MATLAB Code

We provide a function called **cshmeans** that implements the HMEANS algorithm given above. We also have a function called **cskmeans** that checks to see if moving individual observations changes the sum-square error. With both of these functions, the user can specify the initial centers as an input argument. However, if that argument is omitted, then the function will randomly pick the initial cluster centers.

As we stated in the body of the text, there are many MATLAB functions available that the analyst can use to develop classifiers using Bayes decision theory. These are any of the functions in the Statistics Toolbox that estimates a probability density function using the parametric approach: **normfit**, **expfit**, **gamfit**, **unifit**, **betafit**, and **weibfit**. These functions return

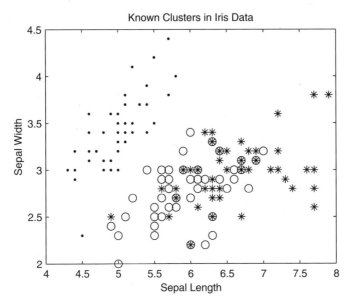

**FIGURE 9.19**
This is a scatterplot of the *first two features* of the `iris` data. The three classes are represented by different plotting symbols. From this, we expect that the *k*-means algorithm should find the cluster in the upper left corner, but have trouble separating the other two clusters. Note that because this represents the first two features, some of the symbols (circles and asterisks) are on top of each other.

the appropriate distribution parameters estimated from the sample. For the nonparametric approach, one can use any of the techniques from Chapter 8: histograms, frequency polygons, kernel methods, finite mixtures or adaptive mixtures. Also, there is a function in the Statistics Toolbox called `classify`. This performs linear discriminant analysis [Duda, Hart, and Stork, 2001] using Mahalanobis distances. Class labels are assigned based on the distance between the observation and the cases in the training set.

A set of M-files implementing many of the methods described in Ripely [1996] are available for download at

`ftp.mathworks.com/pub/contrib/v5/stats/discrim/`.

There are functions for *k*-means, Bayesian classifiers and logistic discriminant analysis.

The MATLAB Statistics Toolbox has several functions for clustering. In Examples 9.15 through 9.17, we illustrated the use of `pdist`, `squareform`, `linkage`, and `cophenet`. There are other clustering functions that the data analyst might find useful. One is called `cluster`, which is used to divide the

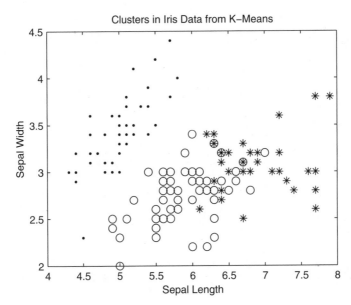

**FIGURE 9.20**
This shows the *first two features* of the clusters found using *k*-means, where all four features were used in the clustering algorithm. As expected, the cluster in the upper left corner is found. The other two clusters do not show the same separation, but the results are reasonable when compared to the true groups shown in Figure 9.19.

output of linkage into clusters. It does this in one of two ways: 1) by finding the natural divisions, or 2) by the user specifying arbitrary clusters. The function **inconsistent** helps the user find natural divisions in the data set by comparing the length of the links in a cluster tree with the lengths of neighboring links. If the link is approximately the same as its neighbors, then it exhibits a high level of consistency. If not, then they are considered to be inconsistent. Inconsistent links might indicate a division of the data. The reader is asked to explore this further in the exercises. Finally, the function **clusterdata** combines the three functions, **pdist**, **linkage**, and **cluster** into one. However, **clusterdata** uses Euclidean distance and single linkage clustering. So, if another cluster methodology is needed, the three separate functions must be used.

**TABLE 9.2**

MATLAB Functions for Statistical Pattern Recognition

| Purpose | MATLAB Function |
|---|---|
| Creating, pruning and displaying classification trees | `csgrowc` |
| | `csprunec` |
| | `cstreec` |
| | `csplotreec` |
| | `cspicktreec` |
| Creating, analyzing and displaying clusters | `clusterdata` |
| | `pdist/squareform` |
| | `linkage` |
| | `cluster` |
| | `cophenet` |
| | `dendrogram` |
| | `cshmeans` |
| | `cskmeans` |
| Statistical pattern recognition using Bayes decision theory | `csrocgen` |
| | `cskernmd` |
| | `cskern2d` |

## 9.7 Further Reading

There are many excellent books on statistical pattern recognition that can be used by students at the graduate level or researchers with a basic knowledge of calculus and linear algebra. The text by Duda and Hart [1973] is a classic book on pattern recognition and includes the foundational theory behind Bayes decision theory, classification and discriminant analysis. It has recently been revised and updated [Duda, Hart, and Stork, 2001]. This second edition contains many new topics, examples, and pseudo-code. Fukunaga [1990] is at the same level and includes similar subjects; however, it goes into more detail on the feature extraction aspects of pattern recognition. Devroye, Gyorfi, and Lugosi [1996] CONTAINS an extensive treatment of the probabilistic theory behind pattern recognition. Ripley [1996] covers pattern recognition from a neural network perspective. This book is recommended for both students and researchers as a standard reference. An excellent book that discusses all aspects of statistical pattern recognition is the text by Webb [1999]. This is suitable for advanced undergraduate students and professionals. The author explains the techniques in a way that is understandable, and he provides enough theory to explain the methodology, but does not overwhelm the reader with it.

The definitive book on classification trees is the one by Breiman, et al. [1984]. This text provides algorithms for building classification trees using ordered or categorical data, mixtures of data types, and splitting nodes using more than one variable. They also provide the methodology for using trees in regression. A paper by Safavian and Landgrebe [1991] provides a review of methodologies for building and using classification trees. A description of classification trees can also be found in Webb [1999] and Duda, Hart, and Stork [2001].

Many books are available that describe clustering techniques, and we mention a few of them here. The books by Hartigan [1975], Spath [1980], Anderberg [1973], Kaufman and Rousseeuw [1990], and Jain and Dubes [1988] provide treatments of the subject at the graduate level. Most of the texts mentioned above on statistical pattern recognition discuss clustering also. For example, see Duda and Hart [1973], Duda, Hart and Stork [2001], Ripley [1996], or Webb [1999]. For two books that are appropriate at the undergraduate level, we refer the reader to Everitt [1993] and Gordon [1999].

We conclude this chapter with a brief discussion of a technique that combines agglomerative clustering and finite mixtures. This method is called model-based clustering [Fraley, 1998; Fraley and Raftery, 1998]. First, agglomerative clustering is performed, where clusters are merged based on the finite mixture model, rather than the distances. The partitions obtained from the model-based agglomerative clustering provide an initialization (number of components, means, variances and weights) to the finite mixtures EM algorithm (with normal components). An approximation to Bayes factors is used to pick the best model.

## Exercises

9.1. Load the **insect** data [Hand, et al., 1994; Lindsey, et al., 1987]. These are three variables measured on each of ten insects from three species. Using the parametric approach and assuming that these data are multivariate normal with different covariances, construct a Bayes classifier. Use the classifier to classify the following vectors as species I, II, or III:

| $X_1$ | $X_2$ | $X_3$ |
|-------|-------|-------|
| 190 | 143 | 52 |
| 174 | 131 | 50 |
| 218 | 126 | 49 |
| 130 | 131 | 51 |
| 138 | 127 | 52 |
| 211 | 129 | 49 |

9.2. The **household** [Hand, et al., 1994; Aitchison, 1986] data set contains the expenditures for housing, food, other goods, and services (four expenditures) for households comprised of single people. Apply the clustering methods of Section 9.5 to see if there are two groups in the data, one for single women and one for single men. To check your results, the first 20 cases correspond to single men, and the last 20 cases are for single women.

9.3. Grow a classification tree for the **household** data, using the class labels as given in problem 9.2. Is the tree consistent with the results from the clustering?

9.4. The **measure** [Hand, et. al., 1994] data contain 20 measurements of chest, waist and hip data. Half of the measured individuals are women and half are men. Use cluster analysis to see if there is evidence of two groups.

9.5. Use the online **help** to find out more about the MATLAB Statistics Toolbox functions **cluster** and **inconsistent**. Use these with the data and clusters of Examples 9.15 through 9.17 to extract the clusters.

9.6. Apply the cross-validation procedure and ROC curve analysis of Example 9.8 to the **tibetan** data. Designate Type A skulls as the target class and Type B skulls as the non-target class.

9.7. Use the **bank** data along with the independent test sample approach to estimate the probability of correctly classifying patterns (see Example 9.7). The file contains two matrices, one corresponding to features taken from 100 forged Swiss bank notes and the other comprising features from 100 genuine Swiss bank notes [Flury and Riedwyl, 1988]. There are six features: length of the bill, left width of the bill, right width of the bill, width of the bottom margin, width of the top margin and length of the image diagonal. Compare classifiers obtained from: 1) the parametric approach, assuming the class-conditionals are multivariate normal with different covariances, and 2) the nonparametric approach, estimating the class-conditional probabilities using the product kernel. Which classifier performs better based on the estimated probability of correct classification?

9.8. Apply the cross-validation procedure and ROC curve analysis of Example 9.8 to the **bank** data. The target class corresponds to the forged bills. Obtain ROC curves for a classifier built using: 1) the parametric approach, assuming the class-conditionals are multivariate normal with different covariances, and 2) the nonparametric approach, estimating the class-conditional probabilities using the product kernel. Which classifier performs better, based on the ROC curve analysis?

9.9. For the **bank** data, obtain a classification tree. Use the independent test sample approach to pick a final pruned tree.

9.10. Apply $k$-means clustering to the complete **bank** data, without class labels. Apply the hierarchical clustering methods to the data. Is there significant evidence of two groups?

9.11. Do a Monte Carlo study of the probability of misclassification. Generate $n$ random variables using the class-conditional probabilities and the priors from Example 9.3. Estimate the probability of misclassification based on the data. Note that you will have to do some probability density estimation here. Record the probability of error for this trial. Repeat for $M$ Monte Carlo trials. Plot a histogram of the errors. What can you say about the probability of error based on this Monte Carlo experiment?

9.12. The **flea** data set [Hand, et al., 1994; Lubischew, 1962] contains measurements on three species of flea beetle: *Chaetocnema concinna*, *Chaetocnema heikertingeri*, and *Chaetocnema heptapotamica*. The features for classification are the maximal width of aedeagus in the forepart (microns) and the front angle of the aedeagus (units are 7.5 degrees). Build a classifier for these data using a Bayes classifier. For the Bayes classifier, experiment with different methods of estimating the class-conditional probability densities. Construct ROC curves and use them to compare the classifiers.

9.13. Build a classification tree using the **flea** data. Based on a three-term multivariate normal finite mixture model for these data, obtain an estimate of the model. Using the estimated model, generate an independent test sample to pick the best tree in the sequence of subtrees.

9.14. The **k-*nearest neighbor rule*** assigns patterns **x** to the class that is the most common amongst its $k$ nearest neighbors. To fix the notation, let $k_m$ represent the number of cases belonging to class $\omega_m$ out of the $k$ nearest neighbors to **x**. We classify **x** as belonging to class $\omega_m$, if $k_m \geq k_i$, for $i = 1, ..., J$. Write a MATLAB function that implements this classifier.

9.15. Repeat Example 9.7 using all of the features for **versicolor** and **virginica**. What is your estimated probability of correct classification?

9.16. Apply the method of Example 9.7 to the **virginica** and **setosa** classes.

# Chapter 10

## Nonparametric Regression

### 10.1 Introduction

In Chapter 7, we briefly introduced the concepts of linear regression and showed how cross-validation can be used to determine a model that provides a good fit to the data. We return to linear regression in this section to introduce nonparametric regression and smoothing. We first revisit classical linear regression and provide more information on how to analyze and visualize the results of the model. We also examine more of the capabilities available in MATLAB for this type of analysis. In Section 10.2, we present a method for scatterplot smoothing called loess. Kernel methods for nonparametric regression are discussed in Section 10.3, and regression trees are presented in Section 10.4.

Recall from Chapter 7 that one model for linear regression is

$$Y = \beta_0 + \beta_1 X + \beta_2 X^2 + \ldots + \beta_d X^d + \varepsilon. \tag{10.1}$$

We follow the terminology of Draper and Smith [1981], where the *'linear'* refers to the fact that the model is linear with respect to the coefficients, $\beta_j$. It is not that we are restricted to fitting only straight lines to the data. In fact, the model given in Equation 10.1 can be expanded to include multiple predictors $X_j, j = 1, \ldots k$. An example of this type of model is

$$Y = \beta_0 + \beta_1 X_1 + \ldots + \beta_k X_k + \varepsilon. \tag{10.2}$$

In parametric linear regression, we can model the relationship using any combination of predictor variables, order (or degree) of the variables, etc. and use the least squares approach to estimate the parameters. Note that it is called *'parametric'* because we are assuming an explicit model for the relationship between the predictors and the response.

To make our notation consistent, we present the matrix formulation of linear regression for the model in Equation 10.1. Let $\mathbf{Y}$ be an $n \times 1$ vector of observed values for the response variable and let **X** represent a matrix of observed values for the predictor variables, where each row of **X** corresponds to one observation and powers of that observation. Specifically, **X** is of dimension $n \times (d+1)$. We have $d+1$ columns to accommodate a constant term in the model. Thus, the first column of **X** is a column of ones. The number of columns in **X** depends on the chosen parametric model (the number of predictor variables, cross terms and degree) that is used. Then we can write the model in matrix form as

$$\mathbf{Y} = \mathbf{X}\boldsymbol{\beta} + \boldsymbol{\varepsilon},\tag{10.3}$$

where $\boldsymbol{\beta}$ is a $(d+1) \times 1$ vector of parameters to be estimated and $\boldsymbol{\varepsilon}$ is an $n \times 1$ vector of errors, such that

$$E[\boldsymbol{\varepsilon}] = \mathbf{0}$$
$$V(\boldsymbol{\varepsilon}) = \sigma^2 \mathbf{I}.$$

The least squares solution for the parameters can be found by solving the so-called 'normal equations' given by

$$\hat{\boldsymbol{\beta}} = (\mathbf{X}^T\mathbf{X})^{-1}\mathbf{X}^T\mathbf{Y}.\tag{10.4}$$

The solutions formed by the parameter estimate $\hat{\boldsymbol{\beta}}$ obtained using Equation 10.4 is valid in that it is the solution that minimizes the error sum-of-squares $\boldsymbol{\varepsilon}^T\boldsymbol{\varepsilon}$, regardless of the distribution of the errors. However, normality assumptions (for the errors) must be satisfied if one is conducting hypothesis testing or constructing confidence intervals that depend on these estimates.

### Example 10.1

In this example, we explore two ways to perform least squares regression in MATLAB. The first way is to use Equation 10.4 to explicitly calculate the inverse. The data in this example were used by Longley [1967] to verify the computer calculations from a least squares fit to data. They can be downloaded from **http://www.itl.nist.gov/div898**. The data set contains 6 predictor variables so the model follows that in Equation 10.2:

$$y = \beta_0 + \beta_1 x_1 + \beta_2 x_2 + \beta_3 x_3 + \beta_4 x_4 + \beta_5 x_5 + \beta_6 x_6 + \varepsilon.$$

We added a column of ones to the original data to allow for a constant term in the model. The following sequence of MATLAB code obtains the parameter estimates using Equation 10.4

```
load longley
bhat1 = inv(X'*X)*X'*Y;
```

The results are

```
-3482258.65,   15.06,   -0.04,   -2.02,   -1.03,   -0.05,
    1829.15
```

A more efficient way to get the estimates is using MATLAB's backslash operator '\'. Not only is the backslash more efficient, it is better conditioned, so it is less prone to numerical problems. When we try it on the `longley` data, we see that the parameter estimates match. The command

```
bhat = X\Y;
```

yields the same parameter estimates. In some more difficult situations, the backslash operator can be more accurate numerically.
❑

   Recall that the purpose of regression is to estimate the relationship between the independent or predictor variable $X_j$ and the dependent or response variable $Y$. Once we have such a model, we can use it to predict a value of $y$ for a given $x$. We obtain the model by finding the values of the parameters that minimize the sum of the squared errors.

   Once we have our model, it is important to look at the resultant predictions to see if any of the assumptions are violated, and how the model is a good fit to the data for all values of $X$. For example, the least squares method assumes that the errors are normally distributed with the same variance. To determine whether or not these assumptions are reasonable, we can look at the difference between the observed $Y_i$ and the predicted value $\hat{Y}_i$ that we obtain from the fitted model. These differences are called the residuals and are defined as

$$\hat{\varepsilon}_i = Y_i - \hat{Y}_i; \qquad i = 1, \ldots, n, \qquad (10.5)$$

where $Y_i$ is the observed response at $X_i$ and $\hat{Y}_i$ is the corresponding prediction at $X_i$ using the model. The residuals can be thought of as the observed errors.

   We can use the visualization techniques of Chapter 5 to make plots of the residuals to see if the assumptions are violated. For example, we can check the assumption of normality by plotting the residuals against the quantiles of a normal distribution in a q-q plot. If the points fall (roughly) on a straight line, then the normality assumption seems reasonable. Other possibilities include a histogram (if $n$ is large), box plots, etc., to see if the distribution of the residuals looks approximately normal.

   Another and more common method of examining the residuals using graphics is to construct a scatterplot of the residuals against the fitted values. Here the vertical axis units are given by the residuals $\hat{\varepsilon}_i$, and the fitted values $\hat{Y}_i$ are shown on the horizontal axis. If the assumptions are correct for the model, then we would expect a horizontal band of points with no patterns or trends. We do not plot the residuals versus the observed values $Y_i$, because they are correlated [Draper and Smith, 1981], while the $\hat{\varepsilon}_i$ and $\hat{Y}_i$ are not. We can also plot the residuals against the $X_i$, called a ***residual dependence plot*** [Clevelend, 1993]. If this scatterplot still shows a continued relationship between the residuals (the remaining variation not explained by the model) and the predictor variable, then the model is inadequate and adding additional columns in the **X** matrix is indicated. These ideas are explored further in the exercises.

## Example 10.2

The purpose of this example is to illustrate another method in MATLAB for fitting polynomials to data, as well as to show what happens when the model is not adequate. We use the function **polyfit** to fit polynomials of various degrees to data where we have one predictor and one response. Recall that the function **polyfit** takes three arguments: a vector of measured values of the predictor, a vector of response measurements and the degree of the polynomial. One of the outputs from the function is a vector of estimated parameters. Note that MATLAB reports the coefficients in descending powers: $\hat{\beta}_d, \ldots, \hat{\beta}_0$. We use the **filip** data in this example, which can be downloaded from **http://www.itl.nist.gov/div898**. Like the **longley** data, this data set is used as a standard to verify the results of least squares regression. The model for these data are

$$y = \beta_0 + \beta_1 x + \beta_2 x^2 + \ldots + \beta_{10} x^{10} + \varepsilon.$$

We first load up the data and then naively fit a straight line. We suspect that this model will not be a good representation of the relationship between $x$ and $y$.

```
load filip
% This loads up two vectors: x and y.
[p1,s] = polyfit(x,y,1);
% Get the curve from this fit.
yhat1 = polyval(p1,x);
plot(x,y,'k.',x,yhat1,'k')
```

By looking at **p1** we see that the estimates for the parameters are a $y$-intercept of 1.06 and a slope of 0.03. A scatterplot of the data points, along with the estimated line are shown in Figure 10.1. Not surprisingly, we see that the model is not adequate. Next, we try a polynomial of degree $d = 10$.

```
[p10,s] = polyfit(x,y,10);
% Get the curve from this fit.
yhat10 = polyval(p10,x);
```

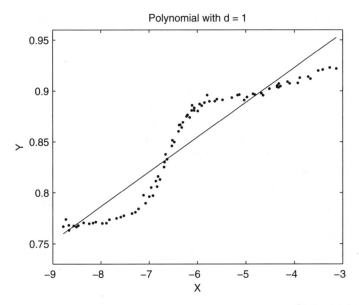

**FIGURE 10.1**
This shows a scatterplot of the **filip** data, along with the resulting line obtained using a polynomial of degree one as the model. It is obvious that this model does not result in an adequate fit.

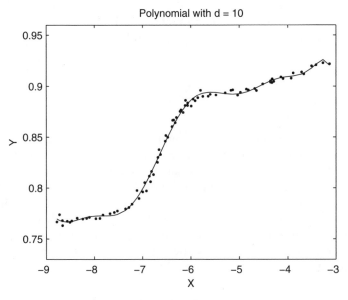

**FIGURE 10.2**
In this figure, we show the scatterplot for the **filip** data along with a curve using a polynomial of degree ten as the model.

```
plot(x,y,'k.',x,yhat10,'k')
```

The curve obtained from this model is shown in Figure 10.2, and we see that it is a much better fit. The reader will be asked to explore these data further in the exercises.
❑

The standard MATLAB program (Version 6) has added an interface that can be used to fit curves. It is only available for 2-D data (i.e., fitting $Y$ as a function of one predictor variable $X$). It enables the user to perform many of the tasks of curve fitting (e.g., choosing the degree of the polynomial, plotting the residuals, annotating the graph, etc.) through one graphical interface. The **Basic Fitting** interface is enabled through the **Figure** window **Tools** menu. To activate this graphical interface, plot a 2-D curve using the **plot** command (or something equivalent) and click on **Basic Fitting** from the **Figure** window **Tools** menu. The MATLAB Statistics Toolbox has an interactive graphical tool called **polytool** that allows the user to see what happens when the degree of the polynomial that is used to fit the data is changed.

---

## 10.2 Smoothing

The previous discussion on classical regression illustrates the situation where the analyst assumes a parametric form for a model and then uses least squares to estimate the required parameters. We now describe a nonparametric approach, where the model is more general and is given by

$$Y = \sum_{j=1}^{d} f(X_j) + \varepsilon. \tag{10.6}$$

Here, each $f(X_j)$ will be a smooth function and allows for non-linear functions of the dependent variables. In this section, we restrict our attention to the case where we have only two variables: one predictor and one response. In Equation 10.6, we are using a random design where the values of the predictor are randomly chosen. An alternative formulation is the fixed design, in which case the design points are fixed, and they would be denoted by $x_i$. In this book, we will be treating the random design case for the most part.

The function $f(X_j)$ is often called the regression or smoothing function. We are searching for a function that minimizes

$$E[(Y - f(X))^2]. \tag{10.7}$$

It is known from introductory statistics texts that the function which minimizes Equation 10.7 is

$$E[Y|X = x].$$

Note that if we are in the parametric regression setting, then we are assuming a parametric form for the smoothing function such as

$$f(X) = \beta_0 + \beta_1 X.$$

If we do not make any assumptions about the form for $f(X_j)$, then we should use nonparametric regression techniques.

The nonparametric regression method covered in this section is called a scatterplot smooth because it helps to visually convey the relationship between $X$ and $Y$ by graphically summarizing the middle of the data using a smooth function of the points. Besides helping to visualize the relationship, it also provides an estimate or prediction for given values of $x$. The smoothing method we present here is called *loess*, and we discuss the basic version for one predictor variable. This is followed by a version of loess that is made robust by using the bisquare function to re-weight points based upon the magnitude of their residuals. Finally, we show how to use loess to get upper and lower smooths to visualize the spread of the data.

## Loess

Before deciding on what model to use, it is a good idea to look at a scatterplot of the data for insight on how to model the relationship between the variables, as was discussed in Chapter 7. Sometimes, it is difficult to construct a simple parametric formula for the relationship, so smoothing a scatterplot can help the analyst understand how the variables depend on each other. Loess is a method that employs locally weighted regression to smooth a scatterplot and also provides a nonparametric model of the relationship between two variables. It was originally described in Cleveland [1979], and further extensions can be found in Cleveland and McGill [1984] and Cleveland [1993].

The curve obtained from a loess model is governed by two parameters, $\alpha$ and $\lambda$. The parameter $\alpha$ is a smoothing parameter. We restrict our attention to values of $\alpha$ between zero and one, where high values for $\alpha$ yield smoother curves. Cleveland [1993] addresses the case where $\alpha$ is greater than one. The second parameter $\lambda$ determines the degree of the local regression. Usually, a first or second degree polynomial is used, so $\lambda = 1$ or $\lambda = 2$. How to set these parameters will be explored in the exercises.

The general idea behind loess is the following. To get a value of the curve $\hat{y}$ at a given point $x$, we first determine a local neighborhood of $x$ based on $\alpha$.

All points in this neighborhood are weighted according to their distance from $x$, with points closer to $x$ receiving larger weight. The estimate $\hat{y}$ at $x$ is obtained by fitting a linear or quadratic polynomial using the weighted points in the neighborhood. This is repeated for a uniform grid of points $x$ in the domain to get the desired curve.

We describe below the steps for obtaining a loess curve [Hastie and Tibshirani, 1990]. The steps of the loess procedure are illustrated in Figures 10.3 through 10.6.

*PROCEDURE - LOESS CURVE CONSTRUCTION*

1. Let $x_i$ denote a set of $n$ values for a predictor variable and let $y_i$ represent the corresponding response.

2. Choose a value for $\alpha$ such that $0 < \alpha < 1$. Let $k = \lfloor \alpha n \rfloor$, where $k$ is the greatest integer less than or equal to $\alpha n$.

3. For each $x_0$, find the $k$ points $x_i$ that are closest to $x_0$. These $x_i$ comprise a neighborhood of $x_0$, and this set is denoted by $N(x_0)$.

4. Compute the distance of the $x_i$ in $N(x_0)$ that is furthest away from $x_0$ using

$$\Delta(x_0) = \max_{x_i \in N_0} |x_0 - x_i| .$$

5. Assign a weight to each point $(x_i, y_i)$, $x_i$ in $N(x_0)$, using the tri-cube weight function

$$w_i(x_0) = W\left( \frac{|x_0 - x_i|}{\Delta(x_0)} \right),$$

with

$$W(u) = \begin{cases} (1 - u^3)^3; & 0 \le u < 1 \\ 0; & \text{otherwise.} \end{cases}$$

6. Obtain the value $\hat{y}$ of the curve at the point $x_0$ using a weighted least squares fit of the points $x_i$ in the neighborhood $N(x_0)$. (See Equations 10.8 through 10.11.)

7. Repeat steps 3 through 6 for all $x_0$ of interest.

In step 6, one can fit either a straight line to the weighted points $(x_i, y_i)$, $x_i$ in $N(x_0)$, or a quadratic polynomial can be used. If a line is used as the local model, then $\lambda = 1$. The values of $\beta_0$ and $\beta_1$ are found such that the following is minimized

$$\sum_{i=1}^{k} w_i(x_0)(y_i - \beta_0 - \beta_1 x_i)^2, \tag{10.8}$$

for $(x_i, y_i)$, $x_i$ in $N(x_0)$. Letting $\hat{\beta}_0$ and $\hat{\beta}_1$ be the values that minimize Equation 10.8, the loess fit at $x_0$ is given by

$$\hat{y}(x_0) = \hat{\beta}_0 + \hat{\beta}_1 x_0. \tag{10.9}$$

When $\lambda = 2$, then we fit a quadratic polynomial using weighted least-squares using only those points in $N(x_0)$. In this case, we find the values for the $\beta_i$ that minimize

$$\sum_{i=1}^{k} w_i(x_0)(y_i - \beta_0 - \beta_1 x_i - \beta_2 x_i^2)^2. \tag{10.10}$$

As before, if $\hat{\beta}_0$, $\hat{\beta}_1$, and $\hat{\beta}_2$ minimize Equation 10.10, then the loess fit at $x_0$ is

$$\hat{y}(x_0) = \hat{\beta}_0 + \hat{\beta}_1 x_0 + \hat{\beta}_2 x_0^2. \tag{10.11}$$

For more information on weighted least squares see Draper and Smith, [1981].

## Example 10.3

In this example, we use a data set that was analyzed in Cleveland and McGill [1984]. These data represent two variables comprising daily measurements of ozone and wind speed in New York City. These quantities were measured on 111 days between May and September 1973. We are interested in understanding the relationship between ozone (the response variable) and wind speed (the predictor variable). The next lines of MATLAB code load the data set and display the scatterplot shown in Figure 10.3.

```
load environ
% Do a scatterplot of the data to see the relationship.
plot(wind,ozone,'k.')
xlabel('Wind Speed (MPH)'),ylabel('Ozone (PPB)')
```

It is difficult to determine the parametric relationship between the variables from the scatterplot, so the loess approach is used. We illustrate how to use the loess procedure to find the estimate of the ozone for a given wind speed of 10 MPH.

```
n = length(wind); % Find the number of data points.
x0 = 10; % Find the estimate at this point.
```

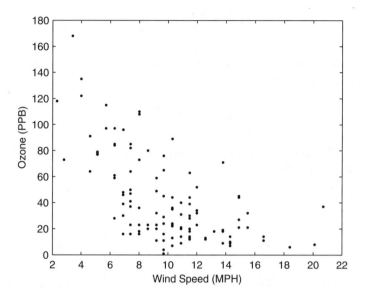

**FIGURE 10.3**
This shows a scatterplot of ozone and wind speed. It is difficult to tell from this plot what type of relationship exists between these two variables. Instead of using a parametric model, we will try the nonparametric approach.

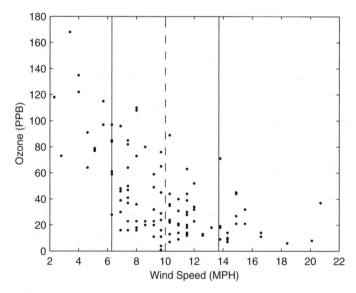

**FIGURE 10.4**
This shows the neighborhood (solid line) of the point $x_0 = 10$ (dashed line).

```
alpha = 2/3;
lambda = 1;
k = floor(alpha*n);
```

Now that we have the parameters for loess, the next step is to find the neighborhood at $x_0 = 10$.

```
% First step is to get the neighborhood.
dist = abs(x0 - wind);
[sdist,ind] = sort(dist);
% Get the points in the neighborhood.
Nx = wind(ind(1:k));
Ny = ozone(ind(1:k));
delxo = sdist(k);   % Maximum distance of neighborhood
```

The neighborhood of $x_0$ is shown in Figure 10.4, where the dashed line indicates the point of interest $x_0$ and the solid line indicates the limit of the local region. All points within this neighborhood receive weights based on their distance from $x_0 = 10$ as shown below.

```
% Delete the ones outside the neighborhood.
sdist((k+1):n) = [];
% These are the arguments to the weight function.
u = sdist/delxo;
% Get the weights for all points in the neighborhood.
w = (1 - u.^3).^3;
```

Using only those points in the neighborhood, we use weighted least squares to get the estimate at $x_0$.

```
% Now using only those points in the neighborhood,
% do a weighted least squares fit of degree 1.
% We will follow the procedure in 'polyfit'.
x = Nx(:); y = Ny(:); w = w(:);
W = diag(w);% get weight matrix
A = vander(x);% get right matrix for X
A(:,1:length(x)-lambda-1) = [];
V = A'*W*A;
Y = A'*W*y;
[Q,R] = qr(V,0);
p = R\(Q'*Y);
p = p';% to fit MATLAB convention
% This is the polynomial model for the local fit.
% To get the value at that point, use polyval.
yhat0 = polyval(p,x0);
```

In Figure 10.5, we show the local fit in the neighborhood of $x_0$. We include a function called **csloess** that will determine the smooth for all points in a given vector. We illustrate its use below.

```
% Now call the loess procedure and plot the result.
% Get a domain over which to evaluate the curve.
x0 = linspace(min(wind),max(wind),50);
yhat = csloess(wind,ozone,x0,alpha,lambda);
% Plot the results.
plot(wind,ozone,'k.',x0,yhat,'k')
xlabel('Wind Speed (MPH)'),ylabel('Ozone (PPB)')
```

The resulting scatterplot with loess smooth is shown in Figure 10.6. The final curve is obtained by linearly interpolating between the estimates from loess.
❏

As we will see in the exercises, fitting curves is an iterative process. Different values for the parameters $\alpha$ and $\lambda$ should be used to obtain various loess curves. Then the scatterplot with superimposed loess curve and residuals plots can be examined to determine whether or not the model adequately describes the relationship.

### Robust Loess Smoothing

Loess is not robust, because it relies on the method of least squares. A method is called *robust* if it performs well when the associated underlying assumptions (e.g., normality) are not satisfied [Kotz and Johnson, Vol. 8, 1988]. There are many ways in which assumptions can be violated. A common one is the presence of *outliers* or extreme values in the response data. These are points in the sample that deviate from the pattern of the other observations. Least squares regression is vulnerable to outliers, and it takes only one extreme value to unduly influence the result. This is easily seen in Figure 10.7, where there is an outlier in the upper left corner. The dashed line is obtained using least squares with the outlier present, and the solid line is obtained with the outlier removed. It is obvious that the outlier affects the slope of the line and would change the predictions one gets from the model.

Cleveland [1993, 1979] and Cleveland and McGill [1984] present a method for smoothing a scatterplot using a robust version of loess. This technique uses the bisquare method [Hoaglin, Mosteller, and Tukey, 1983; Mosteller and Tukey, 1977; Huber, 1973; Andrews, 1974] to add robustness to the weighted least squares step in loess. The idea behind the bisquare is to re-weight points based on their residuals. If the residual for a given point in the neighborhood is large (i.e., it has a large deviation from the model), then the weight for that point should be decreased, since large residuals tend to indicate outlying observations. On the other hand, if the point has a small residual, then it should be weighted more heavily.

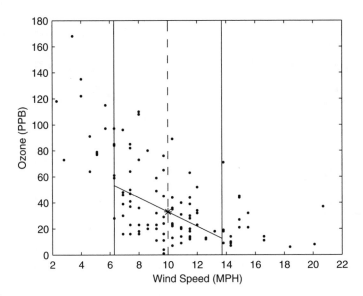

**FIGURE 10.5**
This shows the local fit at $x_0 = 10$ using weighted least squares. Here $\lambda = 1$ and $\alpha = 2/3$.

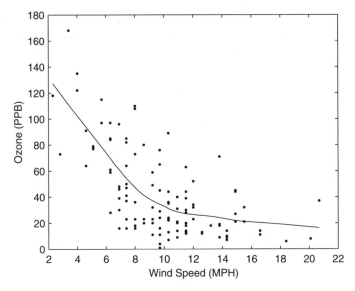

**FIGURE 10.6**
This shows the scatterplot of ozone and wind speed along with the accompanying loess smooth.

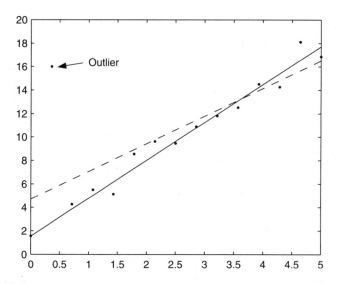

**FIGURE 10.7**
This is an example of what can happen with the least squares method when an outlier is
present. The dashed line is the fit with the outlier present, and the solid line is the fit with
the outlier removed. The slope of the line is changed when the outlier is used to fit the model.

Before showing how the bisquare method can be incorporated into loess,
we first describe the general bisquare least squares procedure. First a linear
regression is used to fit the data, and the residuals $\hat{\varepsilon}_i$ are calculated from

$$\hat{\varepsilon}_i = Y_i - \hat{Y}_i. \tag{10.12}$$

The residuals are used to determine the weights from the bisquare function
given by

$$B(u) = \begin{cases} (1 - u^2)^2; & |u| < 1 \\ 0; & \text{otherwise.} \end{cases} \tag{10.13}$$

The robustness weights are obtained from

$$r_i = B\left(\frac{\hat{\varepsilon}_i}{6\hat{q}_{0.5}}\right), \tag{10.14}$$

where $\hat{q}_{0.5}$ is the median of $|\hat{\varepsilon}_i|$. A weighted least squares regression is performed using $r_i$ as the weights.

To add bisquare to loess, we first fit the loess smooth, using the same procedure as before. We then calculate the residuals using Equation 10.12 and determine the robust weights from Equation 10.14. The loess procedure is repeated using weighted least squares, but the weights are now $r_i w_i(x_0)$. Note that the points used in the fit are the ones in the neighborhood of $x_0$. This is an iterative process and is repeated until the loess curve converges or stops changing. Cleveland and McGill [1984] suggest that two or three iterations are sufficient to get a reasonable model.

*PROCEDURE - ROBUST LOESS*

1. Fit the data using the loess procedure with weights $w_i$,
2. Calculate the residuals, $\hat{\varepsilon}_i = y_i - \hat{y}_i$ for each observation.
3. Determine the median of the absolute value of the residuals, $\hat{q}_{0.5}$.
4. Find the robustness weight from

$$r_i = B\left(\frac{\hat{\varepsilon}_i}{6\hat{q}_{0.5}}\right),$$

using the bisquare function in Equation 10.13.

5. Repeat the loess procedure using weights of $r_i w_i$.
6. Repeat steps 2 through 5 until the loess curve converges.

In essence, the robust loess iteratively adjusts the weights based on the residuals. We illustrate the robust loess procedure in the next example.

## Example 10.4
We return to the **filip** data in this example. We create some outliers in the data by adding noise to five of the points.

```
load filip
% Make several of the points outliers by adding noise.
n = length(x);
ind = unidrnd(n,1,5);% pick 5 points to make outliers
y(ind) = y(ind) + 0.1*randn(size(y(ind)));
```

A function that implements the robust version of loess is included with the text. It is called **csloessr** and takes the following input arguments: the observed values of the predictor variable, the observed values of the response variable, the values of $x_0$, $\alpha$ and $\lambda$. We now use this function to get the loess curve.

```
% Get the x values where we want to evaluate the curve.
xo = linspace(min(x),max(x),25);
% Use robust loess to get the smooth.
alpha = 0.5;
deg = 1;
yhat = csloessr(x,y,xo,alpha,deg);
```

The resulting smooth is shown in Figure 10.8. Note that the loess curve is not affected by the presence of the outliers.
❑

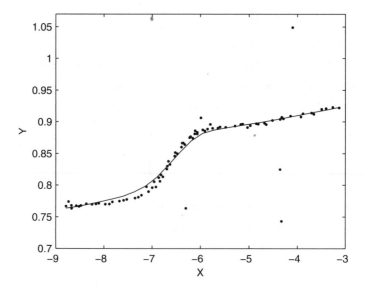

**FIGURE 10.8**
This shows a scatterplot of the `filip` data, where five of the responses deviate from the rest of the data. The curve is obtained using the robust version of loess, and we see that the curve is not affected by the presence of the outliers.

## Upper and Lower Smooths

The loess smoothing method provides a model of the middle of the distribution of $Y$ given $X$. This can be extended to give us upper and lower smooths [Cleveland and McGill, 1984], where the distance between the upper and lower smooths indicates the spread. The procedure for obtaining the upper and lower smooths follows.

*PROCEDURE - UPPER AND LOWER SMOOTHS (LOESS)*

1. Compute the fitted values $\hat{y}_i$ using loess or robust loess.

2. Calculate the residuals $\hat{\varepsilon}_i = y_i - \hat{y}_i$.

3. Find the positive residuals $\hat{\varepsilon}_i^+$ and the corresponding $x_i$ and $\hat{y}_i$ values. Denote these pairs as $(x_i^+, \hat{y}_i^+)$.

4. Find the negative residuals $\hat{\varepsilon}_i^-$ and the corresponding $x_i$ and $\hat{y}_i$ values. Denote these pairs as $(x_i^-, \hat{y}_i^-)$.

5. Smooth the $(x_i^+, \hat{\varepsilon}_i^+)$ and add the fitted values from that smooth to $\hat{y}_i^+$. This is the upper smoothing.

6. Smooth the $(x_i^-, \hat{\varepsilon}_i^-)$ and add the fitted values from this smooth to $\hat{y}_i^-$. This is the lower smoothing.

## Example 10.5

In this example, we generate some data to show how to get the upper and lower loess smooths. These data are obtained by adding noise to a sine wave. We then use the function called **csloessenv** that comes with the Computational Statistics Toolbox. The inputs to this function are the same as the other loess functions.

```
% Generate some x and y values.
x = linspace(0, 4 * pi,100);
y = sin(x) + 0.75*randn(size(x));
% Use loess to get the upper and lower smooths.
[yhat,ylo,xlo,yup,xup]=csloessenv(x,y,x,0.5,1,0);
% Plot the smooths and the data.
plot(x,y,'k.',x,yhat,'k',xlo,ylo,'k',xup,yup,'k')
```

The resulting middle, upper and lower smooths are shown in Figure 10.9, and we see that the smooths do somewhat follow a sine wave. It is also interesting to note that the upper and lower smooths indicate the symmetry of the noise and the constancy of the spread.
❑

## 10.3 Kernel Methods

This section follows the treatment of kernel smoothing methods given in Wand and Jones [1995]. We first discussed kernel methods in Chapter 8, where we applied them to the problem of estimating a probability density function in a nonparametric setting. We now present a class of smoothing

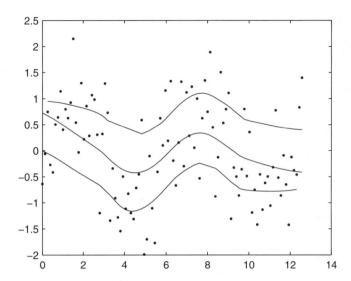

**FIGURE 10.9**
The data for this example are generated by adding noise to a sine wave. The middle curve is the usual loess smooth, while the other curves are obtained using the upper and lower loess smooths.

methods based on kernel estimators that are similar in spirit to loess, in that they fit the data in a local manner. These are called *local polynomial kernel estimators*. We first define these estimators in general and then present two special cases: the *Nadaraya-Watson estimator* and the *local linear kernel estimator*.

With local polynomial kernel estimators, we obtain an estimate $\hat{y}_0$ at a point $x_0$ by fitting a $d$-th degree polynomial using weighted least squares. As with loess, we want to weight the points based on their distance to $x_0$. Those points that are closer should have greater weight, while points further away have less weight. To accomplish this, we use weights that are given by the height of a kernel function that is centered at $x_0$.

As with probability density estimation, the kernel has a bandwidth or smoothing parameter represented by $h$. This controls the degree of influence points will have on the local fit. If $h$ is small, then the curve will be wiggly, because the estimate will depend heavily on points closest to $x_0$. In this case, the model is trying to fit to local values (i.e., our 'neighborhood' is small), and we have over fitting. Larger values for $h$ means that points further away will have similar influence as points that are close to $x_0$ (i.e., the 'neighborhood' is large). With a large enough $h$, we would be fitting the line to the whole data set. These ideas are investigated in the exercises.

We now give the expression for the local polynomial kernel estimator. Let $d$ represent the degree of the polynomial that we fit at a point $x$. We obtain the estimate $\hat{y} = \hat{f}(x)$ by fitting the polynomial

$$\beta_0 + \beta_1(X_i - x) + \ldots + \beta_d(X_i - x)^d \tag{10.15}$$

using the points $(X_i, Y_i)$ and utilizing the weighted least squares procedure. The weights are given by the kernel function

$$K_h(X_i - x) = \frac{1}{h}K\left(\frac{X_i - x}{h}\right). \tag{10.16}$$

The value of the estimate at a point $x$ is $\hat{\beta}_0$, where the $\hat{\beta}_i$ minimize

$$\sum_{i=1}^{n} K_h(X_i - x)(Y_i - \beta_0 - \beta_1(X_i - x) - \ldots - \beta_d(X_i - x)^d)^2. \tag{10.17}$$

Because the points that are used to estimate the model are all centered at $x$ (see Equation 10.15), the estimate at $x$ is obtained by setting the argument in the model equal to zero. Thus, the only parameter left is the constant term $\beta_0$.

The attentive reader will note that the argument of the $K_h$ is backwards from what we had in probability density estimation using kernels. There, the kernels were centered at the random variables $X_i$. We follow the notation of Wand and Jones [1995] that shows explicitly that we are centering the kernels at the points $x$ where we want to obtain the estimated value of the function.

We can write this weighted least squares procedure using matrix notation. According to standard weighted least squares theory [Draper and Smith, 1981], the solution can be written as

$$\hat{\beta} = (\mathbf{X}_x^T \mathbf{W}_x \mathbf{X}_x)^{-1} \mathbf{X}_x^T \mathbf{W}_x \mathbf{Y}, \tag{10.18}$$

where $\mathbf{Y}$ is the $n \times 1$ vector of responses,

$$\mathbf{X}_x = \begin{bmatrix} 1 & X_1 - x & \ldots & (X_1 - \overset{*}{x})^d \\ \vdots & \vdots & \ldots & \vdots \\ 1 & X_n - x & \ldots & (X_n - x)^d \end{bmatrix}, \tag{10.19}$$

and $\mathbf{W}_x$ is an $n \times n$ matrix with the weights along the diagonal. These weights are given by

$$w_{ii}(x) = K_h(X_i - x).$$     (10.20)

Some of these weights might be zero depending on the kernel that is used. The estimator $\hat{y} = \hat{f}(x)$ is the intercept coefficient $\beta_0$ of the local fit, so we can obtain the value from

$$\hat{f}(x) = \mathbf{e}_1^T (\mathbf{X}_x^T \mathbf{W}_x \mathbf{X}_x)^{-1} \mathbf{X}_x^T \mathbf{W}_x \mathbf{Y}$$     (10.21)

where $\mathbf{e}_1^T$ is a vector of dimension $(d+1) \times 1$ with a one in the first place and zeroes everywhere else.

## Nadaraya-Watson Estimator

Some explicit expressions exist when $d = 0$ and $d = 1$. When $d$ is zero, we fit a constant function locally at a given point $x$. This estimator was developed separately by Nadaraya [1964] and Watson [1964]. The Nadaraya-Watson estimator is given below.

*NADARAYA-WATSON KERNEL ESTIMATOR:*

$$\hat{f}_{NW}(x) = \frac{\sum_{i=1}^{n} K_h(X_i - x) Y_i}{\sum_{i=1}^{n} K_h(X_i - x)}.$$     (10.22)

Note that this is for the case of a random design. When the design points are fixed, then the $X_i$ is replaced by $x_i$, but otherwise the expression is the same [Wand and Jones, 1995].

There is an alternative estimator that can be used in the *fixed design* case. This is called the Priestley-Chao kernel estimator [Simonoff, 1996].

*PRIESTLEY-CHAO KERNEL ESTIMATOR:*

$$\hat{f}_{PC}(x) = \frac{1}{h} \sum_{i=1}^{n} (x_i - x_{i-1}) K\left(\frac{x - x_i}{h}\right) y_i,$$     (10.23)

where the $x_i$, $i = 1, ..., n$, represent a fixed set of ordered nonrandom numbers. The Nadarya-Watson estimator is illustrated in Example 10.6, while the Priestley-Chao estimator is saved for the exercises.

## Example 10.6

We show how to implement the Nadarya-Watson estimator in MATLAB. As in the previous example, we generate data that follows a sine wave with added noise.

```
% Generate some noisy data.
x = linspace(0, 4 * pi,100);
y = sin(x) + 0.75*randn(size(x));
```

The next step is to create a MATLAB `inline` function so we can evaluate the weights. Note that we are using the normal kernel.

```
% Create an inline function to evaluate the weights.
mystrg='(2*pi*h^2)^(-1/2)*exp(-0.5*((x - mu)/h).^2)';
wfun = inline(mystrg);
```

We now get the estimates at each value of $x$.

```
% Set up the space to store the estimated values.
% We will get the estimate at all values of x.
yhatnw = zeros(size(x));
n = length(x);
% Set the window width.
h = 1;
% find smooth at each value in x
for i = 1:n
 w = wfun(h,x(i),x);
 yhatnw(i) = sum(w.*y)/sum(w);
end
```

The smooth from the Nadarya-Watson estimator is shown in Figure 10.10.
❑


### Local Linear Kernel Estimator

When we fit a straight line at a point $x$, then we are using a local linear estimator. This corresponds to the case where $d = 1$, so our estimate is obtained as the solutions $\hat{\beta}_0$ and $\hat{\beta}_1$ that minimize the following,

$$\sum_{i=1}^{n} K_h(X_i - x)(Y_i - \beta_0 - \beta_1(X_i - x))^2.$$

We give an explicit formula for the estimator below.

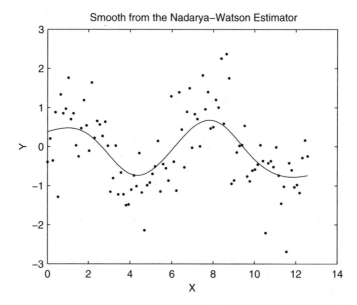

**FIGURE 10.10**
This figure shows the smooth obtained from the Nadarya-Watson estimator with $h = 1$.


*LOCAL LINEAR KERNEL ESTIMATOR:*

$$\hat{f}_{LL}(x) = \frac{1}{n}\sum_{i=1}^{n}\frac{\{\hat{s}_2(x) - \hat{s}_1(x)(X_i - x)\}K_h(X_i - x)Y_i}{\hat{s}_2(x)\hat{s}_0(x) - \hat{s}_1(x)^2}, \qquad (10.24)$$

where

$$\hat{s}_r(x) = \frac{1}{n}\sum_{i=1}^{n}(X_i - x)^r K_h(X_i - x).$$

As before, the fixed design case is obtained by replacing the random variable $X_i$ with the fixed point $x_i$.

When using the kernel smoothing methods, problems can arise near the boundary or extreme edges of the sample. This happens because the kernel window at the boundaries has missing data. In other words, we have weights from the kernel, but no data to associate with them. Wand and Jones [1995] show that the local linear estimator behaves well in most cases, even at the boundaries. If the Nadaraya-Watson estimator is used, then modified kernels are needed [Scott, 1992; Wand and Jones, 1995].

## Example 10.7

The local linear estimator is applied to the same generated sine wave data. The entire procedure is implemented below and the resulting smooth is shown in Figure 10.11. Note that the curve seems to behave well at the boundary.

```
% Generate some data.
x = linspace(0, 4 * pi,100);
y = sin(x) + 0.75*randn(size(x));
h = 1;
deg = 1;
% Set up inline function to get the weights.
mystrg = ...
    '(2*pi*h^2)^(-1/2)*exp(-0.5*((x - mu)/h).^2)';
wfun = inline(mystrg);
% Set up space to store the estimates.
yhatlin = zeros(size(x));
n = length(x);
% Find smooth at each value in x.
for i = 1:n
 w = wfun(h,x(i),x);
 xc = x-x(i);
 s2 = sum(xc.^2.*w)/n;
 s1 = sum(xc.*w)/n;
 s0 = sum(w)/n;
 yhatlin(i) = sum(((s2-s1*xc).*w.*y)/(s2*s0-s1^2))/n;
end
```

□

## 10.4 Regression Trees

The tree-based approach to nonparametric regression is useful when one is trying to understand the structure or interaction among the predictor variables. As we stated earlier, one of the main uses of modeling the relationship between variables is to be able to make predictions given future measurements of the predictor variables. Regression trees accomplish this purpose, but they also provide insight into the structural relationships and the possible importance of the variables. Much of the information about classification

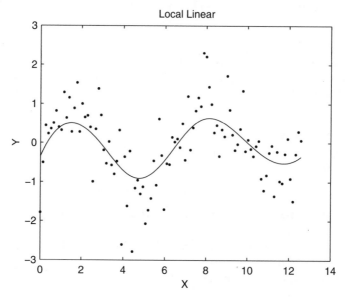

**FIGURE 10.11**
This figure shows the smooth obtained from the local linear estimator.

trees applies in the regression case, so the reader is encouraged to read Chapter 9 first, where the procedure is covered in more detail.

In this section, we move to the multivariate situation where we have a response variable $Y$ along with a set of predictors $X = (X_1, ..., X_d)$. Using a procedure similar to classification trees, we will examine all predictor variables for a best split, such that the two groups are homogeneous with respect to the response variable $Y$. The procedure examines all possible splits and chooses the split that yields the smallest within-group variance in the two groups. The result is a binary tree, where the predicted responses are given by the average value of the response in the corresponding terminal node. To predict the value of a response given an observed set of predictors $x = (x_1, ..., x_d)$, we drop $x$ down the tree, and assign to $\hat{y}$ the value of the terminal node that it falls into. Thus, we are estimating the function using a piecewise constant surface.

Before we go into the details of how to construct regression trees, we provide the notation that will be used.

*NOTATION: REGRESSION TREES*

    $d(x)$ represents the prediction rule that takes on real values. Here $d$ will be our regression tree.

**L** is the learning sample of size $n$. Each case in the learning sample comprises a set of measured predictors and the associated response.

$\mathbf{L}_v$, $v = 1, ..., V$ is the $v$-th partition of the learning sample **L** in cross-validation. This set of cases is used to calculate the prediction error in $d^{(v)}(\mathbf{x})$.

$\mathbf{L}^{(v)} = \mathbf{L} - \mathbf{L}_v$ is the set of cases used to grow a sequence of subtrees.

$(\mathbf{x}_i, y_i)$ denotes one case, where $\mathbf{x}_i = (x_{1_i}, ..., x_{d_i})$ and $i = 1, ..., n$.

$R^*(d)$ is the true mean squared error of predictor $d(\mathbf{x})$.

$\hat{R}^{TS}(d)$ is the estimate of the mean squared error of $d$ using the independent test sample method.

$\hat{R}^{CV}(d)$ denotes the estimate of the mean squared error of $d$ using cross-validation.

$T$ is the regression tree.

$T_{max}$ is an overly large tree that is grown.

$T_{max}^{(v)}$ is an overly large tree grown using the set $L^{(v)}$.

$T_k$ is one of the nested subtrees from the pruning procedure.

$t$ is a node in the tree $T$.

$t_L$ and $t_R$ are the left and right child nodes.

$\widehat{T}$ is the set of terminal nodes in tree $T$.

$|\widehat{T}|$ is the number of terminal nodes in tree $T$.

$n(t)$ represents the number of cases that are in node $t$.

$\bar{y}(t)$ is the average response of the cases that fall into node $t$.

$R(t)$ represents the weighted within-node sum-of-squares at node $t$.

$R(T)$ is the average within-node sum-of-squares for the tree $T$.

$\Delta R(s, t)$ denotes the change in the within-node sum-of-squares at node $t$ using split $s$.

To construct a regression tree, we proceed in a manner similar to classification trees. We seek to partition the space for the predictor values using a sequence of binary splits so that the resulting nodes are better in some sense than the parent node. Once we grow the tree, we use the minimum error complexity pruning procedure to obtain a sequence of nested trees with decreasing complexity. Once we have the sequence of subtrees, independent test samples or cross-validation can be used to select the best tree.

## Growing a Regression Tree

We need a criterion that measures node impurity in order to grow a regression tree. We measure this impurity using the squared difference between the predicted response from the tree and the observed response. First, note that the predicted response when a case falls into node $t$ is given by the average of the responses that are contained in that node,

$$\bar{y}(t) = \frac{1}{n(t)} \sum_{x_i \in t} y_i.$$  (10.25)

The squared error in node $t$ is given by

$$R(t) = \frac{1}{n} \sum_{x_i \in t} (y_i - \bar{y}(t))^2.$$  (10.26)

Note that Equation 10.26 is the average error with respect to the entire learning sample. If we add up all of the squared errors in all of the terminal nodes, then we obtain the mean squared error for the tree. This is also referred to as the total within-node sum-of-squares, and is given by

$$R(T) = \sum_{t \in \widehat{T}} R(t) = \frac{1}{n} \sum_{t \in \widehat{T}} \sum_{x_i \in t} (y_i - \bar{y}(t))^2.$$  (10.27)

The regression tree is obtained by iteratively splitting nodes so that the decrease in $R(T)$ is maximized. Thus, for a split $s$ and node $t$, we calculate the change in the mean squared error as

$$\Delta R(s, t) = R(t) - R(t_L) - R(t_R),$$  (10.28)

and we look for the split $s$ that yields the largest $\Delta R(s, t)$.

We could grow the tree until each node is pure in the sense that all responses in a node are the same, but that is an unrealistic condition. Breiman et al. [1984] recommend growing the tree until the number of cases in a terminal node is five.

## Example 10.8

We show how to grow a regression tree using a simple example with generated data. As with classification trees, we do not provide all of the details of how this is implemented in MATLAB. The interested reader is referred to Appendix D for the source code. We use bivariate data such that the response in each region is constant (with no added noise). We are using this simple toy example to illustrate the concept of a regression tree. In the next example, we will add noise to make the problem a little more realistic.

```
% Generate bivariate data.
X(1:50,1) = unifrnd(0,1,50,1);
X(1:50,2) = unifrnd(0.5,1,50,1);
y(1:50) = 2;
X(51:100,1) = unifrnd(-1,0,50,1);
X(51:100,2) = unifrnd(-0.5,1,50,1);
y(51:100) = 3;
X(101:150,1) = unifrnd(-1,0,50,1);
X(101:150,2) = unifrnd(-1,-0.5,50,1);
y(101:150) = 10;
X(151:200,1) = unifrnd(0,1,50,1);
X(151:200,2) = unifrnd(-1,0.5,50,1);
y(151:200) = -10;
```

These data are shown in Figure 10.12. The next step is to use the function **csgrowr** to get a tree. Since there is no noise in the responses, the tree should be small.

```
% This will be the maximum number in nodes.
% This is high to ensure a small tree for simplicity.
maxn = 75;
% Now grow the tree.
tree = csgrowr(X,y,maxn);
csplotreer(tree); % plots the tree
```

The tree is shown in Figure 10.13 and the partition view is given in Figure 10.14. Notice that the response at each node is exactly right because there is no noise. We see that the first split is at $x_1$, where values of $x_1$ less than 0.034 go to the left branch, as expected. Each resulting node from this split is partitioned based on $x_2$. The response of each terminal node is given in Figure 10.13, and we see that the tree does yield the correct response. ❑.

## Pruning a Regression Tree

Once we grow a large tree, we can prune it back using the same procedure that was presented in Chapter 9. Here, however, we define an error-complexity measure as follows

$$R_\alpha(T) = R(t) + \alpha|\widehat{T}| \tag{10.29}$$

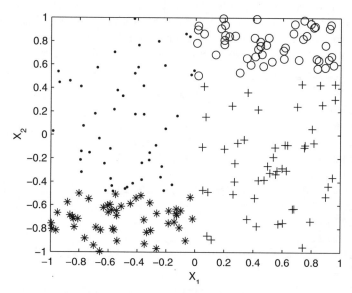

**FIGURE 10.12**
This shows the bivariate data used in Example 10.8. The observations in the upper right corner have response $y = 2$ ('o'); the points in the upper left corner have response $y = 3$ ('.'); the points in the lower left corner have response $y = 10$ ('*'); and the observations in the lower right corner have response $y = -10$ ('+'). No noise has been added to the responses, so the tree should partition this space perfectly.

From this we obtain a sequence of nested trees

$$T_{max} > T_1 > \ldots > T_K = \{t_1\},$$

where $\{t_1\}$ denotes the root of the tree. Along with the sequence of pruned trees, we have a corresponding sequence of values for $\alpha$, such that

$$0 = \alpha_1 < \alpha_2 < \ldots < \alpha_k < \alpha_{k+1} < \ldots < \alpha_K.$$

Recall that for $\alpha_k \leq \alpha < \alpha_{k+1}$, the tree $T_k$ is the smallest subtree that minimizes $R_\alpha(T)$.

### Selecting a Tree

Once we have the sequence of pruned subtrees, we wish to choose the best tree such that the complexity of the tree and the estimation error $R(T)$ are both minimized. We could obtain minimum estimation error by making the

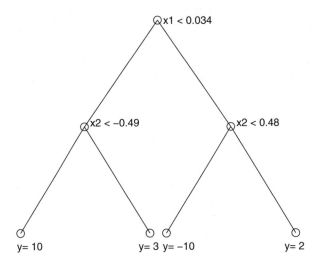

**FIGURE 10.13**
This is the regression tree for Example 10.8.

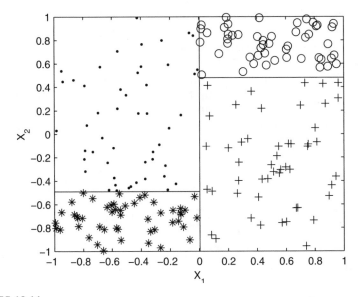

**FIGURE 10.14**
This shows the partition view of the regression tree from Example 10.8. It is easier to see how the space is partitioned. The method first splits the region based on variable $x_1$. The left side of the space is then partitioned at $x_2 = -0.49$, and the right side of the space is partitioned at $x_2 = 0.48$.

tree very large, but this increases the complexity. Thus, we must make a trade-off between these two criteria.

To select the right sized tree, we must have honest estimates of the true error $R^*(T)$. This means that we should use cases that were not used to create the tree to estimate the error. As before, there are two possible ways to accomplish this. One is through the use of independent test samples and the other is cross-validation. We briefly discuss both methods, and the reader is referred to Chapter 9 for more details on the procedures. The independent test sample method is illustrated in Example 10.9.

To obtain an estimate of the error $R^*(T)$ using the independent test sample method, we randomly divide the learning sample $\mathbf{L}$ into two sets $\mathbf{L}_1$ and $\mathbf{L}_2$. The set $\mathbf{L}_1$ is used to grow the large tree and to obtain the sequence of pruned subtrees. We use the set of cases in $\mathbf{L}_2$ to evaluate the performance of each subtree, by presenting the cases to the trees and calculating the error between the actual response and the predicted response. If we let $d_k(\mathbf{x})$ represent the predictor corresponding to tree $T_k$, then the estimated error is

$$\hat{R}^{TS}(T_k) = \frac{1}{n_2} \sum_{(\mathbf{x}_i, y_i) \in \mathbf{L}_2} (y_i - d_k(\mathbf{x}_i))^2 , \tag{10.30}$$

where the number of cases in $\mathbf{L}_2$ is $n_2$.

We first calculate the error given in Equation 10.30 for all subtrees and then find the tree that corresponds to the smallest estimated error. The error is an estimate, so it has some variation associated with it. If we pick the tree with the smallest error, then it is likely that the complexity will be larger than it should be. Therefore, we desire to pick a subtree that has the fewest number of nodes, but is still in keeping with the prediction accuracy of the tree with the smallest error [Breiman, et al. 1984].

First we find the tree that has the smallest error and call the tree $T_0$. We denote its error by $\hat{R}_{min}^{TS}(T_0)$. Then we find the standard error for this estimate, which is given by [Breiman, et al., 1984, p. 226]

$$\hat{SE}(\hat{R}_{min}^{TS}(T_0)) = \frac{1}{\sqrt{n_2}} \left[ \frac{1}{n_2} \sum_{i=1}^{n_2} (y_i - d(\mathbf{x}_i))^4 - (\hat{R}_{min}^{TS}(T_0))^2 \right]^{\frac{1}{2}} . \tag{10.31}$$

We then select the smallest tree $T_k^*$, such that

$$\hat{R}^{TS}(T_k^*) \leq \hat{R}_{min}^{TS}(T_0) + \hat{SE}(\hat{R}_{min}^{TS}(T_0)) . \tag{10.32}$$

Equation 10.32 says that we should pick the tree with minimal complexity that has accuracy equivalent to the tree with the minimum error.

If we are using cross-validation to estimate the prediction error for each tree in the sequence, then we divide the learning sample $\mathbf{L}$ into sets

$\mathbf{L}_1, \ldots, \mathbf{L}_V$. It is best to make sure that the $V$ learning samples are all the same size or nearly so. Another important point mentioned in Breiman, et al. [1984] is that the samples should be kept balanced with respect to the response variable $Y$. They suggest that the cases be put into levels based on the value of their response variable and that stratified random sampling (see Chapter 3) be used to get a balanced sample from each stratum.

We let the $v$-th learning sample be represented by $\mathbf{L}^{(v)} = \mathbf{L} - \mathbf{L}_v$, so that we reserve the set $\mathbf{L}_v$ for estimating the prediction error. We use each learning sample to grow a large tree and to get the corresponding sequence of pruned subtrees. Thus, we have a sequence of trees $T^{(v)}(\alpha)$ that represent the minimum error-complexity trees for given values of $\alpha$.

At the same time, we use the entire learning sample $\mathbf{L}$ to grow the large tree and to get the sequence of subtrees $T_k$ and the corresponding sequence of $\alpha_k$. We would like to use cross-validation to choose the best subtree from this sequence. To that end, we define

$$\alpha'_k = \sqrt{\alpha_k \alpha_{k+1}} , \tag{10.33}$$

and use $d_k^{(v)}(\mathbf{x})$ to denote the predictor corresponding to the tree $T^{(v)}(\alpha'_k)$. The cross-validation estimate for the prediction error is given by

$$\hat{R}^{CV}(T_k(\alpha'_k)) = \frac{1}{n} \sum_{v=1}^{V} \sum_{(\mathbf{x}_i, y_i) \in L_v} (y_i - d_k^{(v)}(\mathbf{x}_i))^2 . \tag{10.34}$$

We use each case from the test sample $\mathbf{L}_v$ with $d_k^{(v)}(\mathbf{x})$ to get a predicted response, and we then calculate the squared difference between the predicted response and the true response. We do this for every test sample and all $n$ cases. From Equation 10.34, we take the average value of these errors to estimate the prediction error for a tree.

We use the same rule as before to choose the best subtree. We first find the tree that has the smallest estimated prediction error. We then choose the tree with the smallest complexity such that its error is within one standard error of the tree with minimum error.

We obtain an estimate of the standard error of the cross-validation estimate of the prediction error using

$$\hat{SE}(\hat{R}^{CV}(T_k)) = \sqrt{\frac{s^2}{n}} , \tag{10.35}$$

where

$$s^2 = \frac{1}{n} \sum_{(\mathbf{x}_i, y_i)} [(y_i - d_k^{(v)}(\mathbf{x}_i))^2 - \hat{R}_{CV}(T_k)]^2 . \qquad (10.36)$$

Once we have the estimated errors from cross-validation, we find the subtree that has the smallest error and denote it by $T_0$. Finally, we select the smallest tree $T_k^*$, such that

$$\hat{R}^{CV}(T_k^*) \le \hat{R}_{min}^{CV}(T_0) + \hat{SE}(\hat{R}_{min}^{CV}(T_0)) \qquad (10.37)$$

Since the procedure is somewhat complicated for cross-validation, we list the procedure below. In Example 10.9, we implement the independent test sample process for growing and selecting a regression tree. The cross-validation case is left as an exercise for the reader.

*PROCEDURE - CROSS-VALIDATION METHOD*

1. Given a learning sample $\mathbf{L}$, obtain a sequence of trees $T_k$ with associated parameters $\alpha_k$.

2. Determine the parameter $\alpha'_k = \sqrt{\alpha_k \alpha_{k+1}}$ for each subtree $T_k$.

3. Partition the learning sample $\mathbf{L}$ into $V$ partitions, $\mathbf{L}_v$. These will be used to estimate the prediction error for trees grown using the remaining cases.

4. Build the sequence of subtrees $T_k^{(v)}$ using the observations in all $\mathbf{L}^{(v)} = \mathbf{L} - \mathbf{L}_v$.

5. Now find the prediction error for the subtrees obtained from the entire learning sample $\mathbf{L}$. For tree $T_k$ and $\alpha'_k$, find all equivalent trees $T_k^{(v)}, v = 1, ..., V$ by choosing trees $T_k^{(v)}$ such that

$$\alpha'_k \in [\alpha_k^{(v)}, \alpha_{k+1}^{(v)}) .$$

6. Take all cases in $\mathbf{L}_v, v = 1, ..., V$ and present them to the trees found in step 5. Calculate the error as the squared difference between the predicted response and the true response.

7. Determine the estimated error for the tree $\hat{R}^{CV}(T_k)$ by taking the average of the errors from step 6.

8. Repeat steps 5 through 7 for all subtrees $T_k$ to find the prediction error for each one.

9. Find the tree $T_0$ that has the minimum error,

$$\hat{R}_{min}^{CV}(T_0) = \min_k \{\hat{R}^{CV}(T_k)\} .$$

10. Determine the standard error for tree $T_0$ using Equation 10.35.

11. For the final model, select the tree that has the fewest number of nodes and whose estimated prediction error is within one standard error (Equation 10.36) of $\hat{R}_{min}^{CV}(T_0)$.

## Example 10.9

We return to the same data that was used in the previous example, where we now add random noise to the responses. We generate the data as follows.

```
X(1:50,1)   = unifrnd(0,1,50,1);
X(1:50,2)   = unifrnd(0.5,1,50,1);
y(1:50)     = 2+sqrt(2)*randn(1,50);
X(51:100,1) = unifrnd(-1,0,50,1);
X(51:100,2) = unifrnd(-0.5,1,50,1);
y(51:100)   = 3+sqrt(2)*randn(1,50);
X(101:150,1) = unifrnd(-1,0,50,1);
X(101:150,2) = unifrnd(-1,-0.5,50,1);
y(101:150)  = 10+sqrt(2)*randn(1,50);
X(151:200,1) = unifrnd(0,1,50,1);
X(151:200,2) = unifrnd(-1,0.5,50,1);
y(151:200)  = -10+sqrt(2)*randn(1,50);
```

The next step is to grow the tree. The $T_{max}$ that we get from this tree should be larger than the one in Example 10.8.

```
% Set the maximum number in the nodes.
maxn = 5;
tree = csgrowr(X,y,maxn);
```

The tree we get has a total of 129 nodes, with 65 terminal nodes. We now get the sequence of nested subtrees using the pruning procedure. We include a function called **cspruner** that implements the process.

```
% Now prune the tree.
treeseq = cspruner(tree);
```

The variable **treeseq** contains a sequence of 41 subtrees. The following code shows how we can get estimates of the error as in Equation 10.30.

```
% Generate an independent test sample.
nprime = 1000;
X(1:250,1)   = unifrnd(0,1,250,1);
X(1:250,2)   = unifrnd(0.5,1,250,1);
y(1:250)     = 2+sqrt(2)*randn(1,250);
X(251:500,1) = unifrnd(-1,0,250,1);
X(251:500,2) = unifrnd(-0.5,1,250,1);
y(251:500)   = 3+sqrt(2)*randn(1,250);
```

```
X(501:750,1) = unifrnd(-1,0,250,1);
X(501:750,2) = unifrnd(-1,-0.5,250,1);
y(501:750) = 10+sqrt(2)*randn(1,250);
X(751:1000,1) = unifrnd(0,1,250,1);
X(751:1000,2) = unifrnd(-1,0.5,250,1);
y(751:1000) = -10+sqrt(2)*randn(1,250);
% For each tree in the sequence,
% find the mean squared error
k = length(treeseq);
msek = zeros(1,k);
numnodes = zeros(1,k);
for i=1:(k-1)
    err = zeros(1,nprime);
    t = treeseq{i};
    for j=1:nprime
        [yhat,node] = cstreer(X(j,:),t);
        err(j) = (y(j)-yhat).^2;
    end
    [term,nt,imp] = getdata(t);
    % find the # of terminal nodes
    numnodes(i) = length(find(term==1));
    % find the mean
    msek(i) = mean(err);
end
t = treeseq{k};
msek(k) = mean((y-t.node(1).yhat).^2);
```

In Figure 10.15, we show a plot of the estimated error against the number of terminal nodes (or the complexity). We can find the tree that corresponds to the minimum error as follows.

```
% Find the subtree corresponding to the minimum MSE.
[msemin,ind] = min(msek);
minnode = numnodes(ind);
```

We see that the tree with the minimum error corresponds to the one with 4 terminal nodes, and it is the 38th tree in the sequence. The minimum error is 5.77. The final step is to estimate the standard error using Equation 10.31.

```
% Find the standard error for that subtree.
t0 = treeseq{ind};
for j = 1:nprime
    [yhat,node] = cstreer(X(j,:),t0);
    err(j) = (y(j)-yhat).^4-msemin^2;
end
se = sqrt(sum(err)/nprime)/sqrt(nprime);
```

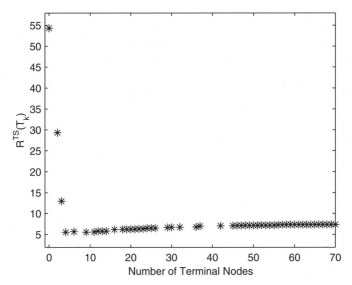

**FIGURE 10.15**
This shows a plot of the estimated error using the independent test sample approach. Note that there is a sharp minimum for $\left| \widehat{T}_k \right| = 4$.

This yields a standard error of 0.97. It turns out that there is no subtree that has smaller complexity (i.e., fewer terminal nodes) and has an error less than $5.77 + 0.97 = 6.74$. In fact, the next tree in the sequence has an error of 13.09. So, our choice for the best tree is the one with 4 terminal nodes. This is not surprising given our results from the previous example.
❑

## 10.5 MATLAB Code

MATLAB does not have any functions for the nonparametric regression techniques presented in this text. The MathWorks, Inc. has a Spline Toolbox that has some of the desired functionality for smoothing using splines. The basic MATLAB package also has some tools for estimating functions using splines (e.g., **spline**, **interp1**, etc.). We did not discuss spline-based smoothing, but references are provided in the next section.

The regression function in the MATLAB Statistics Toolbox is called **regress**. This has more output options than the **polyfit** function. For example, **regress** returns the parameter estimates and residuals, along with corresponding confidence intervals. The **polytool** is an interactive demo available in the MATLAB Statistics Toolbox. It allows the user to explore the effects of changing the degree of the fit.

As mentioned in Chapter 5, the smoothing techniques described in *Visualizing Data* [Cleveland, 1993] have been implemented in MATLAB and are available at **http://www.datatool.com/Dataviz_home.htm** for free download. We provide several functions in the Computational Statistics Toolbox for local polynomial smoothing, loess, regression trees and others. These are listed in Table 10.1.

TABLE 10.1

List of Functions from Chapter 10 Included in the Computational Statistics Toolbox

| Purpose | MATLAB Function |
|---|---|
| These functions are used for loess smoothing. | `csloess` `csloessenv` `csloessr` |
| This function does local polynomial smoothing. | `cslocpoly` |
| These functions are used to work with regression trees. | `csgrowr` `cspruner` `cstreer` `csplotreer` `cspicktreer` |
| This function performs nonparametric regression using kernels. | `csloclin` |

## 10.6 Further Reading

For more information on loess, Cleveland's book *Visualizing Data* [1993] is an excellent resource. It contains many examples and is easy to read and understand. In this book, Cleveland describes many other ways to visualize data, including extensions of loess to multivariate data. The paper by Cleveland and McGill [1984] discusses other smoothing methods such as polar smoothing, sum-difference smooths, and scale-ratio smoothing.

For a more theoretical treatment of smoothing methods, the reader is referred to Simonoff [1996], Wand and Jones [1995], Bowman and Azzalini [1997], Green and Silverman [1994], and Scott [1992]. The text by Loader [1999] describes other methods for local regression and likelihood that are not covered in our book. Nonparametric regression and smoothing are also examined in *Generalized Additive Models* by Hastie and Tibshirani [1990]. This text contains explanations of some other nonparametric regression methods such as splines and multivariate adaptive regression splines.

Other smoothing techniques that we did not discuss in this book, which are commonly used in engineering and operations research, include moving averages and exponential smoothing. These are typically used in applications where the independent variable represents time (or something analogous), and measurements are taken over equally spaced intervals. These smoothing applications are covered in many introductory texts. One possible resource for the interested reader is Wadsworth [1990].

For a discussion of boundary problems with kernel estimators, see Wand and Jones [1995] and Scott [1992]. Both of these references also compare the performance of various kernel estimators for nonparametric regression. When we discussed probability density estimation in Chapter 8, we presented some results from Scott [1992] regarding the integrated squared error that can be expected with various kernel estimators. Since the local kernel estimators are based on density estimation techniques, expressions for the squared error can be derived. Several references provide these, such as Scott [1995], Wand and Jones [1995], and Simonoff [1996].

---

## Exercises

10.1. Generate data according to $y = 4x^3 + 6x^2 - 1 + \varepsilon$, where $\varepsilon$ represents some noise. Instead of adding noise with constant variance, add noise that is variable and depends on the value of the predictor. So, increasing values of the predictor show increasing variance. Do a polynomial fit and plot the residuals versus the fitted values. Do they show that the constant variance assumption is violated? Use MATLAB's **Basic Fitting** tool to explore your options for fitting a model to these data.

10.2. Generate data as in problem 10.1, but use noise with constant variance. Fit a first-degree model to it and plot the residuals versus the observed predictor values $X_i$ (residual dependence plot). Do they show that the model is not adequate? Repeat for $d = 2, 3$.

10.3. Repeat Example 10.1. Construct box plots and histograms of the residuals. Do they indicate normality?

10.4. In some applications, one might need to explore how the spread or scale of $Y$ changes with $X$. One technique that could be used is the following:

    a) determine the fitted values $\hat{Y}_i$;

    b) calculate the residuals $\varepsilon_i = Y_i - \hat{Y}_i$;

    c) plot $|\varepsilon_i|$ against $X_i$; and

    d) smooth using loess [Cleveland and McGill, 1984].

Apply this technique to the **environ** data.

10.5. Use the **filip** data and fit a sequence of polynomials of degree $d = 2, 4, 6, 10$. For each fit, construct a residual dependence plot. What do these show about the adequacy of the models?

10.6. Use the MATLAB Statistics Toolbox graphical user interface **polytool** with the **longley** data. Use the tool to find an adequate model.

10.7. Fit a loess curve to the **environ** data using $\lambda = 1, 2$ and various values for $\alpha$. Compare the curves. What values of the parameters seem to be the best? In making your comparison, look at residual plots and smoothed scatterplots. One thing to look for is excessive structure (wiggliness) in the loess curve that is not supported by the data.

10.8. Write a MATLAB function that implements the Priestley-Chao estimator in Equation 10.23.

10.9. Repeat Example 10.6 for various values of the smoothing parameter *h*. What happens to your curve as *h* goes from very small values to very large ones?

10.10. The **human** data set [Hand, et al., 1994; Mazess, et al., 1984] contains measurements of percent fat and age for 18 normal adults (males and females). Use loess or one of the other smoothing methods to determine how percent fat is related to age.

10.11. The data set called **anaerob** has two variables: oxygen uptake and the expired ventilation [Hand, et al., 1994; Bennett, 1988]. Use loess to describe the relationship between these variables.

10.12. The **brownlee** data contains observations from 21 days of a plant operation for the oxidation of ammonia [Hand, et al., 1994; Brownlee, 1965]. The predictor variables are: $X_1$ is the air flow, $X_2$ is the cooling water inlet temperature (degrees C), and $X_3$ is the percent acid concentration. The response variable $Y$ is the stack loss (the percentage of the ingoing ammonia that escapes). Use a regression tree to determine the relationship between these variables. Pick the best tree using cross-validation.

10.13. The **abrasion** data set has 30 observations, where the two predictor variables are hardness and tensile strength. The response variable is abrasion loss [Hand, et al., 1994; Davies and Goldsmith, 1972]. Construct a regression tree using cross-validation to pick a best tree.

10.14. The data in **helmets** contains measurements of head acceleration (in g) and times after impact (milliseconds) from a simulated motorcycle accident [Hand, et al., 1994; Silverman, 1985]. Do a loess smooth on these data. Include the upper and lower envelopes. Is it necessary to use the robust version?

10.15. Try the kernel methods for nonparametric regression on the **helmets** data.

10.16. Use regression trees on the **boston** data set. Choose the best tree using an independent test sample (taken from the original set) and cross-validation.

# Chapter 11

*Markov Chain Monte Carlo Methods*

## 11.1 Introduction

In many applications of statistical modeling, the data analyst would like to use a more complex model for a data set, but is forced to resort to an oversimplified model in order to use available techniques. Markov chain Monte Carlo (MCMC) methods are simulation-based and enable the statistician or engineer to examine data using realistic statistical models.

We start off with the following example taken from Raftery and Akman [1986] and Roberts [2000] that looks at the possibility that a change-point has occurred in a Poisson process. Raftery and Akman [1986] show that there is evidence for a change-point by determining Bayes factors for the change-point model versus other competing models. These data are a time series that indicate the number of coal mining disasters per year from 1851 to 1962. A plot of the data is shown in Figure 11.8, and it does appear that there has been a reduction in the rate of disasters during that time period. Some questions we might want to answer using the data are:

- What is the most likely year in which the change occurred?
- Did the rate of disasters increase or decrease after the change-point?

Example 11.8, presented later on, answers these questions using Bayesian data analysis and Gibbs sampling.

The main application of the MCMC methods that we present in this chapter is to generate a sample from a distribution. This sample can then be used to estimate various characteristics of the distribution such as moments, quantiles, modes, the density, or other statistics of interest.

In Section 11.2, we provide some background information to help the reader understand the concepts underlying MCMC. Because much of the recent developments and applications of MCMC arise in the area of Bayesian inference, we provide a brief introduction to this topic. This is followed by a discussion of Monte Carlo integration, since one of the applications of

MCMC methods is to obtain estimates of integrals. In Section 11.3, we present several Metropolis-Hastings algorithms, including the random-walk Metropolis sampler and the independence sampler. A widely used special case of the general Metropolis-Hastings method called the Gibbs sampler is covered in Section 11.4. An important consideration with MCMC is whether or not the chain has converged to the desired distribution. So, some convergence diagnostic techniques are discussed in Section 11.5. Sections 11.6 and 11.7 contain references to MATLAB code and references for the theoretical underpinnings of MCMC methods.

## 11.2 Background

### Bayesian Inference

Bayesians represent uncertainty about unknown parameter values by probability distributions and proceed as if parameters were random quantities [Gilks, et al., 1996a]. If we let $D$ represent the data that are observed and $\theta$ represent the model parameters, then to perform any inference, we must know the joint probability distribution $P(D, \theta)$ over all random quantities. Note that we allow $\theta$ to be multi-dimensional. From Chapter 2, we know that the joint distribution can be written as

$$P(D, \theta) = P(\theta)P(D|\theta),$$

where $P(\theta)$ is called the ***prior*** and $P(D|\theta)$ is called the ***likelihood***. Once we observe the data $D$, we can use Bayes' Theorem to get the ***posterior distribution*** as follows

$$P(\theta|D) = \frac{P(\theta)P(D|\theta)}{\int P(\theta)P(D|\theta)d\theta}. \qquad (11.1)$$

Equation 11.1 is the distribution of $\theta$ conditional on the observed data $D$. Since the denominator of Equation 11.1 is not a function of $\theta$ (since we are integrating over $\theta$), we can write the posterior as being proportional to the prior times the likelihood,

$$P(\theta|D) \propto P(\theta)P(D|\theta) = P(\theta)L(\theta;D).$$

We can see from Equation 11.1 that the posterior is a conditional distribution for the model parameters given the observed data. Understanding and using the posterior distribution is at the heart of Bayesian inference, where one is interested in making inferences using various features of the posterior distribution (e.g., moments, quantiles, etc.). These quantities can be written as posterior expectations of functions of the model parameters as follows

$$E[f(\theta)|D] = \frac{\int f(\theta)P(\theta)P(D|\theta)d\theta}{\int P(\theta)P(D|\theta)d\theta}. \tag{11.2}$$

Note that the denominator in Equations 11.1 and 11.2 is a constant of proportionality to make the posterior integrate to one. If the posterior is nonstandard, then this can be very difficult, if not impossible, to obtain. This is especially true when the problem is high dimensional, because there are a lot of parameters to integrate over. Analytically performing the integration in these expressions has been a source of difficulty in applications of Bayesian inference, and often simpler models would have to be used to make the analysis feasible. Monte Carlo integration using MCMC is one answer to this problem.

Because the same problem also arises in frequentist applications, we will change the notation to make it more general. We let $X$ represent a vector of $d$ random variables, with distribution denoted by $\pi(x)$. To a frequentist, $X$ would contain data, and $\pi(x)$ is called a likelihood. For a Bayesian, $X$ would be comprised of model parameters, and $\pi(x)$ would be called a posterior distribution. For both, the goal is to obtain the expectation

$$E[f(X)] = \frac{\int f(x)\pi(x)dx}{\int \pi(x)dx}. \tag{11.3}$$

As we will see, with MCMC methods we only have to know the distribution of $X$ up to the constant of normalization. This means that the denominator in Equation 11.3 can be unknown. It should be noted that in what follows we assume that $X$ can take on values in a $d$-dimensional Euclidean space. The methods can be applied to discrete random variables with appropriate changes.

## Monte Carlo Integration

As stated before, most methods in statistical inference that use simulation can be reduced to the problem of finding integrals. This is a fundamental part of the MCMC methodology, so we provide a short explanation of classical Monte Carlo integration. References that provide more detailed information on this subject are given in the last section of the chapter.

Monte Carlo integration estimates the integral $E[f(X)]$ of Equation 11.3 by obtaining samples $X_t$, $t = 1, ..., n$ from the distribution $\pi(\mathbf{x})$ and calculating

$$E[f(X)] \approx \frac{1}{n}\sum_{t=1}^{n}f(X_t). \qquad (11.4)$$

The notation $t$ is used here because there is an ordering or sequence to the random variables in MCMC methods. We know that when the $X_t$ are independent, then the approximation can be made as accurate as needed by increasing $n$. We will see in the following sections that with MCMC methods, the samples are not independent in most cases. That does not limit their use in finding integrals using approximations such as Equation 11.4. However, care must be taken when determining the variance of the estimate in Equation 11.4 because of dependence [Gentle, 1998; Robert and Casella, 1999]. We illustrate the method of Monte Carlo integration in the next example.

### Example 11.1

For a distribution that is exponential with $\lambda = 1$, we find $E[\sqrt{X}]$ using Equation 11.4. We generate random variables from the required distribution, take the square root of each one and then find the average of these values. This is implemented below in MATLAB.

```
% Generate 500 exponential random
% variables with lambda = 1.
% This is a Statistics Toolbox function.
x = exprnd(1,1,1000);
% Take square root of each one.
xroot = sqrt(x);
% Take the mean - Equation 11.4
exroothat = mean(xroot);
```

From this, we get an estimate of 0.889. We can use MATLAB to find the value using numerical integration.

```
% Now get it using numerical integration
strg = 'sqrt(x).*exp(-x)';
myfun = inline(strg);
% quadl is a MATLAB 6 function.
exroottru = quadl(myfun,0,50);
```

The value we get using numerical integration is 0.886, which closely matches what we got from the Monte Carlo method.

❑

The samples $X_t$ do not have to be independent as long as they are generated using a process that obtains samples from the *'entire'* domain of $\pi(\mathbf{x})$ and in the correct proportions [Gilks, et al., 1996a]. This can be done by constructing a Markov chain that has $\pi(\mathbf{x})$ as its stationary distribution. We now give a brief description of Markov chains.

## Markov Chains

A Markov chain is a sequence of random variables such that the next value or state of the sequence depends only on the previous one. Thus, we are generating a sequence of random variables, $X_0, X_1, \ldots$ such that the next state $X_{t+1}$ with $t \geq 0$ is distributed according to $P(X_{t+1}|X_t)$, which is called the *transition kernel*. A realization of this sequence is also called a Markov chain. We assume that the transition kernel does not depend on $t$, making the chain time-homogeneous.

One issue that must be addressed is how sensitive the chain is to the starting state $X_0$. Given certain conditions [Robert and Casella, 1999], the chain will forget its initial state and will converge to a stationary distribution, which is denoted by $\psi$. As the sequence grows larger, the sample points $X_t$ become dependent samples from $\psi$. The reader interested in knowing the conditions under which this happens and for associated proofs of convergence to the stationary distribution is urged to read the references given in Section 11.7.

Say the chain has been run for $m$ iterations, and we can assume that the sample points $X_t$, $t = m + 1, \ldots, n$ are distributed according to the stationary distribution $\psi$. We can discard the first $m$ iterations and use the remaining $n - m$ samples along with Equation 11.4 to get an estimate of the expectation as follows

$$E[f(X)] \approx \frac{1}{n-m} \sum_{t=m+1}^{n} f(X_t).\tag{11.5}$$

The number of samples $m$ that are discarded is called the *burn-in*. The size of the burn-in period is the subject of current research in MCMC methods. Diagnostic methods to help determine $m$ and $n$ are described in Section 11.5. Geyer [1992] suggests that the burn-in can be between 1% and 2% of $n$, where $n$ is large enough to obtain adequate precision in the estimate given by Equation 11.5.

So now we must answer the question: how large should $n$ be to get the required precision in the estimate? As stated previously, estimating the variance of the estimate given by Equation 11.5 is difficult because the samples are not independent. One way to determine $n$ via simulation is to run several Markov chains in parallel, each with a different starting value. The estimates from Equation 11.5 are compared, and if the variation between them is too great, then the length of the chains should be increased [Gilks, et al., 1996b]. Other methods are given in Roberts [1996], Raftery and Lewis [1996], and in the general references mentioned in Section 11.7.

### Analyzing the Output

We now discuss how the output from the Markov chains can be used in statistical analysis. An analyst might be interested in calculating means, standard deviations, correlations and marginal distributions for components of $X$. If we let $X_{t,j}$ represent the $j$-th component of $X_t$ at the $t$-th step in the chain, then using Equation 11.5, we can obtain the marginal means and variances from

$$\bar{X}_{.j} = \frac{1}{n-m} \sum_{t=m+1}^{n} X_{t,j},$$

and

$$S^2_{.j} = \frac{1}{n-m-1} \sum_{t=m+1}^{n} (X_{t,j} - \bar{X}_{.j})^2.$$

These estimates are simply the componentwise sample mean and sample variance of the sample points $X_t$, $t = m+1, ..., n$. Sample correlations are obtained similarly. Estimates of the marginal distributions can be obtained using the techniques of Chapter 8.

One last problem we must deal with to make Markov chains useful is the stationary distribution $\psi$. We need the ability to construct chains such that the stationary distribution of the chain is the one we are interested in: $\pi(\mathbf{x})$. In the MCMC literature, $\pi(\mathbf{x})$ is often referred to as the *target distribution*. It turns out that this is not difficult and is the subject of the next two sections.

## 11.3 Metropolis-Hastings Algorithms

The Metropolis-Hastings method is a generalization of the Metropolis technique of Metropolis, et al. [1953], which had been used for many years in the physics community. The paper by Hastings [1970] further generalized the technique in the context of statistics. The Metropolis sampler, the independence sampler and the random-walk are all special cases of the Metropolis-

Hastings method. Thus, we cover the general method first, followed by the special cases.

These methods share several properties, but one of the more useful properties is that they can be used in applications where $\pi(\mathbf{x})$ is known up to the constant of proportionality. Another property that makes them useful in a lot of applications is that the analyst does not have to know the conditional distributions, which is the case with the Gibbs sampler. While it can be shown that the Gibbs sampler is a special case of the Metropolis-Hastings algorithm [Robert and Casella, 1999], we include it in the next section because of this difference.

## Metropolis-Hastings Sampler

The Metropolis-Hastings sampler obtains the state of the chain at $t+1$ by sampling a *candidate point Y* from a *proposal distribution* $q(.|X_t)$. Note that this depends only on the previous state $X_t$ and can have any form, subject to regularity conditions [Roberts, 1996]. An example for $q(.|X_t)$ is the multivariate normal with mean $X_t$ and fixed covariance matrix. One thing to keep in mind when selecting $q(.|X_t)$ is that the proposal distribution should be easy to sample from.

The required regularity conditions for $q(.|X_t)$ are irreducibility and aperiodicity [Chib and Greenberg, 1995]. *Irreducibility* means that there is a positive probability that the Markov chain can reach any non-empty set from all starting points. *Aperiodicity* ensures that the chain will not oscillate between different sets of states. These conditions are usually satisfied if the proposal distribution has a positive density on the same support as the target distribution. They can also be satisfied when the target distribution has a restricted support. For example, one could use a uniform distribution around the current point in the chain.

The candidate point is accepted as the next state of the chain with probability given by

$$\alpha(X_t, Y) = \min\left\{1, \frac{\pi(Y)q(X_t|Y)}{\pi(X_t)q(Y|X_t)}\right\}. \tag{11.6}$$

If the point $Y$ is not accepted, then the chain does not move and $X_{t+1} = X_t$. The steps of the algorithm are outlined below. It is important to note that the distribution of interest $\pi(\mathbf{x})$ appears as a ratio, so the constant of proportionality cancels out. This is one of the appealing characteristics of the Metropolis-Hastings sampler, making it appropriate for a wide variety of applications.

*PROCEDURE - METROPOLIS-HASTINGS SAMPLER*

1. Initialize the chain to $X_0$ and set $t = 0$.
2. Generate a candidate point $Y$ from $q(.|X_t)$.
3. Generate $U$ from a uniform $(0, 1)$ distribution.
4. If $U \le \alpha(X_t, Y)$ (Equation 11.6) then set $X_{t+1} = Y$, else set $X_{t+1} = X_t$.
5. Set $t = t + 1$ and repeat steps 2 through 5.

The Metropolis-Hastings procedure is implemented in Example 11.2, where we use it to generate random variables from a standard Cauchy distribution. As we will see, this implementation is one of the special cases of the Metropolis-Hastings sampler described later.

## Example 11.2

We show how the Metropolis-Hastings sampler can be used to generate random variables from a standard Cauchy distribution given by

$$f(x) = \frac{1}{\pi(1 + x^2)}; \qquad -\infty < x < \infty.$$

From this, we see that

$$f(x) \propto \frac{1}{1 + x^2}.$$

We will use the normal as our proposal distribution, with a mean given by the previous value in the chain and a standard deviation given by $\sigma$. We start by setting up **inline** MATLAB functions to evaluate the densities for Equation 11.6.

```
% Set up an inline function to evaluate the Cauchy.
% Note that in both of the functions,
% the constants are canceled.
strg = '1./(1+x.^2)';
cauchy = inline(strg,'x');
% set up an inline function to evaluate the Normal pdf
strg = '1/sig*exp(-0.5*((x-mu)/sig).^2)';
norm = inline(strg,'x','mu','sig');
```

We now generate $n = 10000$ samples in the chain.

```
% Generate 10000 samples in the chain.
% Set up the constants.
n = 10000;
```

```
sig = 2;
x = zeros(1,n);
x(1) = randn(1);% generate the starting point
for i = 2:n
  % generate a candidate from the proposal distribution
  % which is the normal in this case. This will be a
  % normal with mean given by the previous value in the
  % chain and standard deviation of 'sig'
  y = x(i-1) + sig*randn(1);
  % generate a uniform for comparison
  u = rand(1);
  alpha = min([1, cauchy(y)*norm(x(i-1),y,sig)/...
          (cauchy(x(i-1))*norm(y,x(i-1),sig))]);
  if u <= alpha
    x(i) = y;
  else
    x(i) = x(i-1);
  end
end
```

We can plot a density histogram along with the curve corresponding to the true probability density function. We discard the first 500 points for the burn-in period. The plot is shown in Figure 11.1.
❑

## Metropolis Sampler

The Metropolis sampler refers to the original method of Metropolis, et al. [1953], where only symmetric distributions are considered for the proposal distribution. Thus, we have that

$$q(Y|X) = q(X|Y).$$

for all $X$ and $Y$. As before, a common example of a distribution like this is the normal distribution with mean $X$ and fixed covariance. Because the proposal distribution is symmetric, those terms cancel out in the acceptance probability yielding

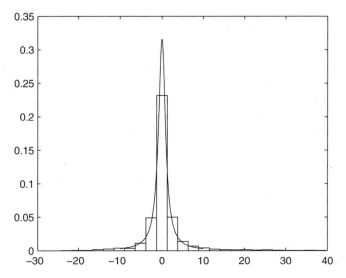

**FIGURE 11.1**
We generated 10,000 variates from the Cauchy distribution using the Metropolis-Hastings sampler. This shows a density histogram of the random variables after discarding the first 500 points. The curve corresponding to the true probability density function is superimposed over the histogram. We see that the random variables do follow the standard Cauchy distribution.

$$\alpha(X_t, Y) = \min\left\{1, \frac{\pi(Y)}{\pi(X_t)}\right\}. \tag{11.7}$$

*PROCEDURE - METROPOLIS SAMPLER*

1. Initialize the chain to $X_0$ and set $t = 0$.
2. Generate a candidate point $Y$ from $q(.|X_t)$.
3. Generate $U$ from a uniform $(0, 1)$ distribution.
4. If $U \le \alpha(X_t, Y)$ (Equation 11.7) then set $X_{t+1} = Y$, else set $X_{t+1} = X_t$.
5. Set $t = t + 1$ and repeat steps 2 through 5.

When the proposal distribution is such that $q(Y|X) = q(|X - Y|)$, then it is called the **random-walk Metropolis**. This amounts to generating a candidate point $Y = X_t + Z$, where $Z$ is an increment random variable from the distribution $q$.

We can gain some insight into how this algorithm works by looking at the conditions for accepting a candidate point as the next sample in the chain. In the symmetric case, the probability of moving is $\pi(Y)/\pi(X_t)$. If $\pi(Y) \geq \pi(X_t)$, then the chain moves to $Y$ because $\alpha(X_t, Y)$ will be equal to 1. This means that a move that climbs up the curve given by the target distribution is always accepted. A move that is worse (i.e., one that goes downhill) is accepted with probability given by $\pi(Y)/\pi(X_t)$. These concepts are illustrated in Figure 11.2. This is the basic algorithm proposed by Metropolis, et al. [1953], and it is the foundation for other optimization algorithms such as simulated annealing [Kirkpatrick, Gelatt, and Vechi, 1983; Aarts and Korst, 1989].

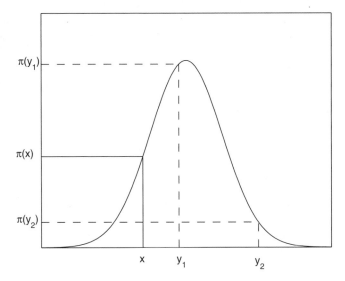

**FIGURE 11.2**
This shows what happens when a candidate point is selected and the proposal distribution is symmetric [Chib and Greenberg, 1995]. In this case, the probability of moving to another point is based on the ratio $\pi(y)/\pi(x)$. If $\pi(y) \geq \pi(x)$, then the chain moves to the candidate point $y$. If $\pi(y) < \pi(x)$, then the chain moves to $y$ with probability $\pi(y)/\pi(x)$. So we see that a move from $x$ to $y_1$ would be automatically accepted, but a move to $y_2$ would be accepted with probability $\pi(y_2)/\pi(x)$.

When implementing any of the Metropolis-Hastings algorithms, it is important to understand how the scale of the proposal distribution affects the efficiency of the algorithm. This is especially apparent with the random-walk version and is illustrated in the next example. If a proposal distribution takes small steps, then the acceptance probability given by Equation 11.7 will be high, yielding a higher rate at which we accept candidate points. The problem here is that the chain will mix slowly, meaning that the chain will take longer to get to the stationary distribution. On the other hand, if the proposal distribution generates large steps, then the chain could move to the tails, resulting in low acceptance probabilities. Again, the chain fails to mix quickly.

## Example 11.3

In this example, we show how to implement the random-walk version of the Metropolis-Hastings sampler [Gilks, et al., 1996a] and use it to generate variates from the standard normal distribution (the target distribution). Of course, we do not have to resort to MCMC methods to generate random variables from this target distribution, but it serves to illustrate the importance of picking the right scale for the proposal distribution. We use the normal as a proposal distribution to generate the candidates for the next value in the chain. The mean of the proposal distribution is given by the current value in the chain $x_t$. We generate three chains with different values for the standard deviation, given by: $\sigma = 0.5, 0.1, 10$. These provide chains that exhibit good mixing, poor mixing due to small step size and poor mixing due to a large step size, respectively. We show below how to generate the three sequences with $n = 500$ variates in each chain.

```
% Get the variances for the proposal distributions.
sig1 = 0.5;
sig2 = 0.1;
sig3 = 10;
% We will generate 500 iterations of the chain.
n = 500;
% Set up the vectors to store the samples.
X1 = zeros(1,n);
X2 = X1;
X3 = X1;
% Get the starting values for the chains.
X1(1) = -10;
X2(1) = 0;
X3(1) = 0;
```

Now that we have everything initialized, we can obtain the chains.

```
% Run the first chain.
for i = 2:n
  % Generate variate from proposal distribution.
  y = randn(1)*sig1 + X1(i-1);
  % Generate variate from uniform.
  u = rand(1);
  % Calculate alpha.
```

```
alpha = normpdf(y,0,1)/normpdf(X1(i-1),0,1);
if u <= alpha
  % Then set the chain to the y.
  X1(i) = y;
else
  X1(i) = X1(i-1);
end
end
% Run second chain.
for i = 2:n
  % Generate variate from proposal distribution.
  y = randn(1)*sig2 + X2(i-1);
  % Generate variate from uniform.
  u = rand(1);
  % Calculate alpha.
  alpha = normpdf(y,0,1)/normpdf(X2(i-1),0,1);
  if u <= alpha
    % Then set the chain to the y.
    X2(i) = y;
  else
    X2(i) = X2(i-1);
  end
end
% Run the third chain.
for i = 2:n
  % Generate variate from proposal distribution.
  y = randn(1)*sig3 + X3(i-1);
  % Generate variate from uniform.
  u = rand(1);
  % Calculate alpha.
  alpha = normpdf(y,0,1)/normpdf(X3(i-1),0,1);
  if u <= alpha
    % Then set the chain to the y.
    X3(i) = y;
  else
    X3(i) = X3(i-1);
  end
end
```

Plots of these sequences are illustrated in Figure 11.3, where we also show horizontal lines at $\pm 2$. These lines are provided as a way to determine if most values in the chain are mixing well (taking on many different values) within two standard deviations of zero, since we are generating standard normal variates. Note that the first one converges quite rapidly and exhibits good mixing in spite of an extreme starting point. The second one with $\sigma = 0.1$ (small steps) is mixing very slowly and does not seem to have converged to

the target distribution in these 500 steps of the chain. The third sequence, where large steps are taken, also seems to be mixing slowly, and it is easy to see that the chain sometimes does not move. This is due to the large steps taken by the proposal distribution.
❑

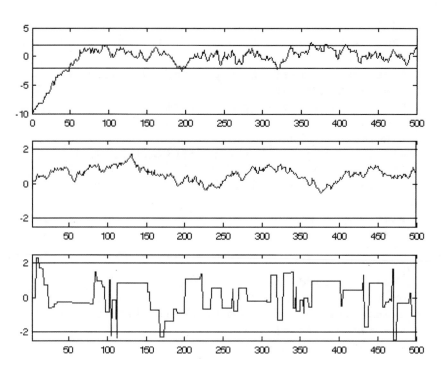

**FIGURE 11.3**
These are the three sequences from Example 11.3. The target distribution is the standard normal. For all three sequences, the proposal distribution is normal with the mean given by the previous element in the sequence. The standard deviations of the proposal distribution are: $\sigma = 0.5, 0.1, 10$. Note that the first sequence approaches the target distribution after the first 50 - 100 iterations. The other two sequences are slow to converge to the target distribution because of slow mixing due to the poor choice of $\sigma$.

### Independence Sampler

The independence sampler was proposed by Tierney [1994]. This method uses a proposal distribution that does not depend on $X$; i.e., it is generated independently of the previous value in the chain. The proposal distribution is of the form $q(Y|X) = q(Y)$, so Equation 11.6 becomes

$$\alpha(X_t, Y) = \min\left\{1, \frac{\pi(Y)q(X_t)}{\pi(X_t)q(Y)}\right\}. \qquad (11.8)$$

This is sometimes written in the literature as

$$\alpha(X_t, Y) = \min\left\{1, \frac{w(Y)}{w(X_t)}\right\}$$

where $w(X) = \pi(X)/q(X)$.

Caution should be used when implementing the independence sampler. In general, this method will not work well unless the proposal distribution $q$ is very similar to the target distribution $\pi$. Gilks, et al. [1996a] show that it is best if $q$ is heavier-tailed than $\pi$. Note also that the resulting sample is still not independent, even though we generate the candidate points independently of the previous value in the chain. This is because the acceptance probability for the next value $X_{t+1}$ depends on the previous one. For more information on the independence sampler and the recommended usage, see Roberts [1996] or Robert and Casella [1999].

### Autoregressive Generating Density

Another choice for a candidate generating density is proposed by Tierney [1994] and described by Chib and Greenberg [1995]. This is represented by an autoregressive process of order 1 and is obtained by generating candidates as follows

$$Y = a + \mathbf{B}(X_t - a) + Z, \qquad (11.9)$$

where $a$ is a vector and $\mathbf{B}$ is a matrix, both of which are conformable in terms of size with $X_t$. The vector $Z$ has a density given by $q$. If $\mathbf{B} = -\mathbf{I}$, then the chains are produced by reflecting about the point $a$, yielding negative correlation between successive values in the sequence. The autoregressive generating density is described in the next example.

### Example 11.4

We show how to use the Metropolis-Hastings sampler with the autoregressive generating density to generate random variables from a target distribution given by a bivariate normal with the following parameters:

$$\mu = \begin{bmatrix} 1 \\ 2 \end{bmatrix} \qquad \Sigma = \begin{bmatrix} 1 & 0.9 \\ 0.9 & 1 \end{bmatrix}.$$

Variates from this distribution can be easily generated using the techniques of Chapter 4, but it serves to illustrate the concepts. In the exercises, the reader is asked to generate a set of random variables using those techniques and compare them to the results obtained in this example. We generate a sequence of $n = 6000$ points and use a burn-in of 4000.

```
% Set up some constants and arrays to store things.
n = 6000;
xar = zeros(n,2); % to store samples
mu = [1;2];  % Parameters - target distribution.
covm = [1 0.9; 0.9 1];
```

We now set up a MATLAB `inline` function to evaluate the required probabilities.

```
% Set up the function to evaluate alpha
% for this problem. Note that the constant
% has been canceled.
strg = 'exp(-0.5*(x-mu)''*inv(covm)*(x-mu))';
norm = inline(strg,'x','mu','covm');
```

The following MATLAB code sets up a random starting point and obtains the elements of the chain.

```
% Generate starting point.
xar(1,:) = randn(1,2);
for i = 2:n
  % Get the next variate in the chain.
  % y is a column vector.
  y = mu - (xar(i-1,:)'-mu) + (-1+2*rand(2,1));
  u = rand(1);
  % Uses inline function 'norm' from above.
  alpha=min([1,norm(y,mu,covm)/...
          norm(xar(i-1,:)',mu,covm)]);
  if u <= alpha
    xar(i,:) = y';
  else
    xar(i,:) = xar(i-1,:);
  end
end
```

A scatterplot of the last 2000 variates is given in Figure 11.4, and it shows that they do follow the target distribution. To check this further, we can get the sample covariance matrix and the sample mean using these points. The result is

$$\hat{\mu} = \begin{bmatrix} 1.04 \\ 2.03 \end{bmatrix} \qquad \hat{\Sigma} = \begin{bmatrix} 1 & 0.899 \\ 0.899 & 1 \end{bmatrix},$$

from which we see that the sample does reflect the target distribution.
❑

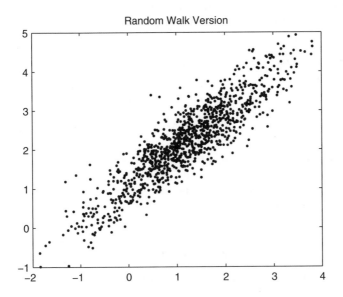

**FIGURE 11.4**
This is a scatterplot of the last 2000 elements of a chain generated using the autoregressive generating density of Example 11.4.

## Example 11.5

This example shows how the Metropolis-Hastings method can be used with an example in Bayesian inference [Roberts, 2000]. This is a genetic linkage example, looking at the genetic linkage of 197 animals. The animals are divided into four categories with frequencies given by

$$Z = (z_1, z_2, z_3, z_4) = (125, 18, 20, 34),$$

with corresponding cell probabilities of

$$\left( \frac{1}{2} + \frac{\theta}{4}, \frac{1}{4}(1-\theta), \frac{1}{4}(1-\theta), \frac{\theta}{4} \right).$$

From this, we get a posterior distribution of $\theta$, given the data $Z$, of

$$P(\theta|Z) = \pi(\theta) \propto (2 + \theta)^{z_1} (1 - \theta)^{z_2 + z_3} \theta^{z_4}.$$

We would like to use this to observe the behavior of the parameter $\theta$ (i.e., what are likely values for $\theta$) given the data. Note that any constants in the denominator in $\pi(\theta)$ have been eliminated because they cancel in the Metropolis-Hastings sampler. We use the random-walk version where the step is generated by the uniform distribution over the interval $(-a, a)$. Note that we set up a MATLAB **inline** function to get the probability of accepting the candidate point.

```
% Set up the preliminaries.
z1 = 125;
z2 = 18;
z3 = 20;
z4 = 34;
n = 1100;
% Step size for the proposal distribution.
a = 0.1;
% Set up the space to store values.
theta = zeros(1,n);
% Get an inline function to evaluate probability.
strg = '((2+th).^z1).*((1-th).^(z2+z3)).*(th.^z4)';
ptheta = inline(strg,'th','z1','z2','z3','z4');
```

We can now generate the chain as shown below.

```
% Use Metropolis-Hastings random-walk
% where y = theta(i-1) + z
% and z is uniform(-a,a).
% Get initial value for theta.
theta(1) = rand(1);
for i = 2:n
 % Generate from proposal distribution.
 y = theta(i-1) - a + 2*a*rand(1);
 % Generate from uniform.
 u = rand(1);
 alpha = min([ ptheta(y,z1,z2,z3,z4)/...
     ptheta(theta(i-1),z1,z2,z3,z4),1]);
 if u <= alpha
   theta(i) = y;
 else
   theta(i) = theta(i-1);
 end
end
```

We set the burn-in period to 100, so only the last 1000 elements are used to produce the density histogram estimate of the posterior density of $\theta$ given in Figure 11.5.

□

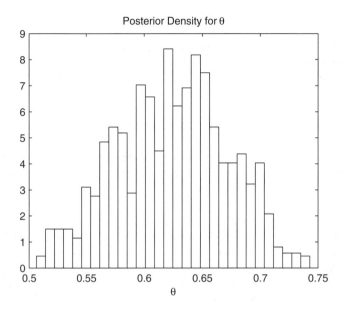

**FIGURE 11.5**
This shows the density histogram estimate of the posterior density of $\theta$ given the observed data.

## 11.4 The Gibbs Sampler

Although the Gibbs sampler can be shown to be a special case of the Metropolis-Hastings algorithm [Gilks, et al., 1996b; Robert and Casella, 1999], we include it in its own section, because it is different in some fundamental ways. The two main differences between the Gibbs sampler and Metropolis-Hastings are:

1) We always accept a candidate point.
2) We must know the full conditional distributions.

In general, the fact that we must know the full conditional distributions makes the algorithm less applicable.

The Gibbs sampler was originally developed by Geman and Geman [1984], where it was applied to image processing and the analysis of Gibbs distributions on a lattice. It was brought into mainstream statistics through the articles of Gelfand and Smith [1990] and Gelfand, et al. [1990].

In describing the Gibbs sampler, we follow the treatment in Casella and George [1992]. Let's assume that we have a joint density that is given by $f(x, y_1, ..., y_d)$, and we would like to understand more about the marginal density. For example, we might want to know the shape, the mean, the variance or some other characteristic of interest.

The marginal density is given by

$$f(x) = \int ... \int f(x, y_1, ..., y_d) dy_1 ... dy_d. \tag{11.10}$$

Equation 11.10 says that to get the marginal distribution, we must integrate over all of the other variables. In many applications, this integration is very difficult (and sometimes impossible) to perform. The Gibbs sampler is a way to get $f(x)$ by simulation. As with the other MCMC methods, we use the Gibbs sampler to generate a sample $X_1, ..., X_m$ from $f(x)$ and then use the sample to estimate the desired characteristic of $f(x)$. Casella and George [1992] note that if $m$ is large enough, then any population characteristic can be calculated with the required degree of accuracy.

To illustrate the Gibbs sampler, we start off by looking at the simpler case where the joint distribution is $f(x_1, x_2)$. Using the notation from the previous sections, $X_t$ is a two element vector with elements given by

$$X_t = (X_{t,1}, X_{t,2})$$

We start the chain with an initial starting point of $X_0 = (X_{0,1}, X_{0,2})$. We then generate a sample from $f(x_1, x_2)$ by sampling from the conditional distributions given by $f(x_1|x_2)$ and $f(x_2|x_1)$. At each iteration, the elements of the random vector are obtained one at a time by alternately generating values from the conditional distributions. We illustrate this in the procedure given below.

*PROCEDURE - GIBBS SAMPLER (BIVARIATE CASE)*

    1. Generate a starting point $X_0 = (X_{0,1}, X_{0,2})$. Set $t = 0$.

    2. Generate a point $X_{t,1}$ from

$$f(X_{t,1}|X_{t,2} = x_{t,2}).$$

    3. Generate a point $X_{t,2}$ from

$$f(X_{t,2}|X_{t+1,1} = x_{t+1,1}).$$

    4. Set $t = t + 1$ and repeat steps 2 through 4.

Note that the conditional distributions are conditioned on the current or most recent values of the other components of $X_t$. Example 11.6 shows how this is done in a simple case taken from Casella and George [1992].

## Example 11.6

To illustrate the Gibbs sampler, we consider the following joint distribution

$$f(x, y) \propto \binom{n}{x} y^{x+\alpha-1} (1-y)^{n-x+\beta-1},$$

where $x = 0, 1, \ldots, n$ and $0 \le y \le 1$. Let's say our goal is to estimate some characteristic of the marginal distribution $f(x)$ of $X$. By ignoring the overall dependence on $n$, $\alpha$ and $\beta$, we find that the conditional distribution $f(x|y)$ is binomial with parameters $n$ and $y$, and the conditional distribution $f(y|x)$ is a beta distribution with parameters $x + \alpha$ and $n - x + \beta$ [Casella and George, 1992]. The MATLAB commands given below use the Gibbs sampler to generate variates from the joint distribution.

```
% Set up preliminaries.
% Here we use k for the chain length, because n
% is used for the number of trials in a binomial.
k = 1000;    % generate a chain of size 1000
m = 500;     % burn-in will be 500
a = 2;    % chosen
b = 4;
x = zeros(1,k);
y = zeros(1,k);
n = 16;
```

We are now ready to generate the elements in the chain. We start off by generating a starting point.

```
% Pick a starting point.
x(1) = binornd(n,0.5,1,1);
y(1) = betarnd(x(1) + a, n - x(1) + b,1,1);
for i = 2:k
 x(i) = binornd(n,y(i-1),1,1);
 y(i) = betarnd(x(i)+a, n-x(i)+b, 1, 1);
end
```

Note that we do not have to worry about whether or not we will accept the next value in the chain. With Gibbs sampling every candidate is accepted. We can estimate the marginal using the following

$$\hat{f}(x) = \frac{1}{k-m} \sum_{i=m+1}^{k} f(x|y_i).$$

This says that we evaluate the probability conditional on the values of $y_i$ that were generated after the burn-in period. This is implemented in MATLAB as follows:

```
% Get the marginal by evaluating the conditional.
% Use MATLAB's Statistics Toolbox.
% Find the P(X=x|Y's)
fhat = zeros(1,17);
for i = 1:17
  fhat(i) = mean(binopdf(i-1,n,y(500:k)));
end
```

The true marginal probability mass function is [Casella and George, 1992]

$$f(x) = \binom{n}{x} \frac{\Gamma(\alpha + \beta)}{\Gamma(\alpha)\Gamma(\beta)} \frac{\Gamma(x + \alpha)\Gamma(n - x + \beta)}{\Gamma(\alpha + \beta + n)},$$

for $x = 0, 1, ..., n$. We plot the estimated probability mass function along with the true marginal in Figure 11.6. This shows that the estimate is very close to the true function.
❑

Casella and George [1992] and Gelfand and Smith [1990] recommend that $K$ different sequences be generated, each one with length $n$. Then the last element of each sequence is used to obtain a sample of size $K$ that is approximately independent for large enough $K$. We do note that there is some disagreement in the literature regarding the utility of running one really long chain to get better convergence to the target distribution or many shorter chains to get independent samples [Gilks, et al., 1996b]. Most researchers in this field observe that one long run would often be used for exploratory analysis and a few moderate size runs is preferred for inferences.

The procedure given below for the general Gibbs sampler is for one chain only. It is easier to understand the basic concepts by looking at one chain, and it is simple to expand the algorithm to multiple chains.

*PROCEDURE - GIBBS SAMPLER*

1. Generate a starting point $X_0 = (X_{0,1}, ..., X_{0,d})$. Set $t = 0$.
2. Generate a point $X_{t,1}$ from

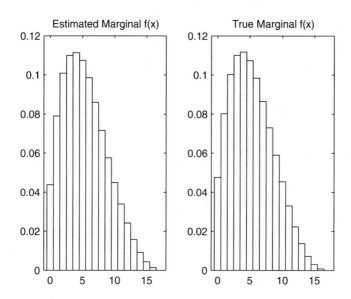

**FIGURE 11.6**
On the left, we have the estimated probability mass function for the marginal distribution
$f(x)$. The mass function on the right is from the true probability mass function. We see that
there is close agreement between the two.

$$f(X_{t,1} | X_{t,2} = x_{t,2}, ..., X_{t,d} = x_{t,d}).$$

Generate a point $X_{t,2}$ from

$$f(X_{t,2} | X_{t+1,1} = x_{t+1,1}, X_{t,3} = x_{t,3}, ..., X_{t,d} = x_{t,d}).$$

. . .

Generate a point $X_{t,d}$ from

$$f(X_{t,d} | X_{t+1,1} = x_{t+1,1}, ..., X_{t+1,d-1} = x_{t+1,d-1}).$$

3. Set $t = t + 1$ and repeat steps 2 through 3.

## Example 11.7
We show another example of Gibbs sampling as applied to bivariate normal
data. Say we have the same model as we had in Example 11.4, where we

wanted to generate samples from a bivariate normal with the following parameters

$$\mu = \begin{bmatrix} \mu_1 \\ \mu_2 \end{bmatrix} = \begin{bmatrix} 1 \\ 2 \end{bmatrix} \qquad \Sigma = \begin{bmatrix} 1 & \rho \\ \rho & 1 \end{bmatrix} = \begin{bmatrix} 1 & 0.9 \\ 0.9 & 1 \end{bmatrix}.$$

From Gelman, et al. [1995] we know that $f(x_1|x_2)$ is univariate normal with mean $\mu_1 + \rho(x_2 - \mu_2)$ and standard deviation $1 - \rho^2$. Similarly, $f(x_2|x_1)$ is univariate normal with mean $\mu_2 + \rho(x_1 - \mu_1)$ and standard deviation $1 - \rho^2$. With this information, we can implement the Gibbs sampler to generate the random variables.

```
% Set up constants and arrays.
n = 6000;
xgibbs = zeros(n,2);
rho = 0.9;
y = [1;2];% This is the mean.
sig = sqrt(1-rho^2);
% Initial point.
xgibbs(1,:) = [10 10];
% Start the chain.
for i = 2:n
    mu = y(1) + rho*(xgibbs(i-1,2)-y(2));
    xgibbs(i,1) = mu + sig*randn(1);
    mu = y(2) + rho*(xgibbs(i,1) - y(1));
    xgibbs(i,2) = mu + sig*randn(1);
end
```

Notice that the next element in the chain is generated based on the current values for $x_1$ and $x_2$. A scatterplot of the last 2000 variates generated with this method is shown in Figure 11.7.
□

We return now to our example described at the beginning of the chapter, where we are investigating the hypothesis that there has been a reduction in coal mining disasters over the years 1851 to 1962. To understand this further, we follow the model given in Roberts [2000]. This model assumes that the number of disasters per year follows a Poisson distribution with a mean rate of $\theta$ until the $k$-th year. After the $k$-th year, the number of disasters is distributed according to the Poisson distribution with a mean rate of $\lambda$. This is represented as

$$Y_i \sim \text{Poisson}(\theta) \qquad i = 1, \dots, k$$
$$Y_i \sim \text{Poisson}(\lambda) \qquad i = k+1, \dots, n,$$

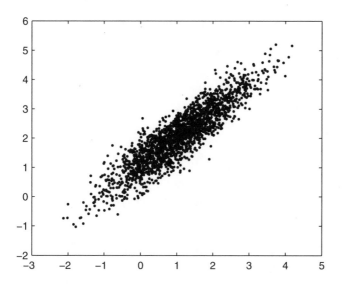

**FIGURE 11.7**
This is a scatterplot of the bivariate normal variates generated using Gibbs sampling. Note that the results are similar to Figure 11.4.

where the notation '~' means '*is distributed as.*'
  A Bayesian model is given by the following

$$\theta \sim \text{Gamma}(a_1, b_1)$$
$$\lambda \sim \text{Gamma}(a_2, b_2)$$
$$b_1 \sim \text{Gamma}(c_1, d_1)$$
$$b_2 \sim \text{Gamma}(c_2, d_2)$$

and the $k$ is discrete uniform over $\{1, ..., 112\}$ (since there are 112 years). Note that $\theta$, $\lambda$ and $k$ are all independent of each other.
  This model leads to the following conditional distributions:

$$\theta \,|\, Y, \lambda, b_1, b_2, k \sim \text{Gamma}\!\left(a_1 + \sum_{i=1}^{k} Y_i, k + b_1\right)$$

$$\lambda \,|\, Y, \theta, b_1, b_2, k \sim \text{Gamma}\!\left(a_2 + \sum_{i=k+1}^{n} Y_i, n - k + b_2\right)$$

$$b_1 | Y, \theta, \lambda, b_2, k \sim \text{Gamma}(a_1 + c_1, \theta + d_1)$$

$$b_2 | Y, \theta, \lambda, b_1, k \sim \text{Gamma}(a_2 + c_2, \lambda + d_2)$$

$$f(k | Y, \theta, \lambda, b_1, b_2) = \frac{L(Y; k, \theta, \lambda)}{\sum_{j=1}^{n} L(Y; j, \theta, \lambda)}$$

The likelihood is given by

$$L(Y; k, \theta, \lambda) = \exp\{k(\lambda - \theta)\}(\theta / \lambda)^{\sum_{i=1}^{k} Y_i} .$$

We use Gibbs sampling to simulate the required distributions and examine the results to explore the change-point model. For example, we could look at the posterior densities of $\theta$, $\lambda$ and $k$ to help us answer the questions posed at the beginning of the chapter.

## Example 11.8

A plot of the time series for the **coal** data is shown in Figure 11.8, where we see graphical evidence supporting the hypothesis that a change-point does occur [Raftery and Akman, 1986] and that there has been a reduction in the rate of coal mine disasters over this time period.

We set up the preliminary data needed to implement Gibbs sampling as follows:

```
% Set up preliminaries.
load coal
% y contains number of disasters.
% year contains the year.
n = length(y);
m = 1100;    % number in chain
% The values for the parameters are the same
% as in Roberts[2000].
a1 = 0.5;
a2 = 0.5;
c1 = 0;
c2 = 0;
d1 = 1;
d2 = 1;
theta = zeros(1,m);
lambda = zeros(1,m);
k = zeros(1,n);
```

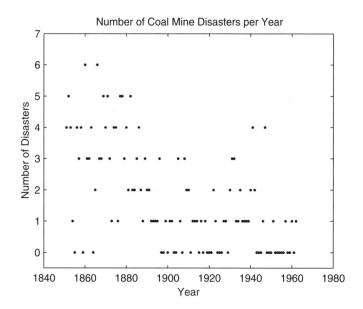

**FIGURE 11.8**

Time series of the **coal** data. It does appear that there was a reduction in the rate of disasters per year, after a certain year. Estimating that year is the focus of this example.

```
% Holds probabilities for k.
like = zeros(1,n);
```

We are now ready to implement the Gibbs sampling. We will run the chain for 1100 iterations and use a burn-in period of 100.

```
% Get starting points.
k(1) = unidrnd(n,1,1);
% Note that k will indicate an index to the year
% that corresponds to a hypothesized change-point.
theta(1) = 1;
lambda(1) = 1;
b1 = 1;
b2 = 1;
% Start the Gibbs Sampler.
for i = 2:m
    kk = k(i-1);
    % Get parameters for generating theta.
    t = a1 + sum(y(1:kk));
    lam = kk + b1;
    % Generate the variate for theta.
```

```
theta(i) = gamrnd(t,1/lam,1,1);
% Get parameters for generating lambda.
t = a2 + sum(y) - sum(y(1:kk));
lam = n-kk+b2;
% Generate the variate for lambda.
lambda(i) = gamrnd(t,1/lam,1,1);
% Generate the parameters b1 and b2.
b1 = gamrnd(a1+c1,1/(theta(i)+d1),1,1);
b2 = gamrnd(a2+c2,1/(lambda(i)+d2),1,1);
% Now get the probabilities for k.
for j = 1:n
    like(j) = exp((lambda(i)-theta(i))*j)*...
        (theta(i)/lambda(i))^sum(y(1:j)));
end
like = like/sum(like);
% Now sample the variate for k.
k(i) = cssample(1:n,like,1);
end
```

The sequences for $\theta$, $\lambda$ and $k$ are shown in Figure 11.9, where we can see that a burn-in period of 100 is reasonable. In Figure 11.10. we plot the frequencies for the estimated posterior distribution using the generated $k$ variates. We see evidence of a posterior mode at $k = 41$, which corresponds to the year 1891. So, we suspect that the change-point most likely occurred around 1891. We can also look at density histograms for the posterior densities for $\theta$ and $\lambda$. These are given in Figure 11.11, and they indicate that the mean rate of disasters did decrease after the change-point.
☐

## 11.5 Convergence Monitoring

The problem of deciding when to stop the chain is an important one and is the topic of current research in MCMC methods. After all, the main purpose of using MCMC is to get a sample from the target distribution and explore its characteristics. If the resulting sequence has not converged to the target distribution, then the estimates and inferences we get from it are suspect.

Most of the methods that have been proposed in the literature are really diagnostic in nature and have the goal of monitoring convergence. Some are appropriate only for Metropolis-Hastings algorithms and some can be applied only to Gibbs samplers. We will discuss in detail one method due to Gelman and Rubin [1992] and Gelman [1996], because it is one of the simplest to understand and to implement. Additionally, it can be used in any of the MCMC algorithms. We also very briefly describe another widely used

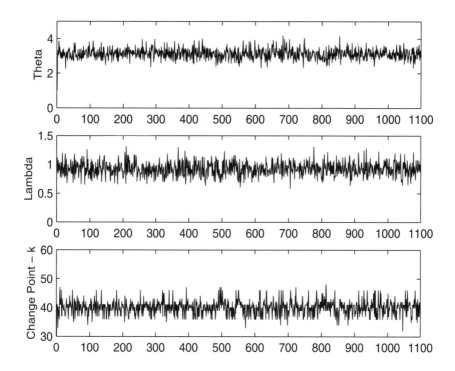

**FIGURE 11.9**
This shows the sequences that were generated using the Gibbs sampler.

method due to Raftery and Lewis [1992, 1996] that can be employed within the MCMC method. Other papers that review and compare the various convergence diagnostics are Cowles and Carlin [1996], Robert [1995] and Brooks [1998]. Some recent research in this area can be found in Canty [1999] and Brooks and Giudici [2000].

## Gelman and Rubin Method

We will use $v$ to represent the characteristic of the target distribution (mean, moments, quantiles, etc.) in which we are interested. One obvious way to monitor convergence to the target distribution is to run multiple sequences of the chain and plot $v$ versus the iteration number. If they do not converge to approximately the same value, then there is a problem. Gelman [1996] points out that lack of convergence can be detected by comparing multiple sequences, but cannot be detected by looking at a single sequence.

The Gelman-Rubin convergence diagnostic is based on running multiple chains. Cowles and Carlin [1996] recommend ten or more chains if the target

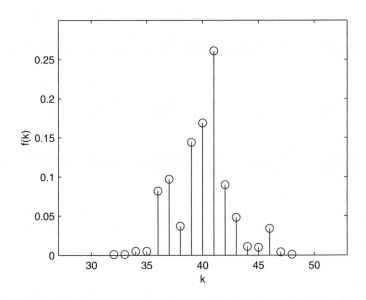

**FIGURE 11.10**
This is the frequency histogram for the random variables *k* generated by the Gibbs sampler of Example 11.8. Note the mode at *k* = 41 corresponding to the year 1891.

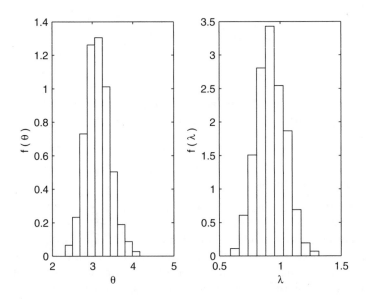

**FIGURE 11.11**
This figure shows density histograms for the posterior distributions for θ and λ, and there seems to be evidence showing that there was a reduction in the mean rate of disasters per year.

distribution is unimodal. The starting points for these chains are chosen to be widely dispersed in the target distribution. This is important for two reasons. First, it will increase the likelihood that most regions of the target distribution are visited in the simulation. Additionally, any convergence problems are more likely to appear with over-dispersed starting points.

The method is based on the idea that the variance within a single chain will be less than the variance in the combined sequences, if convergence has not taken place. The Gelman-Rubin approach monitors the scalar quantities of interest in the analysis (i.e., v ).

We start off with $k$ parallel sequences of length $n$ starting from over-dispersed points in the target distribution. The between-sequence variance $B$ and the within-sequence $W$ are calculated for each scalar summary v . We denote the $j$-th scalar summary in the $i$-th chain by

$$v_{ij}; \qquad i = 1, ..., k, \qquad j = 1, ..., n .$$

Thus, the subscript $j$ represents the position in the chain or sequence and $i$ denotes which sequence it was calculated from.

The between-sequence variance is given as

$$B = \frac{n}{k-1} \sum_{i=1}^{k} (\bar{v}_{i.} - \bar{v}_{..})^2 , \tag{11.11}$$

where

$$\bar{v}_{i.} = \frac{1}{n} \sum_{j=1}^{n} v_{ij} , \tag{11.12}$$

and

$$\bar{v}_{..} = \frac{1}{k} \sum_{i=1}^{k} \bar{v}_{i.} . \tag{11.13}$$

Equation 11.12 is the mean of the $n$ values of the scalar summary in the $i$-th sequence, and Equation 11.13 is the average across sequences.

The within-sequence variance is determined by

$$W = \frac{1}{k} \sum_{i=1}^{k} s_i^2 , \tag{11.14}$$

with

$$s_i^2 = \frac{1}{n-1} \sum_{j=1}^{n} (v_{ij} - \bar{v}_{i.})^2.$$  (11.15)

Note that Equation 11.15 is the sample variance of the scalar summary for the *i*-th sequence, and Equation 11.14 is the average variance for the *k* sequences.

Finally, *W* and *B* are combined to get an overall estimate of the variance of v in the target distribution:

$$\hat{var}(v) = \frac{n-1}{n} W + \frac{1}{n} B.$$  (11.16)

Equation 11.16 is a conservative estimate of the variance of v, if the starting points are over-dispersed [Gelman, 1996]. In other words, it tends to over estimate the variance.

Alternatively, the within-sequence variance given by *W* is an underestimate of the variance of v. This should make sense considering the fact that finite sequences have not had a chance to travel all of the target distribution resulting in less variability for v. As *n* gets large, both $\hat{var}(v)$ and *W* approach the true variance of v, one from above and one from below.

The Gelman-Rubin approach diagnoses convergence by calculating

$$\sqrt{\hat{R}} = \sqrt{\frac{\hat{var}(v)}{W}}.$$  (11.17)

This is the ratio between the upper bound on the standard deviation of v and the lower bound. It estimates the factor by which $\hat{var}(v)$ might be reduced by further iterations. The factor given by Equation 11.17 is called the ***estimated potential scale reduction***. If the potential scale reduction is high, then the analyst is advised to run the chains for more iterations. Gelman [1996] recommends that the sequences be run until $\hat{R}$ for all scalar summaries are less than 1.1 or 1.2.

## Example 11.9

We return to Example 11.3 to illustrate the Gelman-Rubin method for monitoring convergence. Recall that our target distribution is the univariate standard normal. This time our proposal distribution is univariate normal with $\mu = X_t$ and $\sigma = 5$. Our scalar summary v is the mean of the elements of the chain. We implement the Gelman-Rubin method using four chains.

```
% Set up preliminaries.
sig = 5;
```

```
% We will generate 500 iterations of the chain.
n = 5000;
numchain = 4;
% Set up the vectors to store the samples.
% This is 4 chains, 5000 samples.
X = zeros(numchain,n);
% This is 4 sequences (rows) of summaries.
nu = zeros(numchain,n);
% Track the rhat for each iteration:
rhat = zeros(1,n);
% Get the starting values for the chain.
% Use over-dispersed starting points.
X(1,1) = -10;
X(2,1) = 10;
X(3,1) = -5;
X(4,1) = 5;
```

The following implements the chains. Note that each column of our matrices X and nu is one iteration of the chains, and each row contains one of the chains. The X matrix keeps the chains, and the matrix nu is the sequence of scalar summaries for each chain.

```
% Run the chain.
for j = 2:n
    for i = 1:numchain
        % Generate variate from proposal distribution.
        y = randn(1)*sig + X(i,j-1);
        % Generate variate from uniform.
        u = rand(1);
        % Calculate alpha.
        alpha = normpdf(y,0,1)/normpdf(X(i,j-1),0,1);
        if u <= alpha
            % Then set the chain to the y.
            X(i,j) = y;
        else
            X(i,j) = X(i,j-1);
        end
    end
    % Get the scalar summary - means of each row.
    nu(:,j) = mean(X(:,1:j)')';
    rhat(j) = csgelrub(nu(:,1:j));
end
```

The function **csgelrub** will return the estimated $\hat{R}$ for a given set of sequences of scalar summaries. We plot the four sequences for the summary statistics of the chains in Figure 11.12. From these plots, we see that it might be reasonable to assume that the sequences have converged, since they are getting close to the same value in each plot. In Figure 11.13, we show a plot of $\hat{R}$ for each iteration of the sequence. This seems to confirm that the chains are getting close to convergence. Our final value of $\hat{R}$ at the last iteration of the chain is 1.05.

❑

One of the advantages of the Gelman-Rubin method is that the sequential output of the chains does not have to be examined by the analyst. This can be difficult, especially when there are a lot of summary quantities that must be monitored. The Gelman-Rubin method is based on means and variances, so it is especially useful for statistics that approximately follow the normal distribution. Gelman, et al. [1995] recommend that in other cases, extreme quantiles of the between and within sequences should be monitored.

### Raftery and Lewis Method

We briefly describe this method for two reasons. First, it is widely used in applications. Secondly, it is available in MATLAB code through the Econometrics Toolbox (see Section 11.6 for more information) and in Fortran from StatLib. So, the researcher who needs another method besides the one of Gelman and Rubin is encouraged to download these and try them. The article by Raftery and Lewis [1996] is another excellent resource for information on the theoretical basis for the method and for advice on how to use it in practice.

This technique is used to detect convergence of the chain to the target distribution and also provides a way to bound the variance of the estimates obtained from the samples. To use this method, the analyst first runs one chain of the Gibbs sampler for $N_{\text{min}}$. This is the minimum number of iterations needed for the required precision, given that the samples are independent. Using this chain and other quantities as inputs (the quantile to be estimated, the desired accuracy, the probability of getting that accuracy, and a convergence tolerance), the Raftery-Lewis method yields several useful values. Among them are the total number of iterations needed to get the desired level of accuracy and the number of points in the chain that should be discarded (i.e., the burn-in).

---

### 11.6 MATLAB Code

The Statistics Toolbox for MATLAB does not provide functions that implement MCMC methods, but the pieces (i.e., evaluating probability density functions and generating random variables) are there for the analyst to easily code up the required simulations. Also, the examples given in this text can be adapted to fit most applications by simply changing the proposal and target

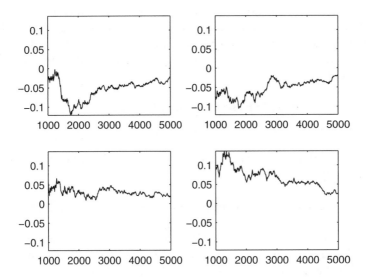

**FIGURE 11.12**
Here are the sequences of summary statistics in Example 11.9. We are tracking the mean of sequences of variables generated by the Metropolis-Hastings sampler. The target distribution is a univariate standard normal. It appears that the sequences are close to converging, since they are all approaching the same value.

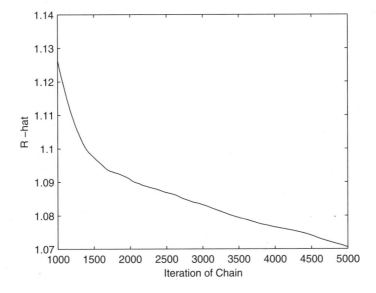

**FIGURE 11.13**
This sequence of values for $\hat{R}$ indicates that it is very close to one, showing near convergence.

distributions. There is an Econometrics Toolbox that contains M-files for the Gibbs sampler and the Raftery-Lewis convergence diagnostic. The software can be freely downloaded at **www.spatial-econometrics.com**. Extensive documentation for the procedures in the Econometrics Toolbox is also available at the website. The Raftery-Lewis method for S-plus and Fortran can be downloaded at:

- S-plus: **http://lib.stat.cmu /S/gibbsit**
- Fortran: **http://lib.stat.cmu/general/gibbsit**.

There are several user-contributed M-files for MCMC available for download at The MathWorks website:

**ftp.mathworks.com/pub/contrib/v5/stats/mcmc/**

For those who do not use MATLAB, another resource for software that will do Gibbs sampling and Bayesian analysis is the BUGS (Bayesian Inference Using Gibbs Sampling) software. The software and manuals can be downloaded at **www.mrc-bsu.cam.ac.uk/bugs/welcome.shtml**.

In the Computational Statistics Toolbox, we provide an M-file function called **csgelrub** that implements the Gelman-Rubin diagnostic. It returns $\hat{R}$ for given sequences of scalar summaries. We also include a function that implements a demo of the Metropolis-Hastings sampler where the target distribution is standard bivariate normal. This runs four chains, and the points are plotted as they are generated so the user can see what happens as the chain grows. The M-file functions pertaining to MCMC that we provide are summarized in Table 11.1.

TABLE 11.1

List of Functions from Chapter 11 Included in the Computational Statistics Toolbox

| Purpose | MATLAB Function |
|---|---|
| Gelman-Rubin convergence diagnostic given sequences of scalar summaries | csgelrub |
| Graphical demonstration of what happens in the Metropolis-Hastings sampler | csmcmcdemo |

## 11.7 Further Reading

For an excellent introduction to Markov chain Monte Carlo methods, we recommend the book *Markov Chain Monte Carlo in Practice* [Gilks, et al., 1996b]. This contains a series of articles written by leading researchers in the area and describes most aspects of MCMC from the theoretical to the practical. For a complete theoretical treatment of MCMC methods and many examples, the reader is referred to Robert and Casella [1999]. This book also contains a description of many of the hybrid MCMC methods that have been developed. The text by Tanner [1996] provides an introduction to computational algorithms for Bayesian and likelihood inference.

Most recent books on random number generation discuss the Metropolis-Hastings sampler and the Gibbs sampler. Gentle [1998] has a good discussion of MCMC methods and includes some examples in MATLAB. Ross [1997] has a chapter on MCMC and also discusses the connection between Metropolis-Hastings and simulated annealing. Ross [2000] also covers the topic of MCMC.

The monograph by Lindley [1995] gives an introduction and review of Bayesian statistics. For an overview of general Markov chain theory, see Tierney [1996], Meyn and Tweedie [1993] or Norris [1997]. If the reader would like more information on Bayesian data analysis, then the book *Bayesian Data Analysis* [Gelman, et al., 1995] is a good place to start. This text also contains some information and examples about the MCMC methods discussed in this chapter. Most of these books also include information on Monte Carlo integration methods, including importance sampling and variance reduction.

Besides simulated annealing, a connection between MCMC methods and the finite mixtures EM algorithm has been discussed in the literature. For more information on this, see Robert and Casella [1999]. There is also another method that, while not strictly an MCMC method, seems to be grouped with them. This is called Sampling Importance Resampling [Rubin, 1987, 1988]. A good introduction to this can be found in Ross [1997], Gentle [1998] and Albert [1993].

**Exercises**

11.1. The von Mises distribution is given by

$$f(x) = \frac{1}{2\pi I_0(b)} e^{b\cos(x)} \qquad -\pi \leq x \leq \pi,$$

where $I_0$ is the modified Bessel function of the first kind and order zero. Letting $b = 3$ and a starting point of 1, use the Metropolis random-walk algorithm to generate 1000 random iterations of the chain. Use the uniform distribution over the interval $(-1, 1)$ to generate steps in the walk. Plot the output from the chain versus the iteration number. Does it look like you need to discard the initial values in the chain for this example? Plot a histogram of the sample [Gentle, 1998].

11.2. Use the Metropolis-Hastings algorithm to generate samples from the beta distribution. Try using the uniform distribution as a candidate distribution. Note that you can simplify by canceling constants.

11.3. Use the Metropolis-Hastings algorithm to generate samples from the gamma distribution. What is a possible candidate distribution? Simplify the ratio by canceling constants.

11.4. Repeat Example 11.3 to generate a sample of standard normal random variables using different starting values and burn-in periods.

11.5. Let's say that $X_{,1}$ and $X_{,2}$ have conditional distributions that are exponential over the interval $(0, B)$, where $B$ is a known positive constant. Thus,

$$f(x_{,1} | x_{,2}) \propto x_{,2} e^{-x_{,2} 2^{x_{,1}}} \qquad 0 < x_{,1} < B < \infty$$

$$f(x_{,2} | x_{,1}) \propto x_{,1} e^{-x_{,1} 1^{x_{,2}}} \qquad 0 < x_{,2} < B < \infty$$

Use Gibbs sampling to generate samples from the marginal distribution $f(x_{,1})$. Choose your own starting values and burn-in period. Estimate the marginal distribution. What is the estimated mean, variance, and skewness coefficient for $f(x_{,1})$? Plot a histogram of the samples obtained after the burn-in period and the sequential output. Start multiple chains from over-dispersed starting points and use the Gelman-Rubin convergence diagnostics for the mean, variance and skewness coefficient [Casella and George, 1992].

11.6. Explore the use of the Metroplis-Hastings algorithm in higher dimensions. Generate 1000 samples for a trivariate normal distribution centered at the origin and covariance equal to the identity matrix. Thus, each coordinate direction should be a univariate standard normal distribution. Use a trivariate normal distribution with covariance matrix $\Sigma = 9 \cdot I$, (i.e., 9's are along the diagonal and 0's everywhere else) and mean given by the current value of the chain $x_t$. Use $x_{0,i} = 10$, $i = 1, ..., 3$ as the starting point of the chain. Plot the sequential output for each coordinate. Construct a histogram for the first coordinate direction. Does it look like a standard normal? What value did you use for the burn-in period? [Gentle, 1998.]

11.7. A joint density is given by

$$f(x_{,1}, x_{,2}, x_{,3}) = C\exp\{-(x_{,1} + x_{,2} + x_{,3} + x_{,1}x_{,2} + x_{,1}x_{,3} + x_{,2}x_{,3})\},$$

where $x_{,i} > 0$. Use one of the techniques from this chapter to simulate samples from this distribution and use them to estimate $E[X_{,1}X_{,2}X_{,3}]$. Start multiple chains and track the estimate to monitor the convergence [Ross, 1997].

11.8. Use Gibbs sampling to generate samples that have the following density

$$f(x_{,1}, x_{,2}, x_{,3}) = kx_{,1}^4 x_{,2}^3 x_{,3}^2 (1 - x_{,1} - x_{,2} - x_{,3})$$

where $x_{,i} > 0$ and $x_{,1} + x_{,2} + x_{,3} < 1$. Let $B(a, b)$ represent a beta distribution with parameters $a$ and $b$. We can write the conditional distributions as

$$X_{,1}|X_{,2}, X_{,3} \sim (1 - X_{,2} - X_{,3})Q \qquad Q \sim B(5, 2)$$
$$X_{,2}|X_{,1}, X_{,3} \sim (1 - X_{,1} - X_{,3})R \qquad R \sim B(4, 2)$$
$$X_{,3}|X_{,1}, X_{,2} \sim (1 - X_{,1} - X_{,2})S \qquad S \sim B(3, 2)$$

where the notation $Q \sim B(a, b)$ means $Q$ is from a beta distribution. Plot the sequential output for each $x_{,i}$ [Arnold, 1993].

11.9. Let's say that we have random samples $Z_1, ..., Z_n$ that are independent and identically distributed from the normal distribution with mean $\theta$ and variance 1. In the notation of Equation 11.1, these constitute the set of observations $D$. We also have a prior distribution on $\theta$ such that

$$P(\theta) \propto \frac{1}{1 + \theta^2},$$

We can write the posterior as follows

$$P(\theta|D) \propto P(\theta)L(\theta;D) = \frac{1}{1 + \theta^2} \times \exp\left\{\frac{-n(\theta - \bar{z})^2}{2}\right\}.$$

Let the true mean be $\theta = 0.06$ and generate a random sample of size $n = 20$ from the normal distribution to obtain the $z_i$. Use Metropolis-Hastings to generate random samples from the posterior distribution and use them to estimate the mean and the variance of the posterior distribution. Start multiple chains and use the Gelman-Rubin diagnostic method to determine when to stop the chains.

11.10. Generate a set of $n = 2000$ random variables for the bivariate distribution given in Example 11.4 using the technique from Chapter 4. Create a scatterplot of these data and compare to the set generated in Example 11.4.

11.11. For the bivariate distribution of Example 11.4, use a random-walk generating density $(Y = X_t + Z)$ where the increment random variable $Z$ is distributed as bivariate uniform. Generate a sequence of 6000 elements and construct a scatterplot of the last 2000 values. Compare to the results of Example 11.4.

11.12. For the bivariate distribution of Example 11.4, use a random-walk generating density $(Y = X_t + Z)$ where the increment random variables $Z$ are bivariate normal with mean zero and covariance

$$\Sigma = \begin{bmatrix} 0.6 & 0 \\ 0 & 0.4 \end{bmatrix}.$$

Generate a sequence of 6000 elements and construct a scatterplot of the last 2000 values. Compare to the results of Example 11.4.

11.13. Use the Metropolis-Hastings sampler to generate random samples from the lognormal distribution

$$f(x) = \frac{1}{x\sqrt{2\pi}} \exp\left\{-\frac{(\ln x)^2}{2}\right\}$$

$$f(x) \propto \frac{1}{x} \exp\left\{-\frac{(\ln x)^2}{2}\right\}.$$

Use the independence sampler and the gamma as a proposal distribution, being careful about the tails. Plot the sample using the density histogram and superimpose the true probability density function to ensure that your random variables are from the desired distribution.

# Chapter 12

## Spatial Statistics

## 12.1 Introduction

We include this final chapter to illustrate an area of data analysis where the methods of computational statistics can be applied. We do not cover this topic in great detail, but we do present some of the areas in spatial statistics that utilize the techniques discussed in the book. These methods include exploratory data analysis and visualization (see Chapter 5), kernel density estimation (see Chapter 8), and Monte Carlo simulation (see Chapter 6).

### What Is Spatial Statistics?

*Spatial statistics* is concerned with statistical methods that explicitly consider the spatial arrangement of the data. Most statisticians and engineers are familiar with time-series data, where the observations are measured at discrete time intervals. We know there is the possibility that the observations that come later in the series are dependent on earlier values. When analyzing such data, we might be interested in investigating the temporal data process that generated the data. This can be thought of as an unobservable curve (that we would like to estimate) that is generated in relation to its own previous values.

Similarly, we can view spatial data as measurements that are observed at discrete locations in a two-dimensional region. As with time series data, the observations might be spatially correlated (in two dimensions), which should be accounted for in the analysis.

Bailey and Gatrell [1995] sum up the definition and purpose of spatial statistics in this way:

> *observational data are available on some process operating in space and methods are sought to describe or explain the behaviour of this process and its possible relationship to other spatial phenomena. The object of the analysis is to increase our basic understanding of the process, assess the evidence in favour of various hypotheses concerning it, or possibly to predict values*

*in areas where observations have not been made. The data with which we
are concerned constitute a sample of observations on the process from which
we attempt to infer its overall behaviour. [Bailey and Gatrell, 1995, p. 7]*

## Types of Spatial Data

Typically, methods in spatial statistics fall into one of three categories that are
based on the type of spatial data that is being analyzed. These types of data
are called: point patterns, geostatistical data, and lattice data. The locations of
the observations might be referenced as points or as areal units. For example,
point locations might be designated by latitude and longitude or by their $x$
and $y$ coordinates. Areal locations could be census tracts, counties, states, etc.

*Spatial point patterns* are data made up of the location of point events. We
are interested in whether or not their relative locations represent a significant
pattern. For example, we might look for patterns such as clustering or regularity. While in some point-pattern data we might have an attribute attached
to an event, we are mainly interested in the locations of the events. Some
examples where spatial statistics methods can be applied to point patterns
are given below.

- We have a data set representing the location of volcanic craters in
  Uganda. It shows a trend in a north-easterly direction, possibly
  representing a major fault. We want to explore and model the
  distribution of the craters using methods for analyzing spatial point
  patterns.

- In another situation, we have two data sets showing thefts in the
  Oklahoma City area in the 1970's. One data set corresponds to those
  committed by Caucasian offenders, and one data set contains information on offences by African-Americans. An analyst might be
  interested in whether there is a difference in the pattern of offences
  committed by each group of offenders.

- Seismologists have data showing the distribution of earthquakes
  in a region. They would like to know if there is any pattern that
  might help them make predictions about future earthquakes.

- Epidemiologists collect data on where diseases occur. They would
  like to determine any patterns that might indicate how the disease
  is passed to other individuals.

With *geostatistical data* (or spatially continuous data), we have a measurement attached to the location of the observed event. The locations can
vary continuously throughout the spatial region, although in practice, measurements (or attributes) are taken at only a finite number of locations. We are
not necessarily interested in the locations themselves. Instead, we want to
understand and model the patterns in the attributes, with the goal of using

the model to predict values of the variable at locations where measurements were not taken. Some examples of geostatistical data analysis include the following:

- Rainfall is recorded at various points in a region. These data could be used to model the rainfall over the entire region.
- Geologists take ore samples at locations in a region. They would like to use these data to estimate the extent of the mineral deposit over the entire region.
- Environmentalists measure the level of a pollutant at locations in a region with the goal of using these data to model and estimate the level of pollutant at other locations in the region.

The third type of spatial data is called **lattice data**. These data are often associated with areas that can be regularly or irregularly spaced. The objective of the analysis of lattice data is to model the spatial pattern in the attributes associated with the fixed areas. Some examples of lattice data are:

- A sociologist has data that comprises socio-economic measures for regions in China. The goal of the analysis might be to describe and to understand any patterns of inequality between the areas.
- Market analysts use socio-economic data from the census to target a promising new area to market their products.
- A political party uses data representing the geographical voting patterns in a previous election to determine a campaign schedule for their candidate.

## Spatial Point Patterns

In this text, we look at techniques for analyzing spatial point patterns only. A spatial point pattern is a set of point locations $s_1, \ldots, s_n$ in a study region $R$. Each point location $s_i$ is a vector containing the coordinates of the $i$-th event,

$$s_i = \begin{bmatrix} s_{i1} \\ s_{i2} \end{bmatrix}.$$

The term *event* can refer to any spatial phenomenon that occurs at a point location. For example, events can be locations of trees growing in a forest, positions of cells in tissue or the incidence of disease at locations in a community. Note that the scale of our study affects the reasonableness of the assumption that the events occur at point locations.

In our analysis of spatial point patterns, we might have to refer to other locations in the study region $R$, where the phenomenon was not observed.

We need a way to distinguish them from the locations where observations were taken, so we refer to these other locations as *points* in the region.

At the simplest level, the data we are analyzing consist only of the coordinate locations of the events. As mentioned before, they could also have an attribute or variable associated with them. For example, this attribute might be the date of onset of the disease, the species of tree that is growing, or the type of crime.

This type of spatial data is sometimes referred to as a *marked point pattern*. In our treatment of spatial point patterns, we assume that the data represent a *mapped point pattern*. This is one where all relevant events in the study region $R$ have been measured. The study region $R$ can be any shape. However, edge effects can be a problem with many methods in spatial statistics. We describe the ramifications of edge effects as they arise with the various techniques. In some cases, edge effects are handled by leaving a specified guard area around the edge of the study region, but still within $R$. The analysis of point patterns is sensitive to the definition of $R$, so one might want to perform the analysis for different guard areas and/or different study regions.

One way we can think of spatial point patterns is in terms of the number of events occurring in an arbitrary sub-region of $R$. We denote the number of events in a sub-region $A$ as $Y(A)$. The spatial process is then represented by the random variables $Y(A)$, $A \subset R$. Since we have a random process, we can look at the behavior in terms of the first-order and second-order properties. These are related to the expected value (i.e., the mean) and the covariance [Bailey and Gatrell, 1995]. The mean and the covariance of $Y(A)$ depend on the number of events in arbitrary sub-regions $A$, and they depend on the size of the areas and the study region $R$. Thus, it is more useful to look at the first- and second-order properties in terms of the limiting behavior per unit area.

The first-order property is described by the intensity $\lambda(s)$. The *intensity* is defined as the mean number of events per unit area at the point $s$. Mathematically, the intensity is given by

$$\lambda(s) = \lim_{ds \to 0} \left\{ \frac{E[Y(ds)]}{ds} \right\}, \tag{12.1}$$

where $ds$ is a small region around the point $s$, and $ds$ is its area. If it is a *stationary point process*, then Equation 12.1 is a constant over the study region. We can then write the intensity as

$$E[Y(A)] = \lambda A, \tag{12.2}$$

where $A$ is the area of the sub-region, and $\lambda$ is the value of the intensity.

To understand the second-order properties of a spatial point process, we need to look at the number of events in pairs of sub-regions of $R$. The second-order property reflects the spatial dependence in the process. We describe this using the second-order intensity $\gamma(s_i, s_j)$. As with the intensity, this is defined using the events per unit area, as follows,

$$\gamma(s_i, s_j) = \lim_{ds_i, ds_j \to 0} \left\{ \frac{E[Y(ds_i)Y(ds_j)]}{ds_i, ds_j} \right\}. \tag{12.3}$$

If the process is stationary, then $\gamma(s_i, s_j) = \gamma(s_i - s_j)$. This means that the second-order intensity depends only on the vector difference of the two points. The process is said to be *second-order* and *isotropic* if the second-order intensity depends only on the distance between $s_i$ and $s_j$. In other words, it does not depend on the direction.

### Complete Spatial Randomness

The benchmark model for spatial point patterns is called *complete spatial randomness* or **CSR**. In this model, events follow a homogeneous Poisson process over the study region. The definition of CSR is given by the following [Diggle, 1983]:

1. The intensity does not vary over the region. Thus, $Y(A)$ follows a Poisson distribution with mean $\lambda A$, where $A$ is the area of $A$ and $\lambda$ is constant.

2. There are no interactions between the events. This means that, for a given $n$, representing the total number of events in $R$, the events are uniformly and independently distributed over the study region.

In a CSR process, an event has the same probability of occurring at any location in $R$, and events neither inhibit nor attract each other. The methods covered in this chapter are mostly concerned with discovering and modeling departures from the CSR model, such as regularity and clustering. Realizations of these three types of spatial point processes are shown in Figures 12.1 through 12.3, so the reader can understand the differences between these point patterns.

In Figure 12.1, we have an example of a spatial point process that follows the CSR model. Note that there does not appear to be systematic regularity or clustering in the process. The point pattern displayed in Figure 12.2 is a realization of a cluster process, where the clusters are obviously present. Finally, in Figure 12.3, we have an example of a spatial point process that exhibits regularity.

In this chapter, we look at methods for exploring and for analyzing spatial point patterns only. We follow the treatment of this subject that is given in Bailey and Gatrell [1995]. In keeping with the focus of this text, we emphasize the simulation and computational approach, rather than the theoretical. In the next section, we look at ways to visualize spatial point patterns using the

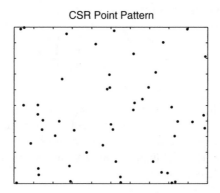

**FIGURE 12.1**
In this figure, we show a realization from a CSR point process.

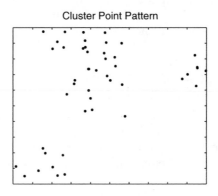

**FIGURE 12.2**
Here we have an example of a spatial point process that exhibits clustering.

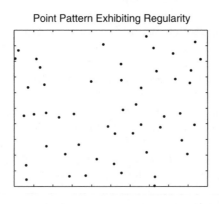

**FIGURE 12.3**
This spatial point process exhibits regularity.

graphical capabilities that come with the basic MATLAB package. Section 12.3 contains information about exploring spatial point patterns and includes methods for estimating first-order and second-order properties of the underlying point process. In Section 12.4, we discuss how to model the observed spatial pattern, with an emphasis on comparing the observed pattern to one that is completely spatially random. Finally, in Section 12.5, we offer some other models for spatial point patterns and discuss how to simulate data from them.

---

## 12.2 Visualizing Spatial Point Processes

The most intuitive way to visualize a spatial point pattern is to plot the data as a dot map. A *dot map* shows the region over which the events are observed, with the events shown using plotting symbols (usually points). When the boundary region is not part of the data set, then the dot map is the same as a scatterplot.

We mentioned briefly in Section 12.1 that some point patterns could have an attribute attached to each event. One way to visualize these attributes is to use different colors or plotting symbols that represent the values of the attribute. Another option is to plot text that specifies the attribute value at the event locations. For example, if the data represent earthquakes, then one could plot the level of the quake at each event location. However, this can be hard to interpret and gets cluttered if there are a lot of observations. Plotting this type of scatterplot is easily done in MATLAB using the **text** function. Its use will be illustrated in the exercises.

In some cases, the demographics of the population (e.g., number of people, age, income, etc.) over the study region is important. For example, if the data represent incidence of disease, then we might expect events to be clustered in regions of high population density. One way to visualize this is to combine the dot map with a surface representing the attribute, similar to what we show in Example 12.4.

We will be using various data sets in this chapter to illustrate spatial statistics for point patterns. We describe them in the next several examples and show how to construct dot maps and boundaries in MATLAB. All of these data sets are analyzed in Bailey and Gatrell [1995].

### Example 12.1
In this first example, we look at data comprised of the crater centers of 120 volcanoes in west Uganda [Tinkler, 1971]. We see from the dot map in Figure 12.4 that there is an indication of a regional trend in the north-easterly direction. The data are contained in the file **uganda**, which contains the

boundary as well as the event locations. The following MATLAB code shows how to obtain a dot map.

```
load uganda
% This loads up x and y vectors corresponding
% to point locations.
% It also loads up a two column matrix
% containing the vertices to the region.
% Plot locations as points.
plot(x,y,'.k')
hold on
% Plot boundary as line.
plot(ugpoly(:,1),ugpoly(:,2),'k')
hold off
title('Volcanic Craters in Uganda')
```

❑

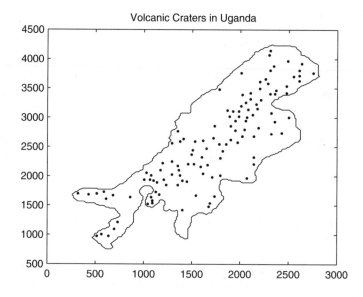

**FIGURE 12.4**
This dot map shows the boundary region for volcanic craters in Uganda.

## Example 12.2

Here we have data for the locations of homes of juvenile offenders living in a housing area in Cardiff, Wales [Herbert, 1980] in 1971. We will use these data in later examples to determine whether they show evidence of clustering or spatial randomness. These data are in the file called **cardiff**. When this is loaded using MATLAB, one also obtains a polygon representing the boundary. The following MATLAB commands construct the dot map using a single call to the **plot** function. The result is shown in Figure 12.5.

```
load cardiff
% This loads up x and y vectors corresponding
% to point locations.It also loads up a two
% column matrix containing the vertices
% to the region.
% Plot locations as points and boundary as line.
% Note: can do as one command:
plot(x,y,'.k',cardpoly(:,1),cardpoly(:,2),'k')
title('Juvenile Offenders in Cardiff')
```

❑

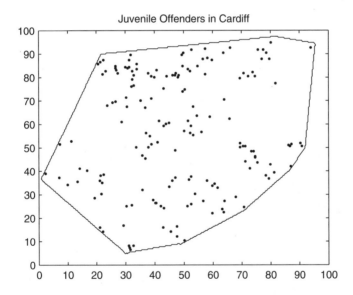

**FIGURE 12.5**
This is the dot map showing the locations of homes of juvenile offenders in Cardiff.

## Example 12.3

These data are the locations where thefts occurred in Oklahoma City in the late 1970's [Bailey and Gatrell, 1995]. There are two data sets: 1) **okwhite** contains the data for Caucasian offenders and 2) **okblack** contains the event locations for thefts committed by African-American offenders. Unlike the previous data sets, these do not have a specific boundary associated with them. We show in this example how to get a boundary for the **okwhite** data

using the MATLAB function `convhull`. This function returns a set of indices
to events in the data set that lie on the convex hull of the locations.

```
load okwhite
% Loads up two vectors: okwhx, okwhy
% These are event locations for the pattern.
% Get the convex hull.
K = convhull(okwhx, okwhy);
% K contains the indices to points on the convex hull.
% Get the events.
cvh = [okwhx(K), okwhy(K)];
plot(okwhx,okwhy,'k.',cvh(:,1),cvh(:,2),'k')
title('Location of Thefts by Caucasian Offenders')
```

A plot of these data and the resulting boundary are shown in Figure 12.6. We
show in one of the exercises how to use a function called `csgetregion`
(included with the Computational Statistics Toolbox) that allows the user to
interactively set the boundary.
❏

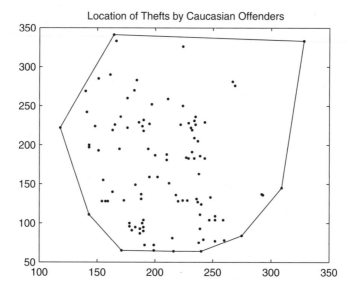

**FIGURE 12.6**
This shows the event locations for locations of thefts in Oklahoma City that were committed
by Caucasians. The boundary is the convex hull.

## 12.3 Exploring First-order and Second-order Properties

In this section, we look at ways to explore spatial point patterns. We see how to apply the density estimation techniques covered in Chapter 8 to estimate the intensity or first-order property of the spatial process. The second-order property can be investigated by using the methods of Chapter 5 to explore the distributions of nearest neighbor distances.

### Estimating the Intensity

One way to summarize the events in a spatial point pattern is to divide the study region into sub-regions of equal area. These are called *quadrats*, which is a name arising from the historical use of square sampling areas used in field sampling. By counting the number of events falling in each of the quadrats, we end up with a histogram or frequency distribution that summarizes the spatial pattern. If the quadrats are non-overlapping and completely cover the spatial region of interest, then the quadrat counts convert the point pattern into area or lattice data. Thus, the methods appropriate for lattice data can be used.

To get an estimate of intensity, we divide the study region using a regular grid, count the number of events that fall into each square and divide each count by the area of the square. We can look at various plots, as shown in Example 12.4, to understand how the intensity of the process changes over the study region.

Note that if edge effects are ignored, then the other methods in Chapter 8, such as frequency polygons or average shifted histograms can also be employed to estimate the first-order effects of a spatial point process.

Not surprisingly, we can apply kernel estimation to get an estimate of the intensity that is smoother than the quadrat method. As before, we let $s$ denote a point in the study region $R$ and $s_1, ..., s_n$ represent the event locations. Then an estimate of the intensity using the kernel method is given by

$$\hat{\lambda}_h(s) = \frac{1}{\delta_h(s)} \sum_{i=1}^{n} \frac{1}{h^2} k\left(\frac{s - s_i}{h}\right), \tag{12.4}$$

where $k$ is the kernel and $h$ is the bandwidth. The kernel is a bivariate probability density function as described in Chapter 8. In Equation 12.4, the edge-correction factor is

$$\delta_h(s) = \int_R \frac{1}{h^2} k\left(\frac{s - u}{h}\right) du. \tag{12.5}$$

Equation 12.5 represents the volume under the scaled kernel centered on **s** which is inside the study region **R**. As with the quadrat method, we can look at how $\hat{\lambda}(s)$ changes to gain insight about the intensity of the point process.

The same considerations, as discussed in Chapter 8, regarding the choice of the kernel and the bandwidth apply here. An overly large $h$ provides an estimate that is very smooth, possibly hiding variation in the intensity. A small bandwidth might indicate more variation than is warranted, making it harder to see the overall pattern in the intensity. A recommended choice for the bandwidth is $h = 0.68n^{-0.2}$, when **R** is the unit square [Diggle, 1981]. This value could be appropriately scaled for the size of the actual study region.

Bailey and Gatrell [1995] recommend the following quartic kernel

$$k(u) = \frac{3}{\pi}(1 - u^T u)^2 \qquad u^T u \le 1.  \tag{12.6}$$

When this is substituted into Equation 12.4, we have the following estimate for the intensity

$$\hat{\lambda}_h(s) = \sum_{d_i \le h} \frac{3}{\pi h^2}\left(1 - \frac{d_i^2}{h^2}\right)^2,  \tag{12.7}$$

where $d_i$ is the distance between point **s** and event location $s_i$ and the correction for edge effects $\delta_h(s)$ has, for simplicity, not been included.

## Example 12.4

In this example, we apply the kernel method as outlined above to estimate the intensity of the **uganda** data. We include a function called **csintenkern** that estimates the intensity of a point pattern using the quartic kernel. For simplicity, this function ignores edge effects. The following MATLAB code shows how to apply this function and how to plot the results. Note that we set the window width to $h = 220$. Other window widths are explored in the exercises. First, we load the data and call the function. The output variable **lamhat** contains the values of the estimated intensity.

```
load uganda
X = [x,y];
h = 220;
[xl,yl,lamhat] = csintenkern(X,ugpoly,h);
```

We use the **pcolor** function to view the estimated intensity. To get a useful color map, we use an inverted gray scale. The estimated intensity is shown in Figure 12.7, where the ridge of higher intensity is visible.

```
pcolor(xl,yl,lamhat)
map = gray(256);
```

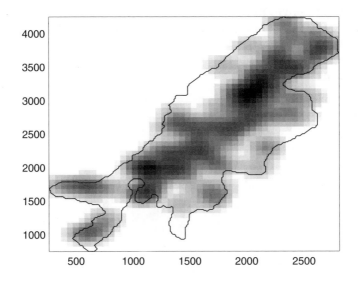

**FIGURE 12.7**
In this figure, we have the estimate of the intensity for the **uganda** crater data. This is obtained using the function **csintkern** with $h = 220$.

```
% Flip the colormap so zero is white and max is black.
map = flipud(map);
colormap(map)
shading flat
hold on
plot(ugpoly(:,1),ugpoly(:,2),'k')
hold off
```

Of course, one could also plot this as a surface. The MATLAB code we provide below shows how to combine a surface plot of the intensity with a dot map below. The axes can be rotated using the toolbar button or the **rotate3d** command to look for an interesting viewpoint.

```
% First plot the surface.
surf(x1,y1,lamhat)
map = gray(256);
map = flipud(map);
colormap(map)
shading flat
% Now plot the dot map underneath the surface.
X(:,3) = -max(lamhat(:))*ones(length(x),1);
ugpoly(:,3) = -max(lamhat(:))*...
```

```
    ones(length(ugpoly(:,1)),1);
hold on
plot3(X(:,1),X(:,2),X(:,3),'.')
plot3(ugpoly(:,1),ugpoly(:,2),ugpoly(:,3),'k')
hold off
axis off
grid off
```

The combination plot of the intensity surface with the dot map is shown in Figure 12.8.
❏

**FIGURE 12.8**
This shows the kernel estimate of the intensity along with a dot map.

## Estimating the Spatial Dependence

We now turn our attention to the problem of exploring the second-order properties of a spatial point pattern. These exploratory methods investigate the second-order properties by studying the distances between events in the study region $R$. We first look at methods based on the nearest neighbor distances between events or between points and events. We then discuss an alternative approach that summarizes the second-order effects over a range of distances.

### Nearest Neighbor Distances - G and F Distributions

The *nearest neighbor event-event distance* is represented by $W$. This is defined as the distance between a randomly chosen event and the nearest neighboring event. The *nearest neighbor point-event distance*, denoted by $X$, is the distance between a randomly selected point in the study region and the nearest event. Note that nearest neighbor distances provide information at small physical scales, which is a reasonable approach if there is variation in the intensity over the region $R$.

It can be shown [Bailey and Gatrell, 1995; Cressie 1993] that, if the CSR model holds for a spatial point process, then the cumulative distribution function for the nearest neighbor event-event distance $W$ is given by

$$G(w) = P(W \le w) = 1 - e^{-\lambda \pi w^2}, \tag{12.8}$$

for $w \ge 0$. The cumulative distribution function for the nearest neighbor point-event distance $X$ is

$$F(x) = P(X \le x) = 1 - e^{-\lambda \pi x^2}, \tag{12.9}$$

with $x \ge 0$.

We can explore the second-order properties of a spatial point pattern by looking at the observed cumulative distribution function of $X$ or $W$. The empirical cumulative distribution function for the event-event distances $W$ is given by

$$\hat{G}(w) = \frac{\#(w_i \le w)}{n}. \tag{12.10}$$

Similarly, the empirical cumulative distribution function for the point-event distances $X$ is

$$\hat{F}(x) = \frac{\#(x_i \le x)}{m}, \tag{12.11}$$

where $m$ is the number of points randomly sampled from the study region.

A plot of $\hat{G}(w)$ and $\hat{F}(x)$ provides possible evidence of inter-event interactions. If there is clustering in the point pattern, then we would expect a lot of short distance neighbors. This means that $\hat{G}(w)$ would climb steeply for smaller values of $w$ and flatten out as the distances get larger. On the other hand, if there is regularity, then there should be more long distance neighbors and $\hat{G}(w)$ would be flat at small distances and climb steeply at larger $w$ or $x$. When we examine a plot of $\hat{F}(x)$, the opposite interpretation holds. For example, if there is an excess of long distances values in $\hat{F}(x)$, then that is evidence for clustering.

We could also plot $\hat{G}(w)$ against $\hat{F}(x)$. If the relationship follows a straight line, then this is evidence that there is no spatial interaction. If there is clustering, then we expect $\hat{G}(w)$ to exceed $\hat{F}(x)$, with the opposite situation occurring if the point pattern exhibits regularity.

From Equation 12.8, we can construct a simpler display for detecting departures from CSR. Under CSR, we would expect a plot of

$$\left\{ \frac{-\log(1 - \hat{G}(w))}{(\hat{\lambda}\pi)} \right\}^{1/2} \qquad (12.12)$$

versus $w$ to be a straight line. In Equation 12.12, we need a suitable estimate for the intensity $\hat{\lambda}$. One possibility is to use $\hat{\lambda} = n/r$, where $r$ is the area of the study region $R$.

So far, we have not addressed the problem of edge effects. Events near the boundary of the region $R$ might have a nearest neighbor that is outside the boundary. Thus, the nearest neighbor distances near the boundary might be biased. One possible solution is to have a guard area inside the perimeter of $R$. We do not compute nearest neighbor distances for points or events in the guard area, but we can use events in the guard area in computing nearest neighbors for points or events inside the rest of $R$. Other solutions for making corrections are discussed in Bailey and Gatrell [1995] and Cressie [1993].

## Example 12.5

The data in **bodmin** represent the locations of granite tors on Bodmin Moor [Pinder and Witherick, 1977; Upton and Fingleton, 1985]. There are 35 locations, along with the boundary. The $x$ and $y$ coordinates for the locations are stored in the **x** and **y** vectors, and the vertices for the region are given in **bodpoly**. The reader is asked in the exercises to plot a dot map of these data. In this example, we use the event locations to illustrate the nearest neighbor distribution functions $\hat{G}(w)$ and $\hat{F}(x)$. First, we show how to get the empirical distribution function for the event-event nearest neighbor distances.

```
load bodmin
% Loads data in x and y and boundary in bodpoly.
% Get the Ghat function first and plot.
X = [x,y];
w = 0:.1:10;
n = length(x);
nw = length(w);
ghat = zeros(1,nw);
% The G function is the nearest neighbor
% distances for each event.
% Find the distances for all points.
dist = pdist(X);
% Convert to a matrix and put large
% numbers on the diagonal.
D = diag(realmax*ones(1,n)) + squareform(dist);
% Find the smallest distances in each row or col.
```

```
mind = min(D);
% Now get the values for ghat.
for i = 1:nw
    ind = find(mind<=w(i));
    ghat(i) = length(ind);
end
ghat = ghat/n;
```

To see whether there is evidence for clustering or regularity, we plot $\hat{G}(w)$ using the following commands.

```
% Plot the Ghat as a function of w. Shows evidence
% of clustering.
figure,plot(w,ghat,'k')
axis([0 10 0 1.1])
xlabel('Event-Event Distances - w'),ylabel('Ghat')
```

We see from Figure 12.9, that the curve climbs steeply at small values of $w$, providing possible evidence for clustering. This indicates that there are many small event-event distances, which is what we would expect for clustering. The reader is asked to explore this further in the exercises by plotting the expression in Equation 12.12 versus $w$. Next, we determine the $\hat{F}(x)$. First we find the nearest neighbor distances for $m = 75$ randomly selected points in the study region.

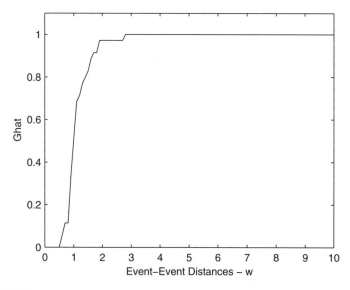

**FIGURE 12.9**
This is the empirical distribution function for the event-event nearest neighbor distances for the **bodmin** data. This provides possible evidence for clustering.

```
xx = w;
m = 75;
nx = length(xx);
fhat = zeros(1,nx);
mind = zeros(1,m);% one for each point m
xt = [0 0; X];
% The F function is the nearest neighbor distances for
% randomly selected points. Generate a point, find its
% closest event.
for i = 1:m
    % Generate a point in the region.
    [xt(1,1), xt(1,2)] = csbinproc(bodpoly(:,1),...
        bodpoly(:,2), 1);
    % Find the distances to all events.
    dist = pdist(xt);
    % The first n in dist are the distance
    % between the point(first row) and all the events.
    % Find the smallest here.
    mind(i) = min(dist(1:n));
end
```

Now that we have the nearest neighbor distances, we can find the empirical distribution function, as follows.

```
% Now get the values for fhat.
for i = 1:nx
    ind = find(mind<=xx(i));
    fhat(i) = length(ind);
end
fhat = fhat/m;
```

We plot the empirical distribution function $\hat{F}(x)$ in Figure 12.10, where it also seems to provide evidence for the cluster model.
❑

### K-Function

The empirical cumulative distribution functions $\hat{G}(w)$ and $\hat{F}(x)$ use distances to the nearest neighbor, so they consider the spatial point pattern over the smallest scales. It would be useful to have some insight about the pattern at several scales. We use an estimate of the $K$-function, which is related to the second-order properties of an isotropic process [Ripley, 1976, 1981]. If the $K$-function is used when there are first-order effects over large scales, then spatial dependence indicated by the $K$-function could be due to first-order effects instead [Bailey and Gatrell, 1995]. If this is the case, the analyst might want to study sub-regions of $R$ where first-order homogeneity is valid.

The $K$-function is defined as

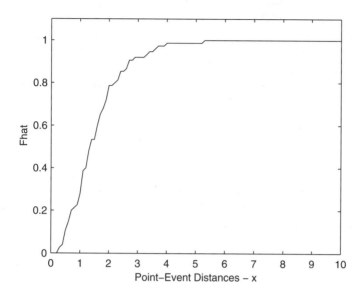

**FIGURE 12.10**
This is the empirical distribution function for the point-event distances of the **bodmin** data.

$$K(d) = \lambda^{-1}E[\# \text{ extra events within distance } d \text{ of an arbitrary event}],$$

where $\lambda$ is a constant representing the intensity over the region and $E[.]$ denotes the expected value.

An edge corrected estimate for the K-function is given by the following

$$\hat{K}(d) = \frac{r}{n^2}\sum_{i \neq j}\sum \frac{I_d(d_{ij})}{w_{ij}}. \tag{12.13}$$

In Equation 12.13, $r$ represents the area of the study region $R$, $n$ is the number of events, $d_{ij}$ is the distance between the $i$-th and $j$-th events, and $I_d$ is an indicator function that takes on the value of one if $d_{ij} \leq d$ and zero otherwise. The $w_{ij}$ in Equation 12.13 is a correction factor for edge effects. If a circle is centered at event $i$ and passes through event $j$, then $w_{ij}$ is the proportion of the circumference of the circle that is in region $R$.

The estimated K-function can be compared to what we would expect if the process that generated the data is completely spatially random. For a CSR spatial point process, the theoretical K-function is

$$K(d) = \pi d^2. \tag{12.14}$$

If our observed process exhibits regularity for a given value of $d$, then we expect that the estimated $K$-function will be less than $\pi d^2$. Alternatively, if the spatial pattern has clustering, then $\hat{K}(d) > \pi d^2$. Plots of the $\hat{K}(d)$ and $K(d)$ under CSR (Equation 12.14) enable us to explore the second-order properties of the spatial process.

Another approach, based on the $K$-function, is to transform $\hat{K}(d)$ using

$$\hat{L}(d) = \sqrt{\frac{\hat{K}(d)}{\pi}} - d. \tag{12.15}$$

Peaks of positive values in a plot of $\hat{L}(d)$ would correspond to clustering, with troughs of negative values indicating regularity, for the corresponding scale $d$. Note that with $\hat{K}(d)$ and $\hat{L}(d)$, we can explore spatial dependence at a range of scales $d$. The quantity

$$L(d) \equiv \sqrt{\frac{K(d)}{\pi}} - d \tag{12.16}$$

is called the *L-function*, and Equation 12.15 is an estimate of it.

**Example 12.6**
In this example, we find $\hat{K}(d)$ and $\hat{L}(d)$ for the **cardiff** data set. We provide a function in the Computational Statistics Toolbox called **cskhat** for estimating the $K$-function and illustrate its use below.

```
load cardiff
% Loads data in x and y and region in cardpoly.
% Get the scales or distances for K_hat.
d = 1:30;
X = [x,y];
% Get the estimate of K_hat.
khat = cskhat(X, cardpoly, 1:30);
```

The next commands show how to plot $\hat{K}(d)$ and the theoretical $K$-function for a random process.

```
% Plot the khat function along with the K-function
% under CSR. Shows clustering because
% khat is above the curve.
plot(d,pi*d.^2,'k',d,khat,'k.')
xlabel('Distances - d')
ylabel('K Function')
```

This plot is given in Figure 12.11, where we see possible evidence for clustering, because the observed $K$-function is above the curve corresponding to a random process. As mentioned previously, we can also plot the function $\hat{L}(d)$. This is shown in Figure 12.12, where we see clustering at all scales.

```
% Get the Lhat function.
% Positive peaks - clustering at all of these scales.
% Clustering shown at d = 10, showing possible
% clustering at that scale.
lhat = sqrt(khat/pi) - d;
plot(d,lhat,'k')
xlabel('Distances - d')
ylabel('Lhat')
```

❑

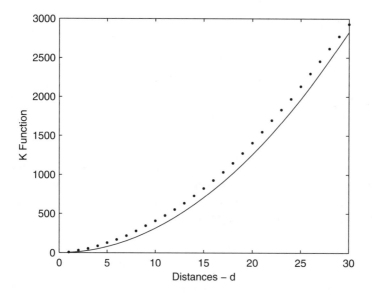

**FIGURE 12.11**
This shows the function $\hat{K}(d)$ for the **cardiff** data. Note that it is above the curve for a random process, indicating possible clustering.

## 12.4 Modeling Spatial Point Processes

When analyzing spatial point patterns, we are mainly interested in discovering patterns such as clustering or regularity versus complete spatial randomness. The exploratory methods of the previous section are meant to provide

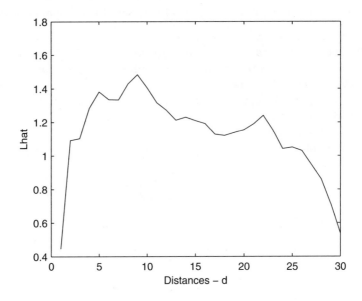

**FIGURE 12.12**
In this plot of $\hat{L}(d)$, we see possible evidence of clustering at all scales.

evidence for a model that might explain the process that generated the spatial point pattern. We now look at ways to use Monte Carlo hypothesis testing to understand the statistical significance of our evidence for departures from CSR. These tests are based on nearest neighbor distances and the $K$-function.

### Nearest Neighbor Distances

Recall that the theoretical cumulative distribution function (under the CSR model) for the nearest neighbor event-event distance $W$ is given by

$$G(w) = P(W \le w) = 1 - e^{-\lambda \pi w^2}; \qquad w \ge 0, \qquad (12.17)$$

and the cumulative distribution function for the nearest neighbor point-event distance $X$ is

$$F(x) = P(X \le x) = 1 - e^{-\lambda \pi x^2}; \qquad x \ge 0. \qquad (12.18)$$

These distributions can be used to implement statistical hypothesis tests that use summary statistics of the observed nearest neighbor distances. The estimated distributions, $\hat{G}(w)$ or $\hat{F}(x)$, can be plotted against the corresponding theoretical distributions under CSR. If the CSR model is valid for the observed spatial point process, then we would expect these plots to follow a straight line. Equations 12.17 and 12.18 assume that no edge effects are present, so it is important to correct for the edge effects when calculating $\hat{G}(w)$ and $\hat{F}(x)$. The reader is referred to Cressie [1993, p. 614] for a description of the edge corrections for $\hat{G}(w)$ and $\hat{F}(x)$. As with the exploratory methods described in the previous section, it is difficult to assess the significance of any departure from CSR that is seen in the plots, even though we might suspect such a departure.

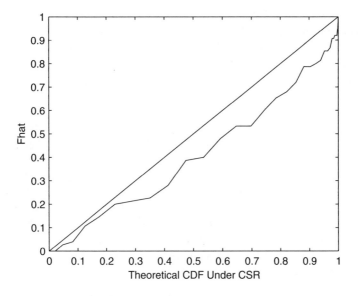

**FIGURE 12.13**
This is the empirical point-event nearest neighbor distribution function $\hat{F}(x)$ for the Bodmin Tors data. Since the curve lies below the 45 degree line, this indicates clustering. Note that edge effects have been ignored.

In the plots discussed in the previous section, we have to judge the general shape of the curve for $\hat{G}(w)$ or $\hat{F}(x)$, which is subjective and not very exact. We now offer another useful way to display these functions. When we plot the empirical distributions for the observed nearest neighbor distances against the theoretical distributions, we expect a straight line, if the point pattern follows a CSR process. In a clustered process, the curve for $\hat{F}(x)$ would lie below the 45 degree line as shown in Figure 12.13 for the **bodmin** data. If the process exhibits regularity, then the empirical distribution function $\hat{F}(x)$ lies above the line. As before, the opposite interpretation holds for the distribution function $\hat{G}(w)$.

We now describe simulation techniques that compare the estimated distribution functions with the distribution under CSR, allowing the analyst to assess the significance of any departure from CSR. These methods are particularly useful, because the edge effects are taken care of by the simulation procedure, so explicit corrections do not need to be made. However, we note that edge-corrected statistics may lead to more powerful tests than those that do not incorporate the edge corrections.

In the procedure explained below, we see that edge effects are accounted for because of the following:

1. The estimated distributions $\hat{G}(w)$ and $\hat{F}(x)$ are obtained for **R** without edge correction.

2. The estimate of the distribution under CSR is obtained via simulation for the particular study region **R**. In other words, we use a procedure that, for a given $n$, yields events that are uniformly and independently distributed over the region. See Section 12.5 for more information.

We describe the method as it applies to the point-event distances $X$, with an analogous approach holding for the event-event distances $W$. In Example 12.7, we illustrate the procedure as it applies to $W$ and leave the other as an exercise for the reader. The simulation estimate for $F(x)$ under CSR is obtained by first generating $B$ spatial point patterns of size $n$ that are independently and uniformly distributed over **R**. The empirical cumulative distribution function is determined for each simulated point pattern, without correcting for edge effects. We denote these by $\hat{F}_b(x)$, $b = 1, ..., B$. Taking the mean of these functions yields an estimate of the distribution of the point-event nearest neighbor distances for a process under CSR,

$$\hat{F}_{CSR}(x) = \frac{1}{B}\sum_{b=1}^{B}\hat{F}_b(x).\qquad(12.19)$$

Letting $\hat{F}_{Obs}(x)$ denote the empirical cumulative distribution function for the observed spatial point pattern, we can plot $\hat{F}_{Obs}(x)$ against $\hat{F}_{CSR}(x)$. If the data follow the CSR model, then the plot should be a straight line. If the data exhibit clustering, then the plot will be above the line. If regularity is present, then the plot will be below the line.

We can assess the significance of the departure from CSR by constructing upper and lower simulation envelopes. These are given by

$$U(x) = \max_b\{\hat{F}_b(x)\},\qquad(12.20)$$

and

$$L(x) = \min_b\{\hat{F}_b(x)\}. \tag{12.21}$$

The significance of the departure from CSR is found using

$$P(\hat{F}_{Obs}(x) > U(x)) = P(\hat{F}_{Obs}(x) < L(x)) = \frac{1}{B+1}. \tag{12.22}$$

For example, if we want to detect clustering that is significant at $\alpha = 0.05$, then (from Equation 12.22) we need 19 simulations. Adding the upper and lower simulation envelopes to the plot of $\hat{F}_{Obs}(x)$ against $\hat{F}_{CSR}(x)$ enables us to determine the significance of the clustering. If $\hat{F}_{Obs}(x)$ is below the upper envelope, then the result showing clustering is significant. Note that Equation 12.22 is for a fixed $x$, so the analyst must look at each point in the curve of $\hat{F}_{Obs}(x)$. In the exercises, we describe an alternative, more powerful test.

*PROCEDURE - MONTE CARLO TEST USING NEAREST NEIGHBOR DISTANCES*

1. Obtain the empirical cumulative distribution function using the observed spatial point pattern, $\hat{F}_{Obs}(x)$ (or $\hat{G}_{Obs}(w)$). Do not correct for edge effects.
2. Simulate a spatial point pattern over the study region of size $n$ from a CSR process.
3. Get the empirical cumulative distribution function $\hat{F}_b(x)$ (or $\hat{G}_b(w)$.) Do not correct for edge effects.
4. Repeat steps 2 and 3, $B$ times, where $B$ is determined from Equation 12.22.
5. Take the average of the $B$ distributions using Equation 12.19 to get the estimated distribution of the nearest neighbor distances under CSR, $\hat{F}_{CSR}(x)$ (or $\hat{G}_{CSR}(w)$).
6. Find the lower and upper simulation envelopes.
7. Plot $\hat{F}_{Obs}(x)$ (or $\hat{G}_{Obs}(w)$) against $\hat{F}_{CSR}(x)$ (or $\hat{G}_{CSR}(w)$).
8. Add plots of the lower and upper simulation envelopes to assess the significance of the test.

## Example 12.7

In this example, we show how to implement the procedure for comparing $\hat{G}_{Obs}(w)$ with an estimate of the empirical distribution function under CSR. We use the **bodmin** data set, so we can compare this with previous results. First we get $\hat{G}_{Obs}(w)$.

```
load bodmin
X = [x,y];
% Note that we are using a smaller range
```

```
% for w than before.
w = 0:.1:6;
nw = length(w);
nx = length(x);
ghatobs = csghat(X,w);
```

The next step is to simulate from a CSR process over the same region and determine the empirical event-event distribution function for each simulation.

```
% Get the simulations.
B = 99;
% Each row is a Ghat from a simulated CSR process.
simul = zeros(B,nw);
for b = 1:B
  [xt,yt] = csbinproc(bodpoly(:,1), bodpoly(:,2), nx);
  simul(b,:) = csghat([xt,yt],w);
end
```

We need to take the average of all of the simulations so we can plot these values along the horizontal axis. The average and the envelopes are easily found in MATLAB. The resulting plot is given in Figure 12.14. Note that there does not seem to be significant evidence for departure from the CSR model using the event-event nearest neighbor distribution function $\hat{G}_{Obs}(w)$.

```
% Get the average.
ghatmu = mean(simul);
% Get the envelopes.
ghatup = max(simul);
ghatlo = min(simul);
plot(ghatmu,ghatobs,'k',ghatmu,ghatup,...
    'k--',ghatmu,ghatlo,'k--')
```

❑

### *K*-Function

We can use a similar approach to formally compare the observed *K*-function with an estimate of the *K*-function under CSR. We determine the upper and lower envelopes as follows

$$U(d) = \max_b\{\hat{K}_b(d)\}, \tag{12.23}$$

and

$$L(d) = \min_b\{\hat{K}_b(d)\}. \tag{12.24}$$

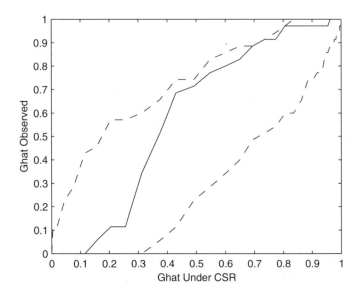

**FIGURE 12.14**
In this figure, we have the upper and lower envelopes for $\hat{G}$ from a CSR process over the **bodmin** region. It does not appear that there is strong evidence for clustering or regularity in the point pattern.

The $\hat{K}_b(d)$ are obtained by simulating spatial point patterns of size $n$ events in $R$ under CSR.

Alternatively, we can use the $L$-function to assess departures from CSR. The upper and lower simulation envelopes for the $L$-function are obtained in the same manner. With the $L$-function, the significance of the peaks or troughs (for fixed $d$) can be assessed using

$$P(\hat{L}_{Obs}(d) > U(d)) = P(\hat{L}_{Obs}(d) < L(d)) = \frac{1}{B+1}. \qquad (12.25)$$

We outline the steps in the following procedure and show how to implement them in Examples 12.8 and 12.9.

*PROCEDURE - MONTE CARLO TEST USING THE K-FUNCTION*

1. Estimate the $K$-function using the observed spatial point pattern to get $\hat{K}_{Obs}(d)$.
2. Simulate a spatial point pattern of size $n$ over the region $R$ from a CSR process.

3. Estimate the *K*-function using the simulated pattern to get $\hat{K}_b(d)$ .

4. Repeat steps 2 and 3, *B* times.

5. Find the upper and lower simulation envelopes using Equations 12.23 and 12.24.

6. Plot $\hat{K}_{Obs}(d)$ and the simulation envelopes.

## Example 12.8

We apply the Monte Carlo test for departure from CSR to the **bodmin** data. We obtain the required simulations using the following steps. First we load up the data and obtain $\hat{K}_{Obs}(d)$ .

```
load bodmin
X = [x,y];
d = 0:.5:10;
nd = length(d);
nx = length(x);
% Now get the Khat for the observed pattern.
khatobs = cskhat(X, bodpoly, d);
```

We are now ready to obtain the *K*-functions for a CSR process through simulation. We use *B* = 20 simulations to obtain the envelopes.

```
% Get the simulations.
B = 20;
% Each row is a Khat from a simulated CSR process.
simul = zeros(B,nd);
for b = 1:B
  [xt,yt] = csbinproc(bodpoly(:,1), bodpoly(:,2), nx);
  simul(b,:) = cskhat([xt,yt],bodpoly, d);
end
```

The envelopes are easily obtained using the MATLAB commands **max** and **min**.

```
% Get the envelopes.
khatup = max(simul);
khatlo = min(simul);
% And plot the results.
plot(d,khatobs,'k',d,khatup,'k--',d,khatlo,'k--')
```

In Figure 12.15, we show the upper and lower envelopes along with the estimated *K*-function $\hat{K}_{Obs}(d)$ . We see from this plot that at the very small scales, there is no evidence for departure from CSR. At some scales there is evidence for clustering and at other scales there is evidence of regularity.

❏

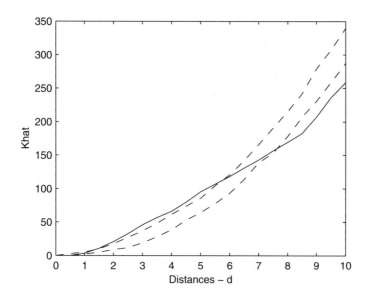

**FIGURE 12.15**
In this figure, we have the results of testing for departures from CSR based on $\hat{K}$ using simulation. We show the upper and lower simulation envelopes for the Bodmin Tor data. At small scales (approximately $d < 2$), the process does not show departure from CSR. This is in agreement with the nearest neighbor results of Figure 12.14. At other scales (approximately $2 < d < 6$), we have evidence for clustering. At higher scales (approximately $7.5 < d$), we see evidence for regularity.

## Example 12.9

In Example 12.6, we estimated the $K$-function for the **cardiff** data. A plot of the associated $L$-function (see Figure 12.12) showed clustering at those scales. We use the simulation approach to determine whether these results are significant. First we get the estimate of the $L$-function as before.

```
load cardiff
X = [x,y];
d = 0:30;
nd = length(d);
nx = length(x);
khatobs = cskhat(X, cardpoly, d);
% Get the lhat function.
lhatobs = sqrt(khatobs/pi) - d;
```

Now we do the same simulations as in the previous example, estimating the $K$-function for each CSR sample. Once we get the $K$-function for the sample, it is easily converted to the $L$-function as shown.

```
% Get the simulations.
B = 20;
% Each row is a Khat from a simulated CSR process.
simul = zeros(B,nd);
for b = 1:B
  [xt,yt] = csbinproc(cardpoly(:,1),...
      cardpoly(:,2), nx);
  temp = cskhat([xt,yt],cardpoly, d);
  simul(b,:) = sqrt(temp/pi) -d;
end
```

We then get the upper and lower simulation envelopes as before. The plot is shown in Figure 12.16. From this, we see that there seems to be compelling evidence that this is a clustered process.

```
% Get the envelopes.
lhatup = max(simul);
lhatlo = min(simul);
plot(d,lhatobs,'k',d,lhatup,'k--',d,lhatlo,'k--')
```

❑

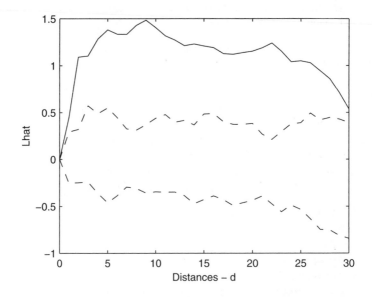

FIGURE 12.16
The upper and lower envelopes were obtained using 20 simulations from a CSR process. Since the $L$-function lies above the upper envelope, the clustering is significant.

## 12.5 Simulating Spatial Point Processes

Once one determines that the model for CSR is not correct, then the analyst should check to see what other model is reasonable. This can be done by simulation as shown in the previous section. Instead of simulating from a CSR process, we can simulate from one that exhibits clustering or regularity. We now discuss other models for spatial point processes and how to simulate them. We include methods for simulating a homogeneous Poisson process with specified intensity, a binomial process, a Poisson cluster process, an inhibition process, and a Strauss process. Before continuing, we note that simulation requires specification of all relevant parameters. To check the adequacy of a model by simulation one has to *"calibrate"* the simulation to the data by estimating the parameters that go into the simulation.

### Homogeneous Poisson Process

We first provide a method for simulating a homogeneous Poisson process with no conditions imposed on the number of events $n$. Unconditionally, a homogeneous Poisson process depends on the intensity $\lambda$. So, in this case, the number of events $n$ changes in each simulated pattern.

We follow the fanning out procedure given in Ross [1997] to generate such a process for a circular region. This technique can be thought of as fanning out from the origin to a radius $r$. The successive radii where events are encountered are simulated by using the fact that the additional area one needs to travel to encounter another event is exponentially distributed with rate $\lambda$. The steps are outlined below.

*PROCEDURE - SIMULATING A POISSON PROCESS*

1. Generate independent exponential variates $X_1, X_{2,\ ...}$, with rate $\lambda$, stopping when

$$N = \min\{n: X_1 + ... + X_n > \pi r^2\}.$$

2. If $N = 1$, then stop, because there are no events in the circular region.

3. If $N > 1$, then for $i = 1, ..., N-1$, find

$$R_i = \sqrt{\frac{X_1 + ... + X_i}{\pi}}.$$

4. Generate $N - 1$ uniform $(0,1)$ variates, $U_1, ..., U_{N-1}$.

5. In polar coordinates, the events are given by $(R_i, 2\pi U_i)$.

Ross [1997] describes a procedure where the region can be somewhat arbitrary. For example, in Cartesian coordinates, the region would be defined between the $x$ axis and a nonnegative function $f(x)$, starting at $x = 0$. A rectangular region with the lower left corner at the origin is an example where this can be applied. For details on the algorithm for an arbitrary region, we refer the reader to Ross [1997]. We show in Example 12.10 how to implement the procedure for a circular region.

### Example 12.10

In this example, we show how to generate a homogeneous Poisson process for a given $\lambda$. This is accomplished using the given MATLAB commands.

```
% Set the lambda.
lambda = 2;
r = 5;
tol = 0;
i=1;
% Generate the exponential random variables.
while tol < pi*r^2
    x(i) = exprnd(1/lambda,1,1);
    tol = sum(x);
    i=i+1;
end
x(end)=[];
N = length(x);
% Get the coordinates for the angles.
th = 2*pi*rand(1,N);
R = zeros(1,N);
% Find the R_i.
for i = 1:N
    R(i) = sqrt(sum(x(1:i))/pi);
end
[Xc,Yc]=pol2cart(th,R);
```

The $x$ and $y$ coordinates for the generated locations are contained in **Xc** and **Yc**. The radius of our circular region is 5, and the intensity is $\lambda = 2$. The result of our sampling scheme is shown in Figure 12.17. We see that the locations are all within the required radius. To verify the intensity, we can estimate it by dividing the number of points in the sample by the area.

```
% estimate the overall intensity
lamhat = length(Xc)/(pi*r^2);
```

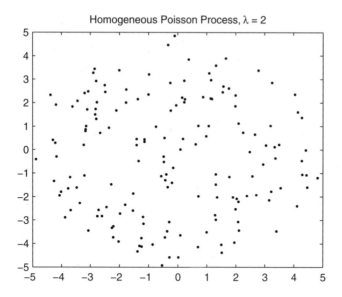

**FIGURE 12.17**
This spatial point pattern was simulated using the procedure for simulating a homogeneous Poisson process with specified intensity.

Our estimated intensity is $\hat{\lambda} = 2.05$.
❑

### Binomial Process

We saw in previous examples that we needed a way to simulate realizations from a CSR process. If we condition on the number of events $n$, then the locations are uniformly and independently distributed over the study region. This type of process is typically called a **binomial process** in the literature [Ripley, 1981]. To distinguish this process from the homogeneous Poisson process, we offer the following:

1. When generating variates from the homogeneous Poisson process, the intensity is specified. Therefore, the number of events in a realization of the process is likely to change for each one generated.
2. When generating variates from a binomial process, the number of events in the region is specified.

To simulate from a binomial process, we first enclose the study region $R$ with a rectangle given by

$$\{(x, y) : x_{min} \leq x \leq x_{max} , y_{min} \leq y \leq y_{max}\} . \qquad (12.26)$$

We can generate the $x$ coordinates for an event location from a uniform distribution over the interval $(x_{min} , x_{max})$. Similarly, we generate the $y$ coordinates from a uniform distribution over the interval $(y_{min} , y_{max})$. If the event is within the study region $R$, then we keep the location. These steps are outlined in the following procedure and are illustrated in Example 12.11.

*PROCEDURE - SIMULATING A BINOMIAL PROCESS*

1. Enclose the study region $R$ in a rectangle, given by Equation 12.26.
2. Obtain a candidate location $s_i$ by generating an $x$ coordinate that is uniformly distributed over $(x_{min} , x_{max})$ and a $y$ coordinate that is uniformly distributed over $(y_{min} , y_{max})$.
3. If $s_i$ is within the study region $R$, then retain the event.
4. Repeat steps 2 through 3 until there are $n$ events in the sample.

## Example 12.11

In this example, we show how to simulate a CSR point pattern using the region given with the **uganda** data set. First we load up the data set and find a rectangular region that bounds $R$.

```
load uganda
% loads up x, y, ugpoly
xp = ugpoly(:,1);
yp = ugpoly(:,2);
n = length(x);
xg = zeros(n,1);
yg = zeros(n,1);
% Find the maximum and the minimum for a 'box' around
% the region. Will generate uniform on this, and throw
% out those points that are not inside the region.
% Find the bounding box.
minx = min(xp);
maxx = max(xp);
miny = min(yp);
maxy = max(yp);
```

Now we are ready to generate the locations, as follows.

```
% Now get the points.
i = 1;
cx = maxx - minx;
cy = maxy - miny;
while i <= n
```

```
    xt = rand(1)*cx + minx;
    yt = rand(1)*cy + miny;
    k = inpolygon(xt, yt, xp, yp);
    if k == 1
      % it is in the region
      xg(i) = xt;
      yg(i) = yt;
      i = i+1;
    end
  end
```

In Figure 12.18, we show a realization of this process. Note that this does look like a CSR process generated these data, unlike the point pattern for the actual crater locations.
❑

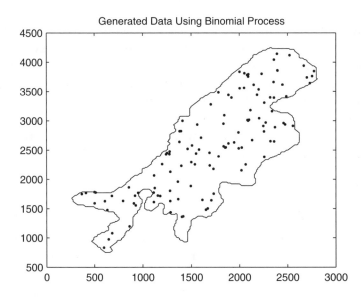

**FIGURE 12.18**
This shows a point pattern generated according to a binomial process.

## Poisson Cluster Process

We can generate a Poisson cluster process by including a spatial clustering mechanism into the model. First, parent events form a homogeneous Poisson process. Each parent gives rise to a random number of offspring according to some probability distribution $f$. The positions of the children relative to their parents are independently distributed according to a bivariate distribution $g$. The events retained in the final pattern are the child events only. The resulting process is isotropic if $g$ is radially symmetric.

To simulate this type of pattern, we first simulate the parents from a homogeneous Poisson process. Note that the parents should be simulated over a region that is larger than the study region. This is to ensure that edge effects are avoided. Parents outside the study region can have offspring that are in $R$, so we want to account for those events. For each parent event, we determine the number of offspring by randomly sampling from $f$. The next step is to locate the number of children around each parent event according to $g$. The steps for this procedure are outlined here.

*PROCEDURE - SIMULATING A POISSON CLUSTER PROCESS*

1. Simulate the desired number of parents over a region that is slightly larger than the study region $R$. The parents are generated according to a CSR process.

2. Generate the number of children for each parent according to a probability distribution $f$. One reasonable choice is to have a Poisson number of children.

3. Generate the locations for each child around the parent according to a bivariate probability distribution $g$. For example, $g$ could be multivariate normal, with the mean given by the parent location.

4. Save only the child events that are within the study region.

In the following example, we apply this procedure to generate a Poisson cluster process over the unit square.

**Example 12.12**
We now show how to generate a Poisson cluster process using MATLAB. We first generate 15 parents from a binomial process over a region that is slightly larger.

```
npar = 15;
% Get the vertices for the regions.
rx = [0 1 1 0 0];
ry = [0 0 1 1 0];
rxp = [-.05 1.05 1.05 -.05 -.05];
ryp = [-.05 -.05 1.05 1.05 -.05];
% Get all of the parents.
[xp,yp] = csbinproc(rxp, ryp, npar);
```

We use a Poisson distribution with mean $\lambda = 15$ to generate the number of children for the parents.

```
lam = 15;
% Get the number of children per parent.
nchild = poissrnd(lam,1,npar);
```

Now we find the locations of the children around the parent using a bivariate normal distribution that is centered at each parent. The covariance of the distribution is given by $\sigma^2 I$, where $I$ is a $2 \times 2$ identity matrix. The value given to the variance $\sigma^2$ would govern the spread of the cluster of children around the parent.

```
X = [];
sig = r*eye(2);
r = 0.05;
% Locate the children.
for i = 1:npar
    xc = randn(nchild(i),2)*sig + ...
        repmat([xp(i) yp(i)],nchild(i),1);
    X = [X; xc];
end
```

To get the final events for our sample, we need to determine which ones are inside the study region $R$. We do this using the MATLAB function **inpolygon**. In Figure 12.19, we show the resulting spatial sample. We provide a function called **csclustproc** that will generate patterns that follow a Poisson cluster process.

```
% Find the ones that are in the region of interest.
ind = find(inpolygon(X(:,1), X(:,2), rx, ry));
% Those are the children for the sample.
x = X(ind,1);
y = X(ind,2);
```

❑

### Inhibition Process

An inhibition process is one that often shows regularity. To simulate this type of process, we include a mechanism in the model that stipulates a minimum distance between two events. We call this distance the *inhibition distance* $\delta$.

One way to obtain such a process is to first generate a homogeneous Poisson process over the region. The events are then thinned by deleting all pairs of events that are closer than $\delta$. Implementing this procedure in MATLAB is left as an exercise.

Another method is to generate a homogeneous Poisson process one event at a time and discard candidate events if they are within distance $\delta$ of any previously retained event. This type of process is sometimes referred to as *Sequential Spatial Inhibition* or SSI [Ripley, 1981]. It is important to keep in mind that if the inhibition distance is too large for the region $R$, then it might

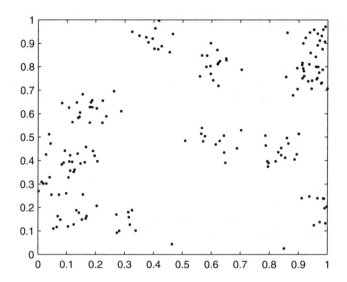

**FIGURE 12.19**
This sample was generated according to a Poisson cluster process.

be difficult (if not impossible) to generate the required number of points. In Example 12.13, we provide the MATLAB code to generate an inhibition spatial point pattern using this procedure.

## Example 12.13

To start the procedure, we set the boundary for the region and the inhibition distance.

```
delta = 0.1;
% Get the vertices for the regions.
rx = [0 2 2 0 0];
ry = [0 0 2 2 0];
n = 100;
```

We generate the initial event from a CSR process. Subsequent events are generated and kept if they are not closer than δ to any existing events.

```
X = zeros(n,2);
% Generate the first event.
X(1,:) = csbinproc(rx,ry,1);
i = 1;
% Generate the other events.
while i<n
```

```
[sx,sy] = csbinproc(rx, ry, 1);
xt = [sx sy ; X(1:i,:)];
% Find the distance between the events
dist = pdist(xt);
% Find the distance between the candidate event
% and the others that have been generated already.
ind = find(dist(1:i) <= delta);
if isempty(ind)
  % Then we keep the event.
  i = i+1;
  X(i,:) = [sx, sy];
end
end
```

To verify that no two events are closer than δ, we find the smallest distance as follows.

```
% Verify that all are no closer than the
% inhibition distance.
dist = pdist(X);
delhat = min(dist);
```

For this spatial point pattern, we get a minimum distance of 0.1008. A point pattern generated according to this procedure is shown in Figure 12.20. ❑

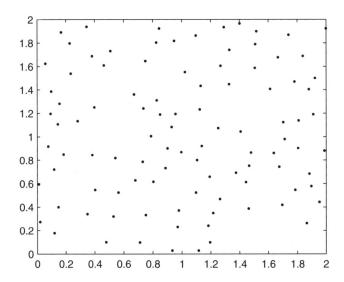

**FIGURE 12.20**
This spatial point pattern was generated under the SSI inhibition process.

## Strauss Process

The Strauss process [Ripley, 1981] is a point pattern where a specified fraction of events is allowed within a distance $\delta$ of any given event. To generate such a pattern, the first event is located uniformly in $R$. Other event locations are generated sequentially, similar to the SSI process. If there are existing events within radius $\delta$ of the candidate location, then it is accepted with probability $c^m$, with $m$ representing the number of events closer than $\delta$. The *inhibition parameter* is given by $c$, which can take on values in the interval $[0, 1]$.

The inhibition parameter specifies the fraction of events allowed within the inhibition distance. If $c = 0$, then the resulting process is the same as SSI. As with the SSI process, care should be taken when specifying the parameters for the process to ensure that the required number of events can be generated. We outline below the steps to generate a spatial point pattern that follows a Strauss process.

*PROCEDURE - SIMULATING A STRAUSS PROCESS*

1. Choose the parameters $n$, $c$, and $\delta$.
2. Generate the first event location $s_1$ uniformly on $R$ (from a CSR process).
3. Generate a candidate location $s_i$ uniformly on $R$.
4. If $m = 0$

    accept the candidate event $s_i$

    Else if $U \le c^m$

    accept the candidate event $s_i$
5. Repeat steps 3 and 4 until there are $n$ locations in the sample.

It should be noted that we are conditioning on the number of points $n$ in the region. So, in this case, we should consider this a conditional Strauss process.

## Example 12.14

We now implement the above procedure in MATLAB. We generate a spatial point pattern of size 100 from a Strauss process over a rectangular region. The inhibition distance is $\delta = 0.1$, and the inhibition parameter is $c = 0.5$. We start by setting these parameters and the boundary of the study region.

```
delta = 0.1;
% Get the vertices for the regions.
rx = [0 1 1 0 0];
ry = [0 0 2 2 0];
% Set number of data points.
n = 100;
% Set the inhibition parameter.
```

```
c = 0.5;
X = zeros(n,2);
% Generate the first point.
X(1,:) = csbinproc(rx,ry,1);
```

The following code is similar to the SSI process, except that we now have a mechanism for accepting points that are closer than the inhibition distance.

```
i = 1;
while i<n
  [sx,sy] = csbinproc(rx, ry, 1);
  xt = [sx sy ; X(1:i,:)];
  % Find the distance between the events.
  dist = pdist(xt);
  % Find the distance between the candidate event
  % and the others that have been generated already.
  ind = find(dist(1:i) <= delta);
  m = length(ind);
  if m == 0
    % Then ok to keep the point - nothing is close.
    i = i+1;
    X(i,:) = [sx, sy];
  elseif rand(1) <= c^m
    % The ok to keep the point.
    i = i+1;
    X(i,:) = [sx, sy];
  end
end
```

A spatial point pattern generated from these commands is shown in Figure 12.21.
❑

---

## 12.6 MATLAB Code

The MathWorks has a Mapping Toolbox for MATLAB, which has some functions for spatial statistics. However, the techniques are mostly applicable to geostatistical data. There is also a user-written Spatial Statistics Toolbox that can be downloaded from the internet at

**http://www.spatial-statistics.com/**

As with the Mapping Toolbox, this has functions mostly for continuous spatial data.

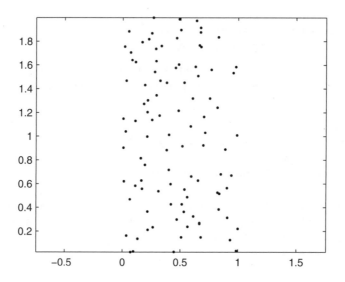

**FIGURE 12.21**
This spatial point pattern was generated from a Strauss process with $\delta = 0.1$ and $c = 0.5$.

We provide functions with the Computational Statistics Toolbox that implement most of the techniques that are described in this chapter. These functions are listed in Table 12.1

**TABLE 12.1**

List of functions from Chapter 12 Included in the Computational Statistics Toolbox

| Purpose | MATLAB Function |
|---|---|
| These functions are used to generate samples from various spatial point processes. | `csbinproc`<br>`csclustproc`<br>`csinhibproc`<br>`cspoissproc`<br>`csstraussproc` |
| This function enables the user to interactively find a study region. | `csgetregion` |
| This is used to estimate the intensity using the quartic kernel. It ignores edge effects. | `csintkern` |
| These functions pertain to the second-order effects of a spatial point pattern. | `csfhat`<br>`csghat`<br>`cskhat` |

## 12.7 Further Reading

For information on the theory for all types of spatial data analysis, we highly recommend Cressie [1993] for a comprehensive treatment of the subject. This text is suitable for scientists and engineers at the graduate level. Those areas that require a higher level of mathematics background are clearly marked. The book has many excellent features, among which are lots of examples that illustrate the concepts and the inclusion of spatial data sets.

We already mentioned the text by Bailey and Gatrell [1995]. This book is another excellent resource for spatial statistics. It includes a discussion of the three types of spatial data (point patterns, geostatistical and lattice data), as well as a fourth type dealing with spatial interaction data. The text has many examples and is easy to understand. For a collection of papers on spatial statistics, we refer the reader to Arlinghaus [1996]. This handbook contains many examples of the application of spatial statistics.

For books that focus mainly on spatial point patterns, we refer the reader to Ripley [1981] and Diggle [1983]. Isaaks and Srivastava [1989] and Journel and Huijbregts [1978] are two texts that discuss geostatistical data. For information on the analysis of lattice data, we recommend Cliff and Ord [1981] and Haining [1993].

---

## Exercises

12.1. We mention in the text that there might be an attribute associated with the spatial point pattern. One way to view this attribute would be to plot the value at each event location rather than the plotting symbol. Load the **okblack** data set. Randomly generate some numbers that would correspond to the dollar amount of the theft at each location. Plot these numbers (attributes) at the locations using the **text** command. Keep in mind that you have to convert the numbers to strings before plotting.

12.2. Repeat the procedure in Example 12.4 using bandwidths of $h = 100, 500$. Plot the estimated intensities. How do they differ from the results in Example 12.4? Which bandwidth is better?

12.3. Using the **bodmin** data, plot a dot map. Does it look like a cluster process is a good model for these events?

12.4. Load the **okwhite** data set. Use the **csgetregion** function to interactively select a boundary. Simply click with the left mouse button at the locations of the vertices for the region. There is no need to close the region. When you are done selecting vertices, right click anywhere in the figure window. The output from this function is a set of vertices for the study region. Plot the event locations and the region.

12.5. Explore the Oklahoma City data sets. Estimate the first-order properties and the second-order properties for both patterns. Do the two sets follow different models?

12.6. Write a MATLAB function that will generate an inhibition process using the thinning approach.

12.7. Repeat Example 12.7 for the point-event nearest neighbor distance distribution. Do you arrive at similar conclusions?

12.8. Repeat Example 12.5. Plot the expression given in Equation 12.12 versus $w$. Does this indicate evidence for departure from CSR?

12.9. The test given in Equation 12.22 suffers from two problems: 1) it is for a fixed $x$, and 2) it is not a powerful test. An alternative would be to use the following test statistic

$$T = \max_x \left| \hat{F}_{Obs}(x) - \hat{F}_{CSR}(x) \right|.$$

Use the Monte Carlo techniques of Chapter 6 to determine whether or not there is significant evidence to reject the null hypothesis (that the point process is CSR). What type of departure from CSR would a large value of $T$ indicate? What type of departure from CSR would a small value of $T$ indicate [Cressie, 1993, p. 636]?

12.10. Generate a realization of a Poisson cluster process. Use your test from problem 12.9 to see if there is significant evidence of clustering.

12.11. Generate a realization of an inhibition process. Apply the nearest-neighbor exploratory graphical techniques ($F$ and $G$ distributions, $K$- and $L$-functions) to see if there is evidence of regularity. Apply the simulation envelope methods to verify that it exhibits regularity.

# Appendix A

## Introduction to MATLAB

### A.1 What Is MATLAB?

MATLAB is a technical computing environment developed by The Math-Works, Inc. for computation and data visualization. It is both an interactive system and a programming language, whose basic data element is an array: scalar, vector, matrix or multi-dimensional array. Besides basic array operations, it offers programming features similar to those of other computing languages (e.g., functions, control flow, etc.).

In this appendix, we provide a brief summary of MATLAB to help the reader understand the algorithms in the text. We do not claim that this introduction is complete, and we urge the reader to learn more about MATLAB from other sources. The documentation that comes with MATLAB is excellent, and the reader should find the tutorials helpful. For a comprehensive overview of MATLAB, we also recommend Hanselman and Littlefield [1998, 2001]. If the reader needs to understand more about the graphics and GUI capabilities in MATLAB, Marchand [1999] is the one to use.

MATLAB will execute on Windows, UNIX, and Linux systems. Here we focus on the Windows version, but most of the information applies to all systems. The main MATLAB software package contains many functions for analyzing data. There are also specialty toolboxes extending the capabilities of MATLAB that are available from The MathWorks and third party vendors. Some toolboxes are also on the internet for free downloading. For more information on these toolboxes, see **http://www.mathworks.com**.

In this text, we use the latest releases of MATLAB (Version 6) and the Statistics Toolbox (Version 3). However, most of the following discussion applies to all versions of MATLAB. We alert the reader to places where they differ.

We assume that readers know how to start MATLAB for their particular platform. When MATLAB is started, you will have a command window with a prompt where you can enter commands. In MATLAB 6, other windows come up (help window, history window, etc.), but we do not cover those here.

## A.2 Getting Help in MATLAB

One useful and important aspect of MATLAB is the **Help** feature. There are many ways to get information about a MATLAB function. Not only does the **Help** provide information about the function, but it also gives references for other related functions. We discuss below the various ways to get help in MATLAB.

- *Command Line:* Typing **help** and then the function name at the command line will, in most cases, tell you everything you need to know about the function. In this text, we do not write about all the capabilities or uses of a function. The reader is strongly encouraged to use command line **help** to find out more. As an example, typing **help plot** at the command line provides lots of useful information about the basic **plot** function. Note that the command line **help** works with the Computational Statistics Toolbox as well.

- *Help Menu:* The **help** files can also be accessed via the usual **Help** menu. This opens up a separate **help** window. Information can be obtained by clicking on links or searching the index (Version 6). In MATLAB 5, you can get a similar window by accessing the **Help Desk** via the **Help** menu.

## A.3 File and Workspace Management

We can enter commands interactively at the command line or save them in an M-file. So, it is important to know some commands for file management. The commands shown in Table A.1 can be used to list, view and delete files.

MATLAB remembers the commands that you enter and all of the values of any variable you create for that session. These variables live in the MATLAB *workspace*. You can recall the variable at any time by typing in the variable name with no punctuation at the end. Note that MATLAB is case sensitive, so **Temp**, **temp**, and **TEMP** represent different variables.

In MATLAB 6, there is a separate command history window. The arrow keys can be used in all versions of MATLAB to recall and edit commands. The up-arrow and down-arrow keys scroll through the commands. The left and right arrows move through the present command. By using these keys, the user can recall commands and edit them using common editing keystrokes.

We can view the contents of the current workspace using the **Workspace Browser**. This is accessed through the **File** menu or the toolbar. All variables in the workspace are listed in the window. The variables can be viewed

TABLE A.1

File Management Commands

| Command | Usage |
|---------|-------|
| `dir, ls` | Shows the files in the present directory. |
| `delete filename` | Deletes *filename*. |
| `cd, pwd` | Show the present directory. |
| `cd dir, chdir` | Changes the directory. In MATLAB 6, there is a pop-up menu on the toolbar that allows the user to change directory. |
| `type filename` | Lists the contents of *filename*. |
| `edit filename` | Brings up *filename* in the editor. |
| `which filename` | Displays the path to *filename*. This can help determine whether a file is part of the standard MATLAB package. |
| `what` | Lists the `.m` files and `.mat` files that are in the current directory. |

TABLE A.2

MATLAB Commands for Workspace Management

| Command | Usage |
|---------|-------|
| `who` | Lists all variables in the workspace. |
| `whos` | Lists all variables in the workspace along with the size in bytes, array dimensions, and object type. |
| `clear` | Removes all variables from the workspace. |
| `clear x y` | Removes variables **x** and **y** from the workspace. |

and edited in a spreadsheet-like window format by double-clicking on the variable name.

The commands contained in Table A.2 help manage the workspace. It is important to be able to get data into MATLAB and to save it. We outline below some of the ways to get data in and out of MATLAB. These are not the only options for file I/O. For example, see **help** on **fprintf**, **fscanf**, and **textread** for more possibilities.

- *Command Line:* The **save** and **load** commands are the main way to perform file I/O in MATLAB. We give some examples of how to use the **save** command. The **load** command works similarly.

| Command | Usage |
|---|---|
| save *filename* | Saves all variables in *filename*.mat. |
| save *filename* var1 var2 | Saves only variables **var1** **var2** in *filename*.mat. |
| save *filename* var1 -ascii | Saves **var1** in ASCII format in *filename*. |

- *File Menu:* There are commands in the **File** menu for saving and loading the workspace.
- *Import Wizard:* In MATLAB 6, there is a spreadsheet-like window for inputting data. To execute the wizard, type **uiimport** at the command line.

## A.4 Punctuation in MATLAB

Table A.3 contains some of the common punctuation characters in MATLAB, and how they are used.

## A.5 Arithmetic Operators

Arithmetic operators ($*$, $/$, $+$, $-$, $\wedge$) in MATLAB follow the convention in linear algebra. If we are multiplying two matrices, **A** and **B**, they must be dimensionally correct. In other words, the number of columns of **A** must be equal to the number of rows of **B**. To multiply, we simply use **A*B**. It is important

**TABLE A.3**

List of MATLAB Punctuation

| Punctuation | Usage |
| --- | --- |
| % | A percent sign denotes a comment line. Information after the % is ignored. |
| , | A comma tells MATLAB to display the results. A blank space works similarly. It also concatenates array elements along a row. |
| ; | A semi-colon suppresses printing the contents of the variable to the screen. It also concatenates array elements along a column. |
| . . . | Three periods denotes the continuation of a statement. Comment statements and variable names cannot be continued with this punctuation. |
| ! | An exclamation tells MATLAB to execute the following as an operating system command. |
| : | The colon specifies a range of numbers. For example, 1:10 means the numbers 1 through 10. A colon in an array dimension accesses all elements in that dimension. |
| . | The period before an operator tells MATLAB to perform the corresponding operation on each element in the array. |

to remember that the default interpretation of an operation is to perform the corresponding array operation.

MATLAB follows the usual order of operations. The precedence can be changed by using parentheses, as in other programming languages.

It is often useful to operate on an array element-by-element. For instance, we might want to square each element of an array. To accomplish this, we add a period before the operator. As an example, to square each element of array **A**, we use **A.^2**. These operators are summarized below in Table A.4.

**TABLE A.4**

List of Element-by-Element Operators in MATLAB

| Operator | Usage |
|:---:|:---:|
| .* | Multiply element-by-element. |
| ./ | Divide element-by-element. |
| .^ | Raise elements to powers. |

## A.6 Data Constructs in MATLAB

### Basic Data Constructs

We do not cover the object-oriented aspects of MATLAB here. Thus, we are concerned mostly with data that are floating point (type **double**) or strings (type **char**). The elements in the arrays will be of these two data types.

The fundamental data element in MATLAB is an array. Arrays can be:

- The $0 \times 0$ empty array created using [ ].
- A $1 \times 1$ scalar array.
- A row vector, which is a $1 \times n$ array.
- A column vector, which is an $n \times 1$ array.
- A matrix with two dimensions, say $m \times n$ or $n \times n$.
- A multi-dimensional array, say $m \times \dots \times n$.

Arrays must always be dimensionally conformal and all elements must be of the same data type. In other words, a $2 \times 3$ matrix must have 3 elements (e.g., numbers) on each of its 2 rows. Table A.5 gives examples of how to access elements of arrays.

### Building Arrays

In most cases, the statistician or engineer will be using outside data in an analysis, so the data would be imported into MATLAB using **load** or some other method described previously. Sometimes, we need to type in simple arrays for testing code or entering parameters, etc. Here we cover some of the ways to build small arrays. Note that this can also be used to concatenate arrays.

Commas or spaces concatenate elements (which can be arrays) as columns. Thus, we get a row vector from the following

```
temp = [1, 4, 5];
```

or we can concatenate two column vectors **a** and **b** into one matrix, as follows

```
temp = [a b];
```

The semi-colon tells MATLAB to concatenate elements as rows. So, we would get a column vector from this command:

```
temp = [1; 4; 5];
```

We note that when concatenating array elements, the sizes must be conformal. The ideas presented here also apply to cell arrays, discussed below.

Before we continue with cell arrays, we cover some of the other useful functions in MATLAB for building arrays. These are summarized here.

| Function | Usage |
|----------|-------|
| **zeros, ones** | These build arrays containing all 0's or all 1's, respectively. |
| **rand, randn** | These build arrays containing uniform (0,1) random variables or standard normal random variables, respectively. See Chapter 4 for more information. |
| **eye** | This creates an identity matrix. |

## Cell Arrays

Cell arrays and structures allow for more flexibility. Cell arrays can have elements that contain any data type (even other cell arrays), and they can be of different sizes. The cell array has an overall structure that is similar to the basic data arrays. For instance, the cells are arranged in dimensions (rows, columns, etc.). If we have a $2 \times 3$ cell array, then each of its 2 rows has to have 3 cells. However, the *content* of the cells can be different sizes and can contain different types of data. One cell might contain **char** data, another **double**, and some can be empty. Mathematical operations are not defined on cell arrays.

In Table A.5, we show some of the common ways to access elements of arrays, which can be cell arrays or basic arrays. With cell arrays, this accesses the cell element, but not the contents of the cells. Curly braces, { }, are used to get to the elements inside the cell. For example, **A{1,1}** would give us the contents of the cell (type **double** or **char**). Whereas, **A(1,1)** is the cell itself

**TABLE A.5**

Examples of Accessing Elements of Arrays

| Notation | Usage |
|---|---|
| **a**(i) | Denotes the *i*-th element (cell) of a row or column vector array (cell array). |
| **A**(:,i) | Accesses the *i*-th column of a matrix or cell array. In this case, the colon in the row dimension tells MATLAB to access all rows. |
| **A**(i,:) | Accesses the *i*-th row of a matrix or cell array. The colon tells MATLAB to gather all of the columns. |
| **A**(1,3,4) | This accesses the element in the first row, third column on the fourth entry of dimension 3 (sometimes called the page). |

and has data type `cell`. The two notations can be combined to access part of the contents of a cell. To get the first two elements of the contents of `A{1,1}`, assuming it contains a vector, we can use

$$A\{1,1\} \ (1:2).$$

Cell arrays are very useful when using strings in plotting functions such as `text`.

Structures are similar to cell arrays in that they allow one to combine collections of dissimilar data into a single variable. Individual structure elements are addressed by names called *fields*. We use the *dot notation* to access the fields. Each element of a structure is called a *record*.

As an example, say we have a structure called **node**, with fields **parent** and **children**. To access the **parent** field of the second node, we use `node(2).parent`. We can get the value of the **child** of the fifth **node** using `node(5).child`. The trees in Chapter 9 and Chapter 10 are programmed using structures.

## A.7 Script Files and Functions

MATLAB programs are saved in M-files. These are text files that contain MATLAB commands, and they are saved with the `.m` extension. Any text editor can be used to create them, but the one that comes with MATLAB is recommended. This editor can be activated using the **File** menu or the toolbar.

When script files are executed, the commands are implemented just as if you typed them in interactively. The commands have access to the workspace and any variables created by the script file are in the workspace when the script finishes executing. To execute a script file, simply type the name of the file at the command line or use the option in the **File** menu.

Script files and functions both have the same **.m** extension. However, a function has a special syntax for the first line. In the general case, this syntax is

```
function [out1,...,outM] = func_name(in1,...,inN)
```

A function does not have to be written with input or output arguments. Whether you have these or not depends on the application and the purpose of the function. The function corresponding to the above syntax would be saved in a file called **func_name.m**. These functions are used in the same way any other MATLAB function is used.

It is important to keep in mind that functions in MATLAB are similar to those in other programming languages. The function has its own workspace. So, communication of information between the function workspace and the main workspace is done via input and output variables.

It is always a good idea to put several comment lines at the beginning of your function. These are returned by the **help** command.

We use a special type of MATLAB function in several examples contained in this book. This is called the **inline** function. This makes a MATLAB **inline** object from a string that represents some mathematical expression or the commands that you want MATLAB to execute. As an optional argument, you can specify the input arguments to the **inline** function object. For example, the variable **gfunc** represents an **inline** object:

```
gfunc = inline('sin(2*pi*f + theta)','f','theta');
```

This calculates the $\sin(2\pi f + \theta)$, based on two input variables: **f** and **theta**. We can now call this function just as we would any MATLAB function.

```
x = 0:.1:4*pi;
thet = pi/2;
ys = gfunc(x, thet);
```

In particular, the **inline** function is useful when you have a simple function and do not want to keep it in a separate file.

## A.8 Control Flow

Most computer languages provide features that allow one to control the flow of execution depending on certain conditions. MATLAB has similar constructs:

- **For** loops
- **While** loops
- **If-else** statements
- **Switch** statement

These should be used sparingly. In most cases, it is more efficient in MATLAB to operate on an entire array rather than looping through it.

### For Loop

The basic syntax for a **for** loop is

```
for i = array
    commands
end
```

Each time through the loop, the loop variable **i** assumes the next value in **array**. The colon notation is usually used to generate a sequence of numbers that **i** will take on. For example,

```
for i = 1:10
```

The commands between the **for** and the **end** statements are executed once for every value in the array. Several **for** loops can be nested, where each loop is closed by **end**.

### While Loop

A **while** loop executes an indefinite number of times. The general syntax is:

```
while expression
    commands
end
```

The commands between the **while** and the **end** are executed as long as **expression** is true. Note that in MATLAB a scalar that is non-zero evaluates to true. Usually a scalar entry is used in the **expression**, but an array can be used also. In the case of arrays, all elements of the resulting array must be true for the commands to execute.

### If-Else Statements

Sometimes, commands must be executed based on a relational test. The **if-else** statement is suitable here. The basic syntax is

```
if expression
   commands
elseif expression
   commands
else
   commands
end
```

Only one **end** is required at the end of the sequence of **if**, **elseif** and **else** statements. Commands are executed only if the corresponding **expression** is true.

### Switch Statement

The **switch** statement is useful if one needs a lot of **if**, **elseif** statements to execute the program. This construct is very similar to that in the C language. The basic syntax is:

```
switch expression
case value1
   commands execute if expression is value1
case value2
   commands execute if expression is value2
...
otherwise
   commands
end
```

**Expression** must be either a scalar or a character string.

---

## A.9 Simple Plotting

For more information on some of the plotting capabilities of MATLAB, the reader is referred to Chapter 5 of this text. Other useful resources are the MATLAB documentation *Using MATLAB Graphics* and *Graphics and GUI's with MATLAB* [Marchand, 1999]. In this appendix, we briefly describe some of the basic uses of **plot** for plotting 2-D graphics and **plot3** for plotting 3-D graphics. The reader is strongly urged to view the **help** file for more information and options for these functions.

When the function **plot** is called, it opens a **Figure** window, if one is not already there, scales the axes to fit the data and plots the points. The default is to plot the points and connect them using straight lines. For example,

```
plot(x,y)
```

plots the values in vector **x** on the horizontal axis and the values in vector **y** on the vertical axis, connected by straight lines. These vectors must be the same size or you will get an error.

Any number of pairs can be used as arguments to **plot**. For instance, the following command plots two curves,

```
plot(x,y1,x,y2)
```

on the same axes. If only one argument is supplied to **plot**, then MATLAB plots the vector versus the index of its values.

The default is a solid line, but MATLAB allows other choices. These are given in Table A.6.

TABLE A.6

Line Styles for Plots

| Notation | Line Type |
|:---:|:---:|
| - | Solid LIne |
| : | Dotted Line |
| -. | Dash-dot Line |
| -- | Dashed line |

If several lines are plotted on one set of axes, then MATLAB plots them as different colors. The predefined colors are listed in Table A.7.

Plotting symbols (e.g., *, **x**, **o**, etc.) can be used for the points. Since the list of plotting symbols is rather long, we refer the reader to the online **help** for **plot** for more information. To plot a curve where both points and a connected curve are displayed, use

```
plot(x, y, x, y, 'b*')
```

This command first plots the points in **x** and **y**, connecting them with straight lines. It then plots the points in **x** and **y** using the symbol * and the color blue.

The **plot3** function works the same as **plot**, except that it takes three vectors for plotting:

```
plot3(x, y, z)
```

TABLE A.7

Line Colors for Plots

| Notation | Color |
|:---:|:---:|
| b | blue |
| g | green |
| r | red |
| c | cyan |
| m | magenta |
| y | yellow |
| k | black |
| w | white |

All of the line styles, colors and plotting symbols apply to **plot3**. Other forms of 3-D plotting (e.g., **surf** and **mesh**) are covered in Chapter 5. Titles and axes labels can be created for all plots using **title, xlabel, ylabel** and **zlabel**.

Before we finish this discussion on simple plotting techniques in MATLAB, we present a way to put several axes or plots in one **figure** window. This is through the use of the **subplot** function. This creates an $m \times n$ matrix of plots (or axes) in the current **figure** window. We provide an example below, where we show how to create two plots side-by-side.

```
% Create the left-most plot.
subplot(1,2,1)
plot(x,y)
% Create the right-most plot
subplot(1,2,2)
plot(x,z)
```

The first two arguments to **subplot** tell MATLAB about the layout of the plots within the **figure** window. The third argument tells MATLAB which plot to work with. The plots are numbered from top to bottom and left to right. The most recent plot that was created or worked on is the one affected by any subsequent plotting commands. To access a previous plot, simply use the **subplot** function again with the proper value for the third argument **p**. You can think of the **subplot** function as a *pointer* that tells MATLAB what set of axes to work with.

Through the use of MATLAB's low-level Handle Graphics functions, the data analyst has complete control over graphical output. We do not present any of that here, because we make limited use of these capabilities. However, we urge the reader to look at the online **help** for **propedit**. This graphical user interface allows the user to change many aspects or properties of the plots.

---

## A.10 Contact Information

For MATLAB product information, please contact:

> The MathWorks, Inc.
> 3 Apple Hill Drive
> Natick, MA, 01760-2098 USA
> Tel: 508-647-7000
> Fax: 508-647-7101
> E-mail: info@mathworks.com
> Web: www.mathworks.com

There are two useful resources that describe new products, programming tips, algorithm development, upcoming events, etc. One is the monthly electronic newsletter called the *MATLAB Digest*. Another is called *MATLAB News & Notes*, published quarterly. You can subscribe to both of these at `www.mathworks.com` or send an email request to

> `subscribe@mathworks.com`

Back issues of these documents are available on-line.

# Appendix B

## *Index of Notation*

### Single Letters

| | |
|---|---|
| $B_k$ | Histogram bin (page 114) |
| $d$ | Dimensionality |
| $h$ | Bin width or smoothing parameter (page 114) |
| $H_0$ | Null hypothesis (page 192) |
| $H_1$ | Alternative hypothesis (page 193) |
| $M_r$ | Sample central moment (page 54) |
| $n$ | Sample size |
| $p$ | Probability |
| $q_p$ | Quantile (page 70) |
| $S^2$ | Sample variance (page 54) |
| $T$ | Statistic (page 61) |
| $T^{(-i)}$ | Jackknife replicate (page 240) |
| $U$ | Uniform (0,1) random variable (page 79) |
| $X$ | A random variable (page 13) |
| $X_{(i)}$ | Order statistic (page 69) |
| $\bar{X}$ | Sample mean (page 53) |
| $\mathbf{x}^* = (x_1^*, ..., x_n^*)$ | Bootstrap sample (page 214) |
| $Z$ | Standard normal random variable (page 31) |

### Other

| | |
|---|---|
| $E[X]$ | Expected value of X (page 21) |
| $f(x)$ | Probability mass or density function (page 14) |

| $F(x)$ | Cumulative distribution function (page 15) |
|---|---|
| | Nearest neighbor point-event cdf (page 479) |
| $f(x, y)$ | Joint probability (mass) function (page 52) |
| $G(w)$ | Nearest neighbor event-event cdf (page 479) |
| $K(d)$ | $K$-function (page 483) |
| $K(t)$ | Kernel (page 280) |
| $L(d)$ | $L$-function (page 484) |
| $L(\theta; x_1, \ldots, x_n)$ | Likelihood function (page 63) |
| $L_R(\mathbf{x})$ | Likelihood ratio (page 331) |
| $P(E)$ | Probability of event $E$ (page 14) |
| $P(E\|F)$ | Conditional probability (page 17) |
| $P(\mathbf{x}\|\omega_j)$ | Class-conditional probability (page 320) |
| $P(\omega_j)$ | Prior probability (page 320) |
| $P(\omega_j\|\mathbf{x})$ | Posterior probability (page 320) |
| $q(. \|X_t)$ | Proposal distribution - MCMC (page 431) |
| $R(f)$ | Roughness (page 265) |
| $V(X)$ | Variance of $X$ (page 21) |

## Greek Symbols

| $\alpha$ | Probability of Type I error (page 195) |
|---|---|
| $\beta$ | Probability of Type II error (page 197) |
| $\alpha(t)$ | Projection vector - grand tour (page 180) |
| $\beta(t)$ | Projection vector - grand tour (page 180) |
| $\alpha(X_t, Y)$ | Acceptance probability - MCMC (page 431) |
| $\hat{\varepsilon}_i$ | Residuals (page 387) |
| $\hat{\theta}^{*b}$ | Bootstrap replicate (page 215) |
| $\lambda(\mathbf{s})$ | Intensity (page 468) |
| $\mu_r$ | $r$-th central moment (page 22) |
| $\mu$ | Mean (page 31) |
| $\nu_k$ | Histogram bin heights (page 114) |
| $\pi(\mathbf{x})$ | Target distribution - MCMC (page 430) |

| | |
|---|---|
| $\rho_{xy}$ | Correlation coefficient (page 56) |
| $\sigma^2$ | Variance (page 31) |
| $\Sigma$ | Covariance matrix (page 41) |
| $\phi(x;\mu, \sigma^2)$ | Standard normal probability density function (page 289) |
| $\Phi$ | Standard normal cdf (page 31) |
| $\psi$ | Stationary distribution - MCMC (page 429) |
| $\omega_j$ | Class $j$ (page 319) |

## Acronyms

| | |
|---|---|
| cdf | Cumulative distribution function |
| CSR | Complete spatial randomness (page 469) |
| EDA | Exploratory data analysis (page 111) |
| IQR | Interquartile range (page 71) |
| ISE | Integrated squared error (page 260) |
| MCMC | Markov chain Monte Carlo (page 425) |
| MIAE | Mean integrated absolute error (page 260) |
| MISE | Mean integrated squared error (page 260) |
| MSE | Mean squared error (page 260) |
| pdf | Probability density function |
| PE | Prediction error (page 236) |
| RSE | Residual squared error (page 234) |
| SE | Standard error (page 62) |

# Appendix C

## Projection Pursuit Indexes

In this appendix, we list several indexes for projection pursuit [Posse, 1995b], and we also provide the M-file source code for the functions included in the Computational Statistics Toolbox.

### C.1 Indexes

Since structure is considered to be departures from normality, these indexes are developed to detect non-normality in the projected data. There are some criteria that we can use to assess the usefulness of projection indexes. These include affine invariance [Huber, 1985], speed of computation, and sensitivity to departure from normality in the core of the distribution rather than the tails. The last criterion ensures that we are pursuing structure and not just outliers.

#### Friedman-Tukey Index

This projection pursuit index [Friedman and Tukey, 1974] is based on inter-point distances and is calculated using the following

$$PI_{FT}(\alpha, \beta) = \sum_{i=1}^{n} \sum_{j=1}^{n} (R^2 - r_{ij}^2)^3 \mathbf{1}(R^2 - r_{ij}^2),$$

where $R = 2.29n^{-1/5}$, $r_{ij}^2 = (z_i^\alpha - z_j^\alpha)^2 + (z_i^\beta - z_j^\beta)^2$, and $\mathbf{1}()$ is the indicator function for positive values,

$$\mathbf{1}(x) = \begin{cases} 1; & x > 0 \\ 0; & x \le 0. \end{cases}$$

This index has been revised from the original to be affine invariant [Swayne, Cook and Buja, 1991] and has computational order $O(n^2)$.

## Entropy Index

This projection pursuit index [Jones and Sibson, 1987] is based on the entropy and is given by

$$PI_E(\alpha, \beta) = \frac{1}{n}\sum_{i=1}^{n}\log\left[\frac{1}{nh_\alpha h_\beta}\sum_{j=1}^{n}\phi_2\left(\frac{(z_i^\alpha - z_j^\alpha)}{h_\alpha}, \frac{(z_i^\beta - z_j^\beta)}{h_\beta}\right)\right] + \log(2\pi e),$$

where $\phi_2$ is the bivariate standard normal density. The bandwidths $h_\gamma, \gamma = \alpha, \beta$ are obtained from

$$h_\gamma = 1.06n^{-1/5}\left(\sum_{i=1}^{n}\left\{z_i^\gamma - \sum_{j=1}^{n}z_j^\gamma/n\right\}^2/(n-1)\right)^{\frac{1}{2}}.$$

This index is also $O(n^2)$.

## Moment Index

This index was developed in Jones and Sibson [1987] and is based on bivariate third and fourth moments. This is very fast to compute, so it is useful for large data sets. However, a problem with this index is that it tends to locate structure in the tails of the distribution. It is given by

$$PI_M(\alpha, \beta) = \frac{1}{12}\left\{\kappa_{30}^2 + 3\kappa_{21}^2 + 3\kappa_{12}^2 + \kappa_{03}^2 + \frac{1}{4}(\kappa_{40}^2 + 4\kappa_{31}^2 + 6\kappa_{22}^2 + 4\kappa_{13}^2 + \kappa_{04}^2)\right\},$$

where

$$\kappa_{30} = \frac{n}{(n-1)(n-2)}\sum_{i=1}^{n}(z_i^\alpha)^3 \qquad \kappa_{03} = \frac{n}{(n-1)(n-2)}\sum_{i=1}^{n}(z_i^\beta)^3$$

$$\kappa_{31} = \frac{n(n+1)}{(n-1)(n-2)(n-3)} \sum_{i=1}^{n} (z_i^{\alpha})^3 z_i^{\beta}$$

$$\kappa_{13} = \frac{n(n+1)}{(n-1)(n-2)(n-3)} \sum_{i=1}^{n} (z_i^{\beta})^3 z_i^{\alpha}$$

$$\kappa_{04} = \frac{n(n+1)}{(n-1)(n-2)(n-3)} \left\{ \sum_{i=1}^{n} (z_i^{\beta})^4 - \frac{3(n-1)^3}{n(n+1)} \right\}$$

$$\kappa_{40} = \frac{n(n+1)}{(n-1)(n-2)(n-3)} \left\{ \sum_{i=1}^{n} (z_i^{\alpha})^4 - \frac{3(n-1)^3}{n(n+1)} \right\}$$

$$\kappa_{22} = \frac{n(n+1)}{(n-1)(n-2)(n-3)} \left\{ \sum_{i=1}^{n} (z_i^{\alpha})^2 (z_i^{\beta})^2 - \frac{(n-1)^3}{n(n+1)} \right\}$$

$$\kappa_{21} = \frac{n}{(n-1)(n-2)} \sum_{i=1}^{n} (z_i^{\alpha})^2 z_i^{\beta} \qquad \kappa_{12} = \frac{n}{(n-1)(n-2)} \sum_{i=1}^{n} (z_i^{\beta})^2 z_i^{\alpha} .$$

## $L^2$ Distances

Several indexes estimate the $L^2$ distance between the density of the projected data and a bivariate standard normal density. The $L^2$ projection indexes use orthonormal polynomial expansions to estimate the marginal densities of the projected data. One of these proposed by Friedman [1987] uses Legendre polynomials with $J$ terms. Note that MATLAB has a function for obtaining these polynomials called **legendre**.

$$PI_{Leg}(\alpha, \beta) = \frac{1}{4} \left\{ \sum_{j=1}^{J} (2j+1) \left( \frac{1}{n} \sum_{i=1}^{n} P_j y_i^{\alpha} \right)^2 \right.$$

$$+ \sum_{k=1}^{J} (2k+1) \left( \frac{1}{n} \sum_{i=1}^{n} P_k y_i^{\beta} \right)^2$$

$$+ \left. \sum_{j=1}^{J} \sum_{k=1}^{J-j} (2j+1)(2k+1) \left( \frac{1}{n} \sum_{i=1}^{n} P_j(y_i^{\alpha}) P_k(y_i^{\beta}) \right)^2 \right\}$$

where $P_a(\ )$ is the Legendre polynomial of order $a$. This index is not affine invariant, so Morton [1989] proposed the following revised index. This is based on a conversion to polar coordinates as follows

$$\rho = (z^\alpha)^2 + (z^\beta)^2 \qquad \theta = \text{atan}\left(\frac{z^\beta}{z^\alpha}\right).$$

We then have the following index where Fourier series and Laguerre polynomials are used:

$$PI_{LF}(\alpha, \beta) = \frac{1}{\pi}\sum_{l=0}^{L}\sum_{k=1}^{K}\left[\left(\frac{1}{n}\sum_{i=1}^{n}L_l(\rho_i)\exp(-\rho_i/2)\cos(k\theta_i)\right)^2\right.$$

$$\left.+\left(\frac{1}{n}\sum_{i=1}^{n}L_l(\rho_i)\exp(-\rho_i/2)\sin(k\theta_i)\right)^2\right] + \frac{1}{2\pi}\sum_{l=0}^{L}\left(\frac{1}{n}\sum_{i=1}^{n}L_l(\rho_i)\exp(-\rho_i/2)\right)^2$$

$$-\frac{1}{2\pi n}\sum_{i=1}^{n}\exp(-\rho_i/2) + \frac{1}{8\pi},$$

where $L_a$ represents the Laguerre polynomial of order $a$. Two more indexes based on the $L^2$ distance using expansions in Hermite polynomials are given in Posse [1995b].

## C.2 MATLAB Source Code

The first function we look at is the one to calculate the chi-square projection pursuit index.

```
function ppi = csppind(x,a,b,n,ck)

% x is the data, a and b are the projection vectors,
% n is the number of data points, and ck is the value
% of the standard normal bivariate cdf for the boxes.
z = zeros(n,2);
ppi = 0;
pk = zeros(1,48);
eta = pi*(0:8)/36;
delang = 45*pi/180;
```

```
delr = sqrt(2*log(6))/5;
angles = 0:delang:(2*pi);
rd = 0:delr:5*delr;
nr = length(rd);
na=length(angles);

for j = 1:9
   % find rotated plane
   aj = a*cos(eta(j))-b*sin(eta(j));
   bj = a*sin(eta(j))+b*cos(eta(j));
   % project data onto this plane
   z(:,1) = x*aj;
   z(:,2) = x*bj;
   % convert to polar coordinates
   [th,r] = cart2pol(z(:,1),z(:,2));
   % find all of the angles that are negative
   ind = find(th<0);
   th(ind) = th(ind)+2*pi;
   % find # points in each box
   for i=1:(nr-1)% loop over each ring
      for k=1:(na-1)% loop over each wedge
         ind = ...
            find(r>rd(i) & r<rd(i+1) & ...
            th>angles(k) & th<angles(k+1));
         pk((i-1)*8+k)=...
            (length(ind)/n-ck((i-1)*8+k))^2...
            /ck((i-1)*8+k);
      end
   end
   % find the number in the outer line of boxes
   for k=1:(na-1)
      ind=...
         find(r>rd(nr) & th>angles(k) & ...
         th<angles(k+1));
      pk(40+k)=(length(ind)/n-(1/48))^2/(1/48);
   end
   ppi = ppi+sum(pk);
end
ppi = ppi/9;
```

Any of the other indexes can be coded in an M-file function and called by the **csppeda** function given below. You would call your function instead of **csppind**.

```
function [as,bs,ppm]=csppeda(Z,c,half,m)

% Z is the sphered data.
```

```
% get the necessary constants
[n,p] = size(Z);
maxiter = 1500;
cs = c;
cstop = 0.00001;
cstop = 0.01;
as = zeros(p,1);% storage for the information
bs = zeros(p,1);
ppm = realmin;

% find the probability of bivariate standard normal
% over each radial box.
% NOTE: the user could put the values in to ck to
% prevent re-calculating each time. We thought the
% reader would be interested in seeing how we did
% it.
% NOTE: MATLAB 5 users should use the function
% quad8 instead of quadl.
fnr = inline('r.*exp(-0.5*r.^2)','r');
ck = ones(1,40);
ck(1:8) = quadl(fnr,0,sqrt(2*log(6))/5)/8;
ck(9:16) = quadl(fnr,sqrt(2*log(6))/5,...
    2*sqrt(2*log(6))/5)/8;
ck(17:24) = quadl(fnr,2*sqrt(2*log(6))/5,...
    3*sqrt(2*log(6))/5)/8;
ck(25:32) = quadl(fnr,3*sqrt(2*log(6))/5,...
    4*sqrt(2*log(6))/5)/8;
ck(33:40) = quadl(fnr,4*sqrt(2*log(6))/5,...
    5*sqrt(2*log(6))/5)/8;

for i=1:m
    % generate a random starting plane
    % this will be the current best plane
    a = randn(p,1);
    mag = sqrt(sum(a.^2));
    astar = a/mag;
    b = randn(p,1);
    bb = b-(astar'*b)*astar;
    mag = sqrt(sum(bb.^2));
    bstar = bb/mag;
    clear a mag b bb
    % find the projection index for this plane
    % this will be the initial value of the index
    ppimax = csppind(Z,astar,bstar,n,ck);

    % keep repeating this search until the value
```

```
% c becomes less than cstop or until the
% number of iterations exceeds maxiter
mi = 0;
% number of iterations without increase in index
h = 0;
c = cs;
while (mi < maxiter) & (c > cstop)
   % generate a p-vector on the unit sphere
   v = randn(p,1);
   mag = sqrt(sum(v.^2));
   v1 = v/mag;
   % find the a1,b1 and a2,b2 planes
   t = astar+c*v1;
   mag = sqrt(sum(t.^2));
   a1 = t/mag;
   t = astar-c*v1;
   mag = sqrt(sum(t.^2));
   a2 = t/mag;
   t = bstar-(a1'*bstar)*a1;
   mag = sqrt(sum(t.^2));
   b1 = t/mag;
   t = bstar-(a2'*bstar)*a2;
   mag = sqrt(sum(t.^2));
   b2 = t/mag;
   ppi1 = csppind(Z,a1,b1,n,ck);
   ppi2 = csppind(Z,a2,b2,n,ck);
   [mp,ip] = max([ppi1,ppi2]);
   if mp > ppimax
      % then reset plane and index to this value
      eval(['astar=a' int2str(ip) ';']);
      eval(['bstar=b' int2str(ip) ';']);
      eval(['ppimax=ppi' int2str(ip) ';']);
   else
      h = h+1;% no increase
   end
   mi = mi+1;
   if h==half% then decrease the neighborhood
      c = c*.5;
      h = 0;
   end
end
if ppimax > ppm
   % save the current projection as a best plane
   as = astar;
   bs = bstar;
   ppm = ppimax;
```

```
        end
    end
```

Finally, we provide the following function for removing the structure from a projection found using PPEDA.

```
function X = csppstrtrem(Z,a,b)

% maximum number of iterations allowed
maxiter = 5;
[n,d] = size(Z);

% find the orthonormal matrix needed via Gram-Schmidt
U = eye(d,d);
U(1,:) = a';% vector for best plane
U(2,:) = b';
for i = 3:d
    for j = 1:(i-1)
        U(i,:) = U(i,:)-(U(j,:)*U(i,:)')*U(j,:);
    end
    U(i,:) = U(i,:)/sqrt(sum(U(i,:).^2));
end

% Transform data using the matrix U.
% To match Friedman's treatment: T is d x n.
T = U*Z';
% These should be the 2-d projection that is 'best'.
x1 = T(1,:);
x2 = T(2,:);

% Gaussianize the first two rows of T.
% set of vector of angles
gam = [0,pi/4, pi/8, 3*pi/8];
for m = 1:maxiter
    % gaussianize the data
    for i=1:4
        % rotate about origin
        xp1 = x1*cos(gam(i))+x2*sin(gam(i));
        xp2 = x2*cos(gam(i))-x1*sin(gam(i));
        % Transform to normality
        [m,rnk1] = sort(xp1);   % get the ranks
        [m,rnk2] = sort(xp2);
        arg1 = (rnk1-0.5)/n;% get the arguments
        arg2 = (rnk2-0.5)/n;
        x1 = norminv(arg1,0,1); % transform to normality
        x2 = norminv(arg2,0,1);
    end
```

```
end

% Set the first two rows of T to the
% Gaussianized values.
T(1,:) = x1;
T(2,:) = x2;
X = (U'*T)';
```

# Appendix D

*MATLAB Code*

In this appendix, we provide the MATLAB functions for some of the more complicated techniques covered in this book. This includes code for the bootstrap $BC_a$ confidence interval, the adaptive mixtures algorithm for probability density estimation, classification trees, and regression trees.

## D.1 Bootstrap $BC_a$ Confidence Interval

```
function [blo,bhi,bvals,z0,ahat]=...
    csbootbca(data,fname,B,alpha)
thetahat = feval(fname,data);
[bh,se,bt] = csboot(data,fname,50);
[n,d] = size(data);
bvals = zeros(B,1);
% Loop over each resample and
% calculate the bootstrap replicates.
for i = 1:B
 % generate the indices for the B bootstrap
 % resamples, sampling with
 % replacement using the discrete uniform.
 ind = ceil(n.*rand(n,1));
 % extract the sample from the data
 % each row corresponds to a bootstrap resample
 xstar = data(ind,:);
 % use feval to evaluate the estimate for
 % the i-th resample
 bvals(i) = feval(fname, xstar);
end
numless = length(find(bvals<thetahat));
z0 = norminv(numless/B,0,1);
% find the estimate for acceleration using jackknife
jvals = zeros(n,1);
```

```
for i = 1:n
    % use feval to evaluate the estimate
    % with the i-th observation removed
    % These are the jackknife replications.
    jvals(i) =...
        feval(fname, [data(1:(i-1));data((i+1):n)]);
end
num = (mean(jvals)-jvals).^3;
den = (mean(jvals)-jvals).^2;
ahat = sum(num)/(6*sum(den)^(3/2));
zlo = norminv(alpha/2,0,1);   % this is the z^(a/2)
zup = norminv(1-alpha/2,0,1); % this is the z^(1-a/2)
% Equation 14.10, E & T
arg = z0 + (z0 + zlo)/(1-ahat*(z0+zlo));
alpha1 = normcdf(arg,0,1);
arg = z0 + (z0 + zup)/(1-ahat*(z0+zup));
alpha2 = normcdf(arg,0,1);
k1 = floor(((B+1)*alpha1));
k2 = ceil(((B+1)*alpha2));    % ???
sbval = sort(bvals);
blo = sbval(k1);
bhi = sbval(k2);
```

## D.2 Adaptive Mixtures Density Estimation

First we provide some of the helper functions that are used in **csadpmix**. This first function calculates the estimated posterior probability, given the current estimated model and the new observation.

```
% function post=rpostup(x,pies,mus,vars,nterms)
% This function will return the posterior.

function post = rpostup(x,pies,mus,vars,nterms)
f = exp(-.5*(x-mus(1:nterms)).^2./...
    vars(1:nterms)).*pies(1:nterms);
f = f/sum(f);
post = f;
```

Next we need a function that will update the mixing coefficients, the means and the variances using the posteriors and the new data point.

```
% This function will update all of the parameters for
% the adaptive mixtures density estimation approach
```

```
function [piess,muss,varss]=...
    csrup(x,pies,mus,vars,posterior,nterms,n)
inertvar = 10;
betan = 1/(n);
piess = pies(1:nterms);
muss = mus(1:nterms);
varss = vars(1:nterms);
post = posterior(1:nterms);
% update the mixing coefficients
piess = piess+(post-piess)*betan;
% update the means
muss = muss+betan*post.*(x-muss)./piess;
% update the variances
denom = (1/betan)*piess+inertvar;
varss = varss+post.*((x-muss).^2-varss)./denom;
```

Finally, the following function will set the initial variance for newly created terms.

```
% This function will update the variances
% in the AMDE. Call with nterms-1,
% since new term is based only on previous terms

function newvar = cssetvar(mus,pies,vars,x,nterms)
f=exp(-.5*(x-mus(1:nterms))...
    .^2./vars(1:nterms)).*pies(1:nterms);
f = f/sum(f);
f = f.*vars(1:nterms);
newvar = sum(f);
```

Here is the main MATLAB function **csadpmix** that ties everything together. For brevity, we show only the part of the function that corresponds to the univariate case. View the M-file for the multivariate case.

```
function [pies,mus,vars] = cadpmix(x,maxterms)
n = length(x);
mus = zeros(1,maxterms);
vars = zeros(1,maxterms);
pies = zeros(1,maxterms);
posterior = zeros(1,maxterms);
tc = 1;
% lower bound on new pies
minpie = .00001;
% bound on variance
sievebd = 1000;
% initialize density to first data point
nterms = 1;
```

```
mus(1) = x(1);
% rule of thumb for initial variance - univariate
vars(1) = (std(x))^2/2.5;
pies(1) = 1;
% loop through all of the data points
for i = 2:n
   md = ((x(i)-mus(1:nterms)).^2)./vars(1:nterms);
   if min(md)>tc & nterms<maxterms
     create = 1;
   else
     create = 0;
   end
   if create == 0 % update terms
      posterior(1:nterms)=...
         csrpostup(x(i),pies,mus,vars,nterms);
      [pies(1:nterms),mus(1:nterms),...
         vars(1:nterms)]=csrup(x(i),pies,mus,...
         vars,posterior,nterms,i);
   else  % create a new term
      nterms = nterms+1;
      mus(nterms) = x(i);
      pies(nterms) = max([1/(i),minpie]);
     % update pies
     pies(1:nterms-1)=...
         pies(1:nterms-1)*(1-pies(nterms));
      vars(nterms)=...
         cssetvar(mus,pies,vars,x(i),nterms-1);
   end   % end if statement
   % to prevent spiking of variances
   index = find(vars(1:nterms)<1/(sievebd*nterms));
   vars(index) = ones(size(index))/(sievebd*nterms);
end    % for i loop
% clean up the model - get rid of the 0 terms
mus((nterms+1):maxterms) = [];
pies((nterms+1):maxterms) = [];
vars((nterms+1):maxterms) = [];
```

## D.3 Classification Trees

In the interest of space, we only include (in the text) the MATLAB code for growing a classification tree. All of the functions for working with trees are included with the Computational Statistics Toolbox, and the reader can easily view the source code for more information.

```
function tree = csgrowc(X,maxn,clas,Nk,pies)

[n,dd] = size(X);
if nargin == 4% then estimate the pies
   pies = Nk/n;
end
% The tree will be implemented as a structure.
% get the initial tree - which is the data set itself
tree.pies = pies;
% need for node impurity calcs:
tree.class = clas;
tree.Nk = Nk;
% maximum number to be allowed in the terminal nodes:
tree.maxn = maxn;
% number of nodes in the tree - total:
tree.numnodes = 1;
% vector of terminal nodes:
tree.termnodes = 1;
% 1=terminal node, 0=not terminal:
tree.node.term = 1;
% total number of points in the node:
tree.node.nt = sum(Nk);
tree.node.impurity = impure(pies);
tree.node.misclass = 1-max(pies);
% prob it is node t:
tree.node.pt = 1;
% root node has no parent
tree.node.parent = 0;
% This will be a 2 element vector of
% node numbers to the children.
tree.node.children = [];
% pointer to sibling node:
tree.node.sibling = [];
% the class membership associated with this node:
tree.node.class = [];
% the splitting value:
tree.node.split = [];
% the variable or dimension that will be split:
tree.node.var = [];
% number of points from each class in this node:
tree.node.nkt = Nk;
% joint prob it is class k and it falls into node t
tree.node.pjoint = pies;
% prob it is class k given node t
tree.node.pclass = pies;
% the root node contains all of the data:
```

```
tree.node.data = X;

% Now get started on growing the very large tree.
% first we have to extract the number of terminal nodes
% that qualify for splitting.
% get the data needed to decide to split the node
[term,nt,imp]=getdata(tree);
% find all of the nodes that qualify for splitting
ind = find( (term==1) & (imp>0) & (nt>maxn) );
% now start splitting
while ~isempty(ind)
  for i=1:length(ind)% check all of them
    % get split
    [split,dim]=...
          splitnode(tree.node(ind(i)).data,...
          tree.node(ind(i)).impurity,...
          tree.class,tree.Nk,tree.pies);
    % split the node
    tree = addnode(tree,ind(i),dim,split);
  end  % end for loop
  [term,nt,imp]=getdata(tree);
  tree.termnodes = find(term==1);
  ind = find( (term==1) & (imp>0) & (nt>maxn) );
  length(tree.termnodes);
  itmp = find(term==1);
end  % end while loop
```

## D.4 Regression Trees

Below is the function for growing a regression tree. The complete set of functions needed for working with regression trees is included with the Computational Statistics Toolbox.

```
function tree = csgrowr(X,y,maxn)

n = length(y);
% The tree will be implemented as a structure
tree.maxn = maxn;
tree.n = n;
tree.numnodes = 1;
tree.termnodes = 1;
tree.node.term = 1;
tree.node.nt = n;
```

```
tree.node.impurity = sqrer(y,tree.n);
tree.node.parent = 0;
tree.node.children = [];
tree.node.sibling = [];
tree.node.yhat = mean(y);
tree.node.split = [];
tree.node.var = [];
tree.node.x = X;
tree.node.y = y;

% Now get started on growing the tree very large
[term,nt,imp]=getdata(tree);
% find all of the nodes that qualify for splitting
ind = find( (term==1) & (imp>0) & (nt>maxn) );
% now start splitting
while ~isempty(ind)
   for i=1:length(ind)
      % get split
      [split,dim]=splitnoder(...
         tree.node(ind(i)).x,...
         tree.node(ind(i)).y,...
         tree.node(ind(i)).impurity,...
         tree.n);
      % split the node
      tree = addnoder(tree,ind(i),dim,split);
   end  % end for loop
   [term,nt,imp]=getdata(tree);
   tree.termnodes = find(term==1);
   ind = find( (term==1) & (imp>0) & (nt>maxn) );
end  % end while loop
```

# Appendix E

## MATLAB *Statistics Toolbox*

The following tables list the functions that are available in the MATLAB Statistics Toolbox, Version 3.0. This toolbox is available for purchase from The MathWorks, Inc.

### TABLE E.1

Functions for Parameter Estimation (**fit**) and Distribution Statistics - Mean and Variance (**stat**)

| Function | Purpose |
|---|---|
| betafit, betastat | Beta distribution. |
| binofit, binostat | Binomial distribution. |
| expfit, expstat | Exponential distribution. |
| fstat | F distribution |
| gamfit, gamstat | Gamma distribution. |
| geostat | Geometric distribution |
| hygestat | Hypergeometric distribution |
| lognstat | Lognormal distribution |
| mle | Maximum likelihood parameter estimation. |
| nbinstat | Negative binomial distribution |
| ncfstat | Noncentral F distribution |
| nctstat | Noncentral t distribution |
| ncx2stat | Noncentral Chi-square distribution |
| normfit, normstat | Normal distribution. |
| poissfit, poisstat | Poisson distribution. |
| raylfit | Rayleigh distribution. |
| tstat | T distribution |
| unidstat | Discrete uniform distribution |
| unifit, unifstat | Uniform distribution. |
| weibfit, weibstat | Weibull distribution. |

## TABLE E.2

Probability Density Functions (**pdf**) and Cumulative Distribution
Functions (**cdf**)

| Function | Purpose |
| --- | --- |
| `betapdf, betacdf` | Beta distribution |
| `binopdf, binocdf` | Binomial distribution |
| `chi2pdf, chi2cdf` | Chi-square distribution |
| `exppdf, expcdf` | Exponential distribution |
| `fpdf, fcdf` | F distribution |
| `gampdf, gamcdf` | Gamma distribution |
| `geopdf, geocdf` | Geometric distribution |
| `hygepdf, hygecdf` | Hypergeometric distribution |
| `lognpdf, logncdf` | Log normal distribution |
| `nbinpdf, nbincdf` | Negative binomial distribution |
| `ncfpdf, ncfcdf` | Noncentral F distribution |
| `nctpdf, nctcdf` | Noncentral t distribution |
| `ncx2pdf, ncx2cdf` | Noncentral chi-square distribution |
| `normpdf, normcdf` | Normal distribution |
| `pdf, cdf` | Probability density/Cumulative distribution |
| `poisspdf, poisscdf` | Poisson distribution |
| `raylpdf, raylcdf` | Rayleigh distribution |
| `tpdf, tcdf` | T distribution |
| `unidpdf, unidcdf` | Discrete uniform distribution |
| `unifpdf, unifcdf` | Continuous uniform distribution |
| `weibpdf, weibcdf` | Weibull distribution |

TABLE E.3

Critical Values (**inv**) and Random Number Generation (**rnd**) for Probability Distribution Functions

| Function | Purpose |
|---|---|
| **betainv, betarnd** | Beta distribution |
| **binoinv, binornd** | Binomial distribution |
| **chi2inv, chi2rnd** | Chi-square distribution |
| **expinv, exprnd** | Exponential distribution |
| **finv, frnd** | F distribution |
| **gaminv, gamrnd** | Gamma distribution |
| **geoinv, geornd** | Geometric distribution |
| **hygeinv, hygernd** | Hypergeometric distribution |
| **logninv, lognrnd** | Log normal distribution |
| **nbininv, nbinrnd** | Negative binomial distribution |
| **ncfinv, ncfrnd** | Noncentral F distribution |
| **nctinv, nctrnd** | Noncentral t distribution |
| **ncx2inv, ncx2rnd** | Noncentral chi-square distribution |
| **norminv, normrnd** | Normal distribution |
| **poissinv, poissrnd** | Poisson distribution |
| **raylinv, raylrnd** | Rayleigh distribution |
| **tinv, trnd** | T distribution |
| **unidinv, unidrnd** | Discrete uniform distribution |
| **unifinv, unifrnd** | Continuous uniform distribution |
| **weibinv, weibrnd** | Weibull distribution |
| **icdf** | Specified inverse cdf |

**TABLE E.4**

Descriptive Statistics

| Function | Purpose |
|---|---|
| `bootstrp` | Bootstrap statistics for any function. |
| `corrcoef` | Correlation coefficient - also in standard MATLAB |
| `cov` | Covariance - also in standard MATLAB |
| `crosstab` | Cross tabulation |
| `geomean` | Geometric mean |
| `grpstats` | Summary statistics by group |
| `harmmean` | Harmonic mean |
| `iqr` | Interquartile range |
| `kurtosis` | Kurtosis |
| `mad` | Median absolute deviation |
| `mean` | Sample average - also in standard MATLAB |
| `median` | Second quartile (50th percentile) of a sample - also in standard MATLAB |
| `moment` | Moments of a sample |
| `nanmax, nanmin` | Maximum/minimum - ignoring NaNs |
| `nanmean, nanmedian` | Mean/median - ignoring NaNs |
| `nanstd, namsum` | Standard deviation/sum - ignoring NaNs |
| `prctile` | Percentiles |
| `range` | Range |
| `skewness` | Skewness |
| `std` | Standard deviation - also in standard MATLAB |
| `tabulate` | Frequency table |
| `trimmean` | Trimmed mean |
| `var` | Variance - also in standard MATLAB |

**TABLE E.5**

Linear Models

| Function | Purpose |
|:---:|:---|
| anova1 | One-way analysis of variance |
| anova2 | Two-way analysis of variance |
| anovan | n-way analysis of variance |
| aoctool | Interactive tool for analysis of covariance |
| dummyvar | Dummy-variable coding |
| friedman | Friedman's test |
| glmfit | Generalized linear model fitting |
| kruskalwallis | Kruskal-Wallis test |
| lscov | Least-squares estimates with known covariance matrix |
| manova1 | One-way multivariate analysis of variance |
| manovacluster | Draw clusters of group means for manova1 |
| multcompare | Multiple comparisons of means and other estimates |
| polyconf | Polynomial evaluation and confidence interval estimation |
| polyfit | Least-squares polynomial fitting- also in standard MATLAB |
| polyval | Predicted values for polynomial functions- also in standard MATLAB |
| rcoplot | Residuals case order plot |
| regress | Multivariate linear regression |
| regstats | Regression diagnostics |
| ridge | Ridge regression |
| robustfit | Robust regression model fitting |
| rstool | Multidimensional response surface visualization |
| stepwise | Interactive tool for stepwise regression |
| x2fx | Factor setting matrix (x) to design matrix (fx) |

TABLE E.6

Nonlinear Models

| Function | Purpose |
|----------|---------|
| nlinfit | Nonlinear least-squares data fitting (Newton's Method) |
| nlintool | Interactive graphical tool for prediction in nonlinear models |
| nlpredci | Confidence intervals for prediction |
| nlparci | Confidence intervals for parameters |
| nnls | Non-negative least-squares |

TABLE E.7

Cluster Analysis

| Function | Purpose |
|----------|---------|
| pdist | Pairwise distance between observations |
| squareform | Square matrix formatted distance |
| linkage | Hierarchical cluster information |
| dendrogram | Generate dendrogram plot |
| inconsistent | Inconsistent values of a cluster tree |
| cophenet | Cophenetic coefficient |
| cluster | Construct clusters from linkage output |
| clusterdata | Construct clusters from data |

## TABLE E.8

Design of Experiments (DOE) and Statistical Process Control (SPC)

| Function | Purpose |
|----------|---------|
| cordexch | D-optimal design (coordinate exchange algorithm) |
| daugment | Augment D-optimal design |
| dcovary | D-optimal design with fixed covariates |
| ff2n | Two-level full-factorial design |
| fracfact | Two-level fractional factorial design |
| fullfact | Mixed-level full-factorial design |
| hadamarad | Hadamard matrices (orthogonal arrays) |
| rowexch | D-optimal (row exchange algorithm) |
| capable | Capability indices |
| capaplot | Capability plot |
| ewmaplot | Exponentially weighted moving average plot |
| histfit | Histogram with superimposed normal density |
| normspec | Plot normal density between specification limits |
| schart | S chart for monitoring variability |
| xbarplot | Xbar chart for monitoring the mean |

## TABLE E.9

Multivariate Statistics and Principal Component Analysis

| Function | Purpose |
|----------|---------|
| classify | Linear discriminant analysis |
| mahal | Mahalanobis distance |
| manova1 | One-way multivariate analysis of variance |
| barttest | Bartlett's test for dimensionality |
| pcacov | Principal components from covariance matrix |
| pcares | Residuals from principal components |
| princomp | Principal component analysis from raw data |

**TABLE E.10**

Hypothesis Tests

| Function | Purpose |
|---|---|
| ranksum | Wilcoxon rank sum test (independent samples) |
| signrank | Wilcoxon sign rank test (paired samples) |
| signtest | Sign test (paired samples) |
| ztest | Z test |
| ttest | One sample t test |
| ttest2 | Two sample t test |
| jbtest | Jarque-Bera test of normality |
| kstest | Kolmogorov-Smirnov test for one sample |
| kstest2 | Kolmogorov-Smirnov test for two samples |
| lillietest | Lilliefors test of normality |

**TABLE E.11**

Statistical Plotting

| Function | Purpose |
|---|---|
| cdfplot | Plot of empirical cumulative distribution function |
| fsurfht | Interactive contour plot of a function |
| gline | Point, drag and click line drawing on figures |
| gname | Interactive point labeling in x-y plots |
| gplotmatrix | Matrix of scatter plots grouped by a common variable |
| gscatter | Scatter plot of two variables grouped by a third |
| lsline | Add least-square fit line to scatter plot |
| normplot | Normal probability plot |
| qqplot | Quantile-quantile plot |
| refcurve | Reference polynomial curve |
| refline | Reference line |
| surfht | Interactive contour plot of a data grid |
| weibplot | Weibull probability plot |

TABLE E.12

Statistics Demos

| Function | Purpose |
|---|---|
| aoctool | Interactive tool for analysis of covariance |
| disttool | GUI tool for exploring probability distribution functions |
| glmdemo | Generalized linear model slide show |
| polytool | Interactive graph for prediction of fitted polynomials |
| randtool | GUI tool for generating random numbers |
| rsmdemo | Reaction simulation |
| robustdemo | Interactive tool to compare robust and least squares fits |

TABLE E.13

File-based I/O

| Function | Purpose |
|---|---|
| tblread | Read in data in tabular format |
| tblwrite | Write out data in tabular format in file |
| tdfread | Read in text and numeric data from tab-delimited file |
| caseread | Read in case names |
| casewrite | Write out case names to file |

# Appendix F

## Computational Statistics Toolbox

The Computational Statistics Toolbox can be downloaded from:

> `http://www.infinityassociates.com`
> `http://lib.stat.cmu.edu`.

Please review the **readme** file for installation instructions and information on any recent changes.

TABLE F.1

Chapter 2 Functions: Probability Distributions

| Distribution<br>PDF (p) / CDF (c) | MATLAB Function |
|---|---|
| Beta | csbetap, csbetac |
| Binomial | csbinop, csbinoc |
| Chi-square | cschip, cschic |
| Exponential | csexpop, csexpoc |
| Gamma | csgammp, csgammc |
| Normal - univariate | csnormp, csnormc |
| Normal - multivariate | csevalnorm |
| Poisson | cspoisp, cspoisc |
| Continuous Uniform | csunifp, csunifc |
| Weibull | csweibp, csweibc |

**TABLE F.2**

Chapter 3 Functions: Statistics

| Purpose | MATLAB Function |
| --- | --- |
| These functions are used to obtain parameter estimates for a distribution. | `csbinpar` |
|  | `csexpar` |
|  | `csgampar` |
|  | `cspoipar` |
|  | `csunipar` |
| These functions return the quantiles. | `csbinoq` |
|  | `csexpoq` |
|  | `csunifq` |
|  | `csweibq` |
|  | `csnormq` |
|  | `csquantiles` |
| Other descriptive statistics | `csmomentc` |
|  | `cskewness` |
|  | `cskurtosis` |
|  | `csmoment` |
|  | `csecdf` |

**TABLE F.3**

Chapter 4 Functions: Random Number Generation

| Distribution | MATLAB Function |
| --- | --- |
| Beta | `csbetarnd` |
| Binomial | `csbinrnd` |
| Chi-square | `cschirnd` |
| Discrete Uniform | `csdunrnd` |
| Exponential | `csexprnd` |
| Gamma | `csgamrnd` |
| Multivariate Normal | `csmvrnd` |
| Poisson | `cspoirnd` |
| Points on a sphere | `cssphrnd` |

## TABLE F.4

Chapter 5 Functions: Exploratory Data Analysis

| Purpose | MATLAB Function |
|---|---|
| Star Plot | csstars |
| Stem-and-leaf Plot | csstemleaf |
| Parallel Coordinates Plot | csparallel |
| Q-Q Plot | csqqplot |
| Poissonness Plot | cspoissplot |
| Andrews Curves | csandrews |
| Exponential Probability Plot | csexpoplot |
| Binomial Plot | csbinoplot |
| PPEDA | csppeda<br>csppstrtrem<br>csppind |

## TABLE F.5

Chapter 6 Functions: Bootstrap

| Purpose | MATLAB Function |
|---|---|
| General bootstrap: resampling, estimates of standard error and bias | csboot |
| Constructing bootstrap confidence intervals | csbootint<br>csbooperint<br>csbootbca |

## TABLE F.6

Chapter 7 Functions: Jackknife

| Purpose | MATLAB Function |
|---|---|
| Implements the jackknife and returns the jackknife estimate of standard error and bias | csjack |
| Implements the jackknife-after-bootstrap and returns the jackknife estimate of the error in the bootstrap | csjackboot |

TABLE F.7

Chapter 8 Functions: Probability Density Estimation

| Purpose | MATLAB Function |
|---|---|
| Bivariate histogram | `cshist2d`<br>`cshistden` |
| Frequency polygon | `csfreqpoly` |
| Averaged Shifted Histogram | `csash` |
| Kernel density estimation | `cskernnd`<br>`cskern2d` |
| Create plots | `csdfplot`<br>`csplotuni` |
| Finite and adaptive mixtures | `csfinmix`<br>`csadpmix` |

TABLE F.8

Chapter 9 Functions: Statistical Pattern Recognition

| Purpose | MATLAB Function |
|---|---|
| Creating, pruning and displaying classification trees | `csgrowc`<br>`csprunec`<br>`cstreec`<br>`csplotreec`<br>`cspicktreec` |
| Creating, analyzing and displaying clusters | `cshmeans`<br>`cskmeans` |
| Statistical pattern recognition using Bayes decision theory | `csrocgen`<br>`cskernmd`<br>`cskern2d` |

## TABLE F.9

Chapter 10 Functions: Nonparametric Regression

| Purpose | MATLAB Function |
|---|---|
| Loess smoothing | `csloess` |
| | `csloessenv` |
| | `csloessr` |
| Local polynomial smoothing | `cslocpoly` |
| Functions for regression trees | `csgrowr` |
| | `cspruner` |
| | `cstreer` |
| | `csplotreer` |
| | `cspicktreer` |
| Nonparametric regression using kernels | `csloclin` |

## TABLE F.10

Chapter 11 Functions: Markov Chain Monte Carlo

| Purpose | MATLAB Function |
|---|---|
| Gelman-Rubin convergence diagnostic | `csgelrub` |
| Graphical demonstration of the Metropolis-Hastings sampler | `csmcmcdemo` |

## TABLE F.11

Chapter 12 Functions: Spatial Statistics

| Purpose | MATLAB Function |
|---|---|
| Functions for generating samples from spatial point processes | `csbinproc` |
| | `csclustproc` |
| | `csinhibproc` |
| | `cspoissproc` |
| | `csstraussproc` |
| Interactively find a study region | `csgetregion` |
| Estimate the intensity using the quartic kernel (no edge effects) | `csintkern` |
| Estimating second-order effects of a spatial point pattern | `csfhat` |
| | `csghat` |
| | `cskhat` |

# Appendix G

## Data Sets

In this appendix, we list the data sets that are used in the book. These data are available for download in either text format (.txt) or MATLAB binary format (.mat). They can be downloaded from

- http://lib.stat.cmu.edu
- http://www.infinityassociates.com

**abrasion**
The **abrasion** data set has 30 observations, where the two predictor variables are hardness and tensile strength (**x**). The response variable is abrasion loss (**y**) [Hand, et al., 1994; Davies and Goldsmith, 1972]. The first column of **x** contains the hardness and the second column contains the tensile strength.

**anaerob**
A subject performs an exercise, gradually increasing the level of effort. The data set called **anaerob** has two variables based on this experiment: oxygen uptake and the expired ventilation [Hand, et al., 1994; Bennett, 1988]. The oxygen uptake is contained in the variable **x** and the expired ventilation is in **y**.

**anscombe**
These data were taken from Hand, et al. [1994]. They were originally from Anscombe [1973], where he created these data sets to illustrate the importance of graphical exploratory data analysis. This file contains four sets of **x** and **y** measurements.

**bank**
This file contains two matrices, one corresponding to features taken from 100 forged Swiss bank notes (**forge**) and the other comprising features from 100 genuine Swiss bank notes (**genuine**) [Flury and Riedwyl, 1988]. There are six features: length of the bill, left width of the bill, right width of the bill,

width of the bottom margin, width of the top margin and length of the image diagonal.

### biology

The **biology** data set contains the number of research papers (**numpaps**) for 1534 biologists [Tripathi and Gupta, 1988; Hand, et al., 1994]. The frequencies are given in the variable **freqs**.

### bodmin

These data represent the locations of granite tors on Bodmin Moor [Pinder and Witherick, 1977; Upton and Fingleton, 1985; Bailey and Gatrell, 1995]. The file contains vectors **x** and **y** that correspond to the coordinates of the tors. The two-column matrix **bodpoly** contains the vertices to the region.

### boston

The **boston** data set contains data for 506 census tracts in the Boston area, taken from the 1970 Census [Harrison and Rubinfeld, 1978]. The predictor variables are: (1) per capita crime rate, (2) proportion of residential land zoned for lots over 25,000 sq.ft., (3) proportion of non-retail business acres, (4) Charles River dummy variable (1 if tract bounds river; 0 otherwise), (5) nitric oxides concentration (parts per 10 million), (6) average number of rooms per dwelling, (7) proportion of owner-occupied units built prior to 1940, (8) weighted distances to five Boston employment centers, (9) index of accessibility to radial highways, (10) full-value property-tax rate per \$10,000, (11) pupil-teacher ratio, (12) proportion of African-Americans, and (13) lower status of the population. These are contained in the variable **x**. The response variable **y** represents the median value of owner-occupied homes in \$1000's. These data were downloaded from

> **http://www.stat.washington.edu/raftery/Courses/**
> **Stat572-96/Homework/Hw1/hw1_96/boston_hw1.html**

### brownlee

The **brownlee** data contains observations from 21 days of a plant operation for the oxidation of ammonia [Hand, et al., 1994; Brownlee, 1965]. The predictor variables are: $X_1$ is the air flow, $X_2$ is the cooling water inlet temperature (degrees C), and $X_3$ is the percent acid concentration. The response variable $Y$ is the stack loss (the percentage of the ingoing ammonia that escapes). The matrix **x** contains the observed predictor values and the vector **y** has the corresponding response variables.

### cardiff

This data set has the locations of homes of juvenile offenders in Cardiff, Wales in 1971 [Herbert, 1980]. The file contains vectors **x** and **y** that correspond to the coordinates of the homes. The two-column matrix **cardpoly** contains the vertices to the region.

## cereal
These data were obtained from ratings of eight brands of cereal [Chakrapani and Ehrenberg, 1981; Venables and Ripley, 1994]. The **cereal** file contains a matrix where each row corresponds to an observation and each column represents one of the variables or the percent agreement to statements about the cereal. It also contains a cell array of strings (**labs**) for the type of cereal.

## coal
The **coal** data set contains the number of coal mining disasters (**y**) over 112 years (**year**) [Raftery and Akman, 1986].

## counting
In the **counting** data set, we have the number of scintillations in 72 second intervals arising from the radioactive decay of polonium [Rutherford and Geiger, 1910; Hand, et al., 1994]. There are a total of 10097 scintillations and 2608 intervals. Two vectors, **count** and **freqs**, are included in this file.

## elderly
The **elderly** data set contains the height measurements (in centimeters) of 351 elderly females [Hand, et al., 1994]. The variable that is loaded is called **heights**.

## environ
This data set was analyzed in Cleveland and McGill [1984]. They represent two variables comprising daily measurements of **ozone** and **wind** speed in New York City. These quantities were measured on 111 days between May and September 1973. One might be interested in understanding the relationship between ozone (the response variable) and wind speed (the predictor variable).

## filip
These data are used as a standard to test the results of least squares calculations. The file contains two vectors **x** and **y**.

## flea
The **flea** data set [Hand, et al., 1994; Lubischew, 1962] contains measurements on three species of flea beetle: *Chaetocnema concinna* (**conc**), *Chaetocnema heikertingeri* (**heik**), and *Chaetocnema heptapotamica* (**hept**). The features for classification are the maximal width of aedeagus in the forepart (microns) and the front angle of the aedeagus (units are 7.5 degrees).

## forearm
These data [Hand, et al., 1994; Pearson and Lee, 1903] consist of 140 measurements of the length (in inches) of the forearm of adult males. The vector **x** contains the measurements.

**geyser**
These data represent the waiting times (in minutes) between eruptions of the Old Faithful geyser at Yellowstone National Park [Hand, et al, 1994; Scott, 1992]. This contains one vector called **geyser**.

**helmets**
The data in **helmets** contain measurements of head acceleration (in g) (**accel**) and times after impact (milliseconds) (**time**) from a simulated motorcycle accident [Hand, et al., 1994; Silverman, 1985].

**household**
The **household** [Hand, et al., 1994; Aitchison, 1986] data set contains the expenditures for housing, food, other goods, and services (four expenditures) for households comprised of single people. The observations are for single **women** and single **men**.

**human**
The **human** data set [Hand, et al., 1994; Mazess, et al., 1984] contains measurements of percent fat and age for 18 normal adults (**males** and **females**).

**insect**
In this data set, we have three variables measured on ten insects from each of three species [Hand, et al.,1994]. The variables correspond to the width of the first joint of the first tarsus, the width of the first joint of the second tarsus and the maximal width of the aedeagus. All widths are measured in microns. When **insect** is loaded, you get one $30 \times 3$ matrix called **insect**. Each group of 10 rows belongs to one of the insect species.

**insulate**
The **insulate** data set [Hand, et al., 1994] contains observations corresponding to the average outside temperature in degrees Celsius (first column) and the amount of weekly gas consumption measured in 1000 cubic feet (second column). One data set is before insulation (**befinsul**) and the other corresponds to measurements taken after insulation (**aftinsul**).

**iris**
The **iris** data were collected by Anderson [1935] and were analyzed by Fisher [1936] (and many statisticians since then!). The data consist of 150 observations containing four measurements based on the petals and sepals of three species of iris. The three species are: *Iris setosa, Iris virginica* and *Iris versicolor*. When the **iris** data are loaded, you get three $50 \times 4$ matrices, one corresponding to each species.

**law/lawpop**
The **lawpop** data set [Efron and Tibshirani, 1993] contains the average scores on the LSAT (**lsat**) and the corresponding average undergraduate grade

point average (**gpa**) for the 1973 freshman class at 82 law schools. Note that these data constitute the entire population. The data contained in **law** comprise a random sample of 15 of these classes, where the **lsat** score is in the first column and the **gpa** is in the second column.

### longley
The data in **longley** were used by Longley [1967] to verify the computer calculations from a least squares fit to data. The data set (**X**) contains measurements of 6 predictor variables and a column of ones representing the constant term. The observed responses are contained in **Y**.

### measure
The **measure** [Hand, et. al., 1994] data contain 20 measurements of chest, waist and hip data. Half of the measured individuals are women and half are men.

### moths
The **moths** data represent the number of moths caught in a trap over 24 consecutive nights [Hand, et al., 1994].

### nfl
The **nfl** data [Csorgo and Welsh, 1989; Hand, et al., 1994] contain bivariate measurements of the game time to the first points scored by kicking the ball between the end posts ($X_1$), and the game time to the first points scored by moving the ball into the end zone ($X_2$). The times are in minutes and seconds.

### okblack and okwhite
These data represent locations where thefts occurred in Oklahoma City in the late 1970's [Bailey and Gatrell, 1995]. The file **okwhite** contains the data for Caucasian offenders, and the file **okblack** contains the data for African-American offenders. The boundary for the region is not included with these data.

### peanuts
The **peanuts** data set [Hand, et al., 1994; Draper and Smith, 1981] contains measurements of the average level of alfatoxin (**X**) of a batch of peanuts and the corresponding percentage of non-contaminated peanuts in the batch (**Y**).

### posse
The **posse** file contains several data sets generated for simulation studies in Posse [1995b]. These data sets are called **croix** (a cross), **struct2** (an L-shape), **boite** (a donut), **groupe** (four clusters), **curve** (two curved groups), and **spiral** (a spiral). Each data set has 400 observations in 8-D. These data can be used in PPEDA.

**quakes**
The **quakes** data [Hand, et al., 1994] contain the time in days between successive earthquakes.

**remiss**
The **remiss** data set contains the remission times for 42 leukemia patients. Some of the patients were treated with the drug called 6-mercaptopurine (**mp**), and the rest were part of the control group (**control**) [Hand, et al., 1994; Gehan, 1965].

**snowfall**
The Buffalo **snowfall** data [Scott, 1992] represent the annual snowfall in inches in Buffalo, New York over the years 1910-1972. This file contains one vector called **snowfall**.

**spatial**
These data came from Efron and Tibshirani [1993]. Here we have a set of measurements of 26 neurologically impaired children who took a test of spatial perception called test A.

**steam**
In the **steam** data set, we have a sample representing the average atmospheric temperature (**x**) and the corresponding amount of steam (**y**) used per month [Draper and Smith, 1981]. We get two vectors **x** and **y** when these data are loaded.

**thrombos**
The **thrombos** data set contains measurements of urinary-thromboglobulin excretion in 12 **normal** and 12 **diabetic** patients [van Oost, et al.; 1983; Hand, et al., 1994].

**tibetan**
This file contains the heights of 32 Tibetan skulls [Hand, et al. 1994; Morant, 1923] measured in millimeters. These data comprise two groups of skulls collected in Tibet. One group of 17 skulls comes from graves in Sikkim and nearby areas of Tibet and the other 15 skulls come from a battlefield in Lhasa. The original data contain five measurements for the 32 skulls. When you load this file, you get a $32 \times 5$ matrix called **tibetan**.

**uganda**
This data set contains the locations of crater centers of 120 volcanoes in west Uganda [Tinkler, 1971, Bailey and Gatrell, 1995]. The file has vectors **x** and **y** that correspond to the coordinates of the craters. The two-column matrix **ugpoly** contains the vertices to the region.

**whisky**

In 1961, 16 states owned the retail liquor stores (**state**). In 26 others, the stores were owned by private citizens (**private**). The data contained in **whisky** reflect the price (in dollars) of a fifth of Seagram 7 Crown Whisky from these 42 states. Note that this represents the population, not a sample [Hand, et al., 1994].

# References

Aarts, E. and J. Korst. 1989. *Simulated Annealing and Boltzmann Machines*, New York: John Wiley & Sons.

Aitchison, J. 1986. *The Statistical Analysis of Compositional Data*, London: Chapman and Hall.

Albert, James H. 1993. "Teaching Bayesian statistics using sampling methods and MINITAB," *The American Statistician*. **47**: pp. 182-191.

Anderberg, Michael R. 1973. *Cluster Analysis for Applications*, New York: Academic Press.

Anderson, E. 1935. "The irises of the Gaspe Peninsula," *Bulletin of the American Iris Society*, **59**: pp. 2-5.

Andrews, D. F. 1972. "Plots of high-dimensional data," *Biometrics*, **28**: pp. 125-136.

Andrews, D. F. 1974. "A robust method of multiple linear regression," *Technometrics*, **16**: pp. 523-531.

Andrews, D. F. and A. M. Herzberg. 1985. *Data: A Collection of Problems from Many Fields for the Student and Research Worker*, New York: Springer-Verlag.

Anscombe, F. J. 1973. "Graphs in statistical analysis," *The American Statistician*, **27**: pp. 17-21.

Arlinghaus, S. L. (ed.). 1996. *Practical Handbook of Spatial Statistics*, Boca Raton: CRC Press.

Arnold, Steven F. 1993. "Gibbs sampling," in *Handbook of Statistics, Vol 9, Computational Statistics*, C. R. Rao, ed., The Netherlands: Elsevier Science Publishers, pp. 599-625.

Ash, Robert. 1972. *Real Analysis and Probability*, New York: Academic Press.

Asimov, Daniel. 1985. "The grand tour: a tool for viewing multidimensional data," *SIAM Journal of Scientific and Statistical Computing*, **6**: pp. 128-143.

Bailey, T. C. and A. C. Gatrell. 1995. *Interactive Spatial Data Analysis*, London: Longman Scientific & Technical.

Bain, L. J. and M. Engelhardt. 1992. *Introduction to Probability and Mathematical Statistics, Second Edition*, Boston: PWS-Kent Publishing Company.

Banks, Jerry, John Carson, Barry Nelson, and David Nicol. 2001. *Discrete-Event Simulation, Third Edition*, New York: Prentice Hall.

Bennett, G. W. 1988. "Determination of anaerobic threshold," *Canadian Journal of Statistics*, **16**: pp. 307-310.

Besag, J. and P. J. Diggle. 1977. "Simple Monte Carlo tests for spatial patterns," *Applied Statistics*, **26**: pp. 327-333.

Bickel, Peter J. and Kjell A. Doksum. 2001. *Mathematical Statistics: Basic Ideas and Selected Topics, Vol 1, Second Edition,* New York: Prentice Hall.

Billingsley, Patrick. 1995. *Probability and Measure, 3rd Edition,* New York: John Wiley & Sons.

Bolton, R. J. and W. J. Krzanowski. 1999. "A characterization of principal components for projection pursuit," *The American Statistician,* **53**: pp. 108-109.

Boos, D. D. and J. Zhang. 2000. "Monte Carlo evaluation of resampling-based hypothesis tests," *Journal of the American Statistical Association,* **95**: pp. 486-492.

Bowman, A. W. and A. Azzalini. 1997. *Applied Smoothing Techniques for Data Analysis: The Kernel Approach with S-Plus Illustrations,* Oxford: Oxford University Press.

Breiman, Leo. 1992. *Probability.* Philadelphia: Society for Industrial and Applied Mathematics.

Breiman, Leo, Jerome H. Friedman, Richard A. Olshen and Charles J. Stone. 1984. *Classification and Regression Trees,* New York: Wadsworth, Inc.

Brooks, S. P. 1998. "Markov chain Monte Carlo and its application," *The American Statistician,* **47**: pp. 69-100.

Brooks, S. P. and P. Giudici. 2000. "Markov chain Monte Carlo convergence assessment via two-way analysis of variance," *Journal of Computational and Graphical Statistics,* **9**: pp. 266-285.

Brownlee, K. A. 1965. *Statistical Theory and Methodology in Science and Engineering, Second Edition,* London: John Wiley & Sons.

Cacoullos, T. 1966. "Estimation of a multivariate density," *Annals of the Institute of Statistical Mathematics,* **18**: pp. 178-189.

Canty, A. J. 1999. "Hypothesis tests of convergence in Markov chain Monte Carlo," *Journal of Computational and Graphical Statistics,* **8**: pp. 93-108.

Carr, D., R. Littlefield, W. Nicholson, and J. Littlefield. 1987. "Scatterplot matrix techniques for large N," *Journal of the American Statistical Association,* **82**: p. 424-436.

Carter, R. L. and K. Q. Hill. 1979. *The Criminals' Image of the City,* Oxford: Pergamon Press.

Casella, George and Roger L. Berger. 1990. *Statistical Inference,* New York: Duxbury Press.

Casella, George, and E. I. George. 1992. "An introduction to Gibbs Sampling," *The American Statistician,* **46**: pp. 167-174.

Cencov, N. N. 1962. "Evaluation of an unknown density from observations," *Soviet Mathematics,* **3**: pp. 1559-1562.

Chakrapani, T. K. and A. S. C. Ehrenberg. 1981. "An alternative to factor analysis in marketing research - Part 2: Between group analysis," *Professional Marketing Research Society Journal,* **1**: pp. 32-38.

Chambers, John. 1999. "Computing with data: Concepts and challenges," *The American Statistician,* **53**: pp. 73-84.

Chambers, John and Trevor Hastie. 1992. *Statistical Models in S,* New York: Wadsworth & Brooks/Cole Computer Science Series.

Chernick, M. R. 1999. *Bootstrap Methods: A Practitioner's Guide,* New York: John Wiley & Sons.

Chernoff, Herman. 19 73. "The use of faces to represent points in k-dimensional space graphically," *Journal of the American Statistical Association*, **68**: 361-368.

Chib, S., and E. Greenberg. 1995. "Understanding the Metropolis-Hastings Algorithm," *The American Statistician*, **49**: pp. 327-335.

Cleveland, W. S. 1979. "Robust locally weighted regression and smoothing scatterplots," *Journal of the American Statistical Association*, **74**, pp. 829-836.

Cleveland, W. S. 1993. *Visualizing Data*, New York: Hobart Press.

Cleveland, W. S. and Robert McGill. 1984. "The many faces of a scatterplot," *Journal of the American Statistical Association*, **79**: pp. 807-822.

Cliff, A. D. and J. K. Ord. 1981. *Spatial Processes: Models and Applications*, London: Pion Limited.

Cook, D., A. Buha, J. Cabrera, and C. Hurley. 1995. "Grand tour and projection pursuit," *Journal of Computational and Graphical Statistics*, **4**: pp. 155-172.

Cowles, M. K. and B. P. Carlin. 1996. "Markov chain Monte Carlo convergence diagnostics: a comparative study," *Journal of the American Statistical Association*, **91**: pp. 883–904.

Crawford, Stuart. 1991. "Genetic optimization for exploratory projection pursuit," *Proceedings of the 23rd Symposium on the Interface*, **23**: pp. 318-321.

Cressie, Noel A. C. 1993. *Statistics for Spatial Data, Revised Edition*. New York: John Wiley & Sons.

Csorgo, S. and A. S. Welsh. 1989. "Testing for exponential and Marshall-Olkin distributions," *Journal of Statistical Planning and Inference*, **23**: pp. 278-300.

David, Herbert A. 1981. *Order Statistics, 2nd edition*, New York: John Wiley & Sons.

Dempster, A. P., Laird, N. M., and Rubin, D. B. 1977. "Maximum likelihood from incomplete data via the EM algorithm (with discussion)," *Journal of the Royal Statistical Society: B*, **39**: pp. 1-38.

Deng, L. and D. K. J. Lin. 2000. "Random number generation for the new century," *The American Statistician*, **54**: pp. 145-150.

Devroye, Luc. and L. Gyorfi. 1985. *Nonparametric Density Estimation: the $L_1$ View*, New York: John Wiley & Sons.

Devroye, Luc, Laszlo Gyorfi and Gabor Lugosi. 1996. *A Probabilistic Theory of Pattern Recognition*, New York: Springer-Verlag.

Diggle, Peter J. 1981. "Some graphical methods in the analysis of spatial point patterns," in *Interpreting Multivariate Data*, V. Barnett, ed., New York: John Wiley & Sons, pp. 55-73.

Diggle, Peter J. 1983. *Statistical Analysis of Spatial Point Patterns*, New York: Academic Press.

Diggle, P. J. and R. J. Gratton. 1984. "Monte Carlo methods of inference for implicit statistical models," *Journal of the Royal Statistical Society: B*, **46**: pp. 193–227.

Draper, N. R. and H. Smith. 1981. *Applied Regression Analysis, 2nd Edition*, New York: John Wiley & Sons.

du Toit, S. H. C., A. G. W. Steyn and R. H. Stumpf. 1986. *Graphical Exploratory Data Analysis*, New York: Springer-Verlag.

Duda, Richard O. and Peter E. Hart. 1973. *Pattern Classification and Scene Analysis*, New York: John Wiley & Sons.

Duda, Richard O., Peter E. Hart, and David G. Stork. 2001. *Pattern Classification, Second Edition*, New York: John Wiley & Sons.

Durrett, Richard. 1994. *The Essentials of Probability*, New York: Duxbury Press.

Efron, B. 1979. "Computers and the theory of statistics: thinking the unthinkable," *SIAM Review*, **21**: pp. 460-479.

Efron, B. 1981. "Nonparametric estimates of standard error: the jackknife, the bootstrap and other methods," *Biometrika*, **68**: pp. 589-599.

Efron, B. 1982. *The Jackknife, the Bootstrap, and Other Resampling Plans*, Philadelphia: Society for Industrial and Applied Mathematics.

Efron, B. 1983. "Estimating the error rate of a prediction rule: improvement on cross-validation," *Journal of the American Statistical Association*, **78**: pp. 316-331.

Efron, B. 1985. "Bootstrap confidence intervals for a class of parametric problems," *Biometrika*, **72**: pp. 45–58.

Efron, B. 1986. "How biased is the apparent error rate of a prediction rule?" *Journal of the American Statistical Association*, **81**: pp. 461-470.

Efron, B. 1987. "Better bootstrap confidence intervals' (with discussion)," *Journal of the American Statistical Association*, **82**: pp. 171-200.

Efron, B. 1990. "More efficient bootstrap computations, *Journal of the American Statistical Association*, **85**: pp. 79-89.

Efron, B. 1992. "Jackknife-after-bootstrap standard errors and influence functions," *Journal of the Royal Statistical Society: B*, **54**: pp. 83-127.

Efron, B. and G. Gong. 1983. "A leisurely look at the bootstrap, the jackknife and cross-validation," *The American Statistician*, **37**: pp. 36-48.

Efron, B. and R. J. Tibshirani. 1991. "Statistical data analysis in the computer age," *Science*, **253**: pp. 390-395.

Efron, B. and R. J. Tibshirani. 1993. *An Introduction to the Bootstrap*, London: Chapman and Hall.

Egan, J. P. 1975. *Signal Detection Theory and ROC Analysis*, New York: Academic Press.

Embrechts, P. and A. Herzberg. 1991. "Variations of Andrews' plots," *International Statistical Review*, **59**: pp. 175-194.

Epanechnikov, V. K. 1969. "Non-parametric estimation of a multivariate probability density," *Theory of Probability and its Applications*, **14**: pp. 153-158.

Everitt, Brian S. 1993. *Cluster Analysis, Third Edition*, New York: Edward Arnold Publishing.

Everitt, B. S. and D. J. Hand. 1981. *Finite Mixture Distributions*, London: Chapman and Hall.

Fienberg, S. 1979. "Graphical methods in statistics," *The American Statistician*, **33**: pp. 165-178.

Fisher, R. A. 1936. "The use of multiple measurements in taxonomic problems," *Annals of Eugenics*, **7**: pp. 179-188.

Flick, T., L. Jones, R. Priest, and C. Herman. 1990. "Pattern classification using projection pursuit," *Pattern Recognition*, **23**: pp. 1367-1376.

Flury, B. and H. Riedwyl. 1988. *Multivariate Statistics: A Practical Approach*, London: Chapman and Hall.

Fortner, Brand. 1995. *The Data Handbook: A Guide to Understanding the Organization and Visualization of Technical Data, Second Edition*, New York: Springer-Verlag.

Fortner, Brand and Theodore E. Meyer. 1997. *Number by Colors: A Guide to Using Color to Understand Technical Data*, New York: Springer-Verlag.

Fraley, C. 1998. "Algorithms for model-based Gaussian hierarchical clustering," *SIAM Journal on Scientific Computing*, **20**: pp. 270-281.

Fraley, C. and A. E. Raftery. 1998. "How many clusters? Which clustering method? Answers via model-based cluster analysis," *The Computer Journal*, **41**: pp. 578-588.

Freedman, D. and P. Diaconis. 1981. "On the histogram as a density estimator: $L_2$ theory," *Zeitschrift fur Wahrscheinlichkeitstheorie und verwandte Gebiete*, **57**: pp. 453-476.

Friedman, J. 1987. "Exploratory projection pursuit," *Journal of the American Statistical Association*, **82**: pp. 249-266.

Friedman, J. and W. Stuetzle. 1981. "Projection pursuit regression," *Journal of the American Statistical Association*, **76**: pp. 817-823.

Friedman, J. and John Tukey. 1974. "A projection pursuit algorithm for exploratory data analysis," *IEEE Transactions on Computers*, **23**: pp. 881-889.

Friedman, J., W. Stuetzle, and A. Schroeder. 1984. "Projection pursuit density estimation," *Journal of the American Statistical Association*, **79**: pp. 599-608.

Frigge, M., C. Hoaglin, and B. Iglewicz. 1989. "Some implementations of the boxplot," *The American Statistician*, **43**: pp. 50-54.

Fukunaga, Keinosuke. 1990. *Introduction to Statistical Pattern Recognition, Second Edition*, New York: Academic Press.

Gehan, E. A. 1965. "A generalized Wilcoxon test for comparing arbitrarily single-censored samples," *Biometrika*, **52**: pp. 203-233.

Gelfand, A. E. and A. F. M. Smith. 1990. "Sampling-based approaches to calculating marginal densities," *Journal of the American Statistical Association*, **85**: pp. 398-409.

Gelfand, A. E., S. E. Hills, A. Racine-Poon, and A. F. M. Smith. 1990. "Illustration of Bayesian inference in normal data models using Gibbs sampling," *Journal of the American Statistical Association*, **85**: pp. 972-985.

Gelman, A. 1996. "Inference and monitoring convergence," in *Markov Chain Monte Carlo in Practice*, W. R. Gilks, S. Richardson, and D. T. Spiegelhalter, eds., London: Chapman and Hall, pp. 131-143.

Gelman, A. and D. B. Rubin. 1992. "Inference from iterative simulation using multiple sequences (with discussion)," *Statistical Science*, **7**: pp. 457–511.

Gelman, A., J. B. Carlin, H. S. Stern, and D. B. Rubin. 1995. *Bayesian Data Analysis*, London: Chapman and Hall.

Geman, S. and D. Geman. 1984. "Stochastic relaxation, Gibbs distributions and the Bayesian restoration of images," *IEEE Transactions PAMI*, **6**: pp. 721-741.

Gentle, James E. 1998. *Random Number Generation and Monte Carlo Methods*, New York: Springer-Verlag.

Gentle, James E. 2001. *Computational Statistics*, (in press), New York: Springer-Verlag.

Geyer, C. J. 1992. "Practical Markov chain Monte Carlo," *Statistical Science*, **7**: pp. 473-511.

Gilks, W. R., S. Richardson, and D. J. Spiegelhalter. 1996a. "Introducing Markov chain Monte Carlo," in *Markov Chain Monte Carlo in Practice*, W. R. Gilks, S. Richardson, and D. T. Spiegelhalter, eds., London: Chapman and Hall, pp. 1-19.

Gilks, W. R., S. Richardson, and D. J. Spiegelhalter (eds.). 1996b. *Markov Chain Monte Carlo in Practice*, London: Chapman and Hall.

Gordon, A. D. 1999. *Classification*, London: Chapman and Hall.

Green P. J. and B. W. Silverman. 1994. *Nonparametric Regression and Generalized Linear Models: A Roughness Penalty Approach*, Chapman and Hall.

Haining, Robert. 1993. *Spatial Data Analysis in the Social and Environmental Sciences*, Cambridge: Cambridge University Press.

Hair, Joseph, Rolph Anderson, Ronald Tatham and William Black. 1995. *Multivariate Data Analysis, Fourth Edition*, New York: Prentice Hall.

Hald, A. 1952. *Statistical Theory with Engineering Applications*, New York: John Wiley & Sons.

Hall, P. 1992. *The Bootstrap and Edgeworth Expansion*, New York: Springer-Verlag.

Hall, P. and M. A. Martin. 1988. "On bootstrap resampling and iteration," *Biometrika*, **75**: pp. 661-671.

Hand, D., F. Daly, A. D. Lunn, K. J. McConway and E. Ostrowski. 1994. *A Handbook of Small Data Sets*, London: Chapman and Hall.

Hanley, J. A. and K. O. Hajian-Tilaki. 1997. "Sampling variability of nonparametric estimates of the areas under receiver operating characteristic curves: An update," *Academic Radiology*, **4**: pp. 49-58.

Hanley, J. A. and B. J. McNeil. 1983. "A method of comparing the areas under receiver operating characteristic curves derived from the same cases," *Radiology*, **148**: pp. 839-843.

Hanselman, D. and B. Littlefield. 1998. *Mastering MATLAB 5: A Comprehensive Tutorial and Reference*, New Jersey: Prentice Hall.

Hanselman, D. and B. Littlefield. 2001. *Mastering MATLAB 6: A Comprehensive Tutorial and Reference*, New Jersey: Prentice Hall.

Harrison, D., and D. L. Rubinfeld. 1978. "Hedonic prices and the demand for clean air," *Journal of Environmental Economics and Management*, **5**: pp. 81-102.

Hartigan, J. 1975. *Clustering Algorithms*, New York: Wiley-Interscience.

Hastie, T. J. and R. H. Tibshirani. 1990. *Generalized Additive Models*, London: Chapman and Hall.

Hastings, W. K. 1970. "Monte Carlo sampling methods using Markov chains and their applications," *Biometrika*, **57**: pp. 97-109.

Herbert, D. T. 1980 "The British experience," in *Crime: a Spatial Perspective*, D. E. Georges-Abeyie and K. D. Harries, eds., New York: Columbia University Press.

Hjorth, J. S. U. 1994. *Computer Intensive Statistical Methods: Validation Model Selection and Bootstrap*, London: Chapman and Hall.

Hoaglin, D. C. and D. F. Andrews. 1975. "The reporting of computation-based results in statistics," *The American Statistician*, **29**: pp. 122-126.

Hoaglin, D. and John Tukey. 1985. "Checking the shape of discrete distributions," in *Exploring Data Tables, Trends and Shapes*, D. Hoaglin, F. Mosteller, J. W. Tukey, eds., New York: John Wiley & Sons.

Hoaglin, D. C., F. Mosteller, and J. W. Tukey (eds.). 1983. *Understanding Robust and Exploratory Data Analysis*, New York: John Wiley & Sons.

Hogg, Robert. 1974. "Adaptive robust procedures: a partial review and some suggestions for future applications and theory (with discussion)," *The Journal of the American Statistical Association*, **69**: pp. 909-927.

Hogg, Robert and Allen Craig. 1978. *Introduction to Mathematical Statistics, 4th Edition*, New York: Macmillan Publishing Co.

Hope, A. C. A. 1968. "A simplified Monte Carlo Significance test procedure," *Journal of the Royal Statistical Society, Series B*, **30**: pp. 582-598.

Huber, P. J. 1973. "Robust regression: asymptotics, conjectures, and Monte Carlo," *Annals of Statistics*, **1**: pp. 799-821.

Huber, P. J. 1981. *Robust Statistics*, New York: John Wiley & Sons.

Huber, P. J. 1985. "Projection pursuit (with discussion)," *Annals of Statistics*, **13**: pp. 435-525.

Hunter, J. Stuart. 1988. "The digidot plot," *The American Statistician*, **42**:. pp. 54-54.

Inselberg, Alfred. 1985. "The plane with parallel coordinates," *The Visual Computer*, **1**: pp. 69-91.

Isaaks. E. H. and R. M. Srivastava. 1989. *An Introduction to Applied Geo-statistics*, New York: Oxford University Press.

Izenman, A. J. 1991. 'Recent developments in nonparametric density estimation," *Journal of the American Statistical Association*, **86**: pp. 205-224.

Jackson, J. Edward. 1991. *A User's Guide to Principal Components*, New York: John Wiley & Sons.

Jain, Anil K. and Richard C. Dubes. 1988. *Algorithms for Clustering Data*, New York: Prentice Hall.

Joeckel, K. 1991. "Monte Carlo techniques and hypothesis testing," *The Frontiers of Statistical Computation, Simulation and Modeling, Volume 1 of the Proceedings ICOSCO-I*, pp. 21-41.

Johnson, Mark E. 1987. *Multivariate Statistical Simulation*, New York: John Wiley & Sons.

Jones, M. C. and R. Sibson. 1987. "What is projection pursuit" (with discussion)," *Journal of the Royal Statistical Society, Series A*, **150**: pp. 1–36.

Journel, A. G. and C. J. Huijbregts. 1978. *Mining Geostatistics*, London: Academic Press.

Kalos, Malvin H. and Paula A. Whitlock. 1986. *Monte Carlo Methods, Volume 1: Basics*, New York: Wiley Interscience.

Kaplan, D. T. 1999. *Resampling Stats in MATLAB*, Arlington, VA: Resampling Stats, Inc.

Kaufman, Leonard and Peter J. Rousseeuw. 1990. *Finding Groups in Data: An Introduction to Cluster Analysis*, New York: John Wiley & Sons.

Keating, Jerome, Robert Mason and Pranab Sen. 1993. *Pitman's Measure of Closeness - A Comparison of Statistical Estimators*, New York: SIAM Press.

Kirkpatrick, S., C. D. Gelatt Jr., and M. P. Vecchi. 1983. "Optimization by simulated annealing," *Science*, **220**: pp. 671-680.

Kotz, Samuel and Norman L. Johnson (eds.). 1986. *Encyclopedia of Statistical Sciences*, New York: John Wiley & Sons.

Launer, R., and G. Wilkinson (eds.). 1979. *Robustness in Statistics,* New York: Academic Press.

Lehmann, E. L. 1994. *Testing Statistical Hypotheses,* London: Chapman and Hall.

Lehmann, E. L. and G. Casella. 1998. *Theory of Point Estimation, Second Edition,* New York: Springer-Verlag.

LePage, R. and L. Billard (eds.). 1992. *Exploring the Limits of the Bootstrap,* New York: John Wiley & Sons.

Levy, Paul S. and Stanley Lemeshow. 1999. *Sampling of Populations: Methods and Applications,* New York: John Wiley & Sons.

Li, G. and Z. Chen. 1985. "Projection-pursuit approach to robust dispersion matrices and principal components: primary theory and Monte Carlo," *Journal of the American Statistical Association,* **80**: pp. 759-766.

Lindgren, Bernard W. 1993. *Statistical Theory, Fourth Edition,* London: Chapman and Hall.

Lindley, D. V. 1995. *Bayesian Statistics, A Review,* Philadelphia: Society for Industrial and Applied Mathematics.

Lindsey, J. C., A. M. Herzberg, and D. G. Watts. 1987. "A method for cluster analysis based on projections and quantile-quantile plots," *Biometrics,* **43**: pp. 327-341.

Loader, Clive. 1999. *Local Regression and Likelihood,* New York: Springer-Verlag.

Loh, W. Y. 1987. "Calibrating confidence coefficients," *Journal of the American Statistical Association,* **82**: pp. 155-162.

Longley, J. W. 1967. "An appraisal of least squares programs for the electronic computer from the viewpoint of the user," *Journal of the American Statistical Association,* **62**: pp. 819-841.

Lubischew, A. A. 1962. " On the use of discriminant functions in taxonomy," *Biometrics,* **18**: pp. 455-477.

Lusted, L. B. 1971. "Signal detectability and medical decision-making," *Science,* **171**: pp. 1217-1219.

Marchand, Patrick. 1999. *Graphics and GUI's with MATLAB, Second Edition,* Boca Raton: CRC Press.

Mazess, R. B., W. W. Peppler, and M. Gibbons. 1984. "Total body composition by dualphoton ($^{153}$Gd) absorptiometry," *American Journal of Clinical Nutrition,* **40**: pp. 834-839.

McGill, Robert, John Tukey, and Wayne Larsen. 1978. "Variations of box plots," *The American Statistician,* **32**: pp. 12-16.

McLachlan, G. J. and K. E. Basford. 1988. *Mixture Models: Inference and Applications to Clustering,* New York: Marcel Dekker.

McLachlan, G. J. and T. Krishnan. 1997. *The EM Algorithm and Extensions,* New York: John Wiley & Sons.

McLachlan, G. J. and D. Peel. 2000. *Finite Mixture Models,* New York: John Wiley & Sons.

McNeil, B. J., E. Keeler, and S. J. Adelstein. 1975. "Primer on certain elements of medical decision making," *New England Journal of Medicine,* **293**: pp. 211-215.

Meeker, William and Luis Escobar. 1998. *Statistical Methods for Reliability Data,* New York: John Wiley & Sons.

Metropolis, N., A. W. Rosenbluth, M. N. Rosenbluth, A. H. Teller, and E. Teller. 1953. "Equations of state calculations by fast computing machine," *Journal of Chemistry and Physics*, **21**: pp. 1087-1091.

Meyn, S. P. and R. L. Tweedie. 1993. *Markov Chains and Stochastic Stability*, New York: Springer-Verlag.

Minnotte, M. and R. West. 1998. "The data image: a tool for exploring high dimensional data sets," *Proceedings of the ASA Section on Statistical Graphics*.

Montanari, Angela and Laura Lizzani. 2001. "A projection pursuit approach to variable selection," *Computational Statistics and Data Analysis*, **35**: pp. 463-473.

Montgomery, Douglas C., George C. Runger and Norma F. Hubele. 1998. *Engineering Statistics*, New York: John Wiley & Sons.

Mood, Alexander, Franklin Graybill and Duane Boes. 1974. *Introduction to the Theory of Statistics, Third Edition*, New York: McGraw-Hill Publishing.

Mooney, C. Z. 1997. *Monte Carlo Simulation*, London: Sage Publications.

Mooney, C. Z. and R. D. Duval. 1993. *Bootstrapping: A Nonparametric Approach to Statistical Inference*, London: Sage University Press.

Morant, G. M. 1923. "A first study of the Tibetan skull," *Biometrika*, **14**: pp. 193-260.

Morton, S. 1989. "Interpretable projection pursuit," Technical Report 106, Stanford University, Laboratory for Computational Statistics.

Mosteller, F. and J. W. Tukey. 1977. *Data Analysis and Regression: A Second Course in Statistics*, New York: Addison-Wesley.

Mosteller, F. and D. L. Wallace. *Inference and Disputed Authorship: The Federalist Papers*, New York: Addison-Wesley.

Murdoch, Duncan J. 2000. "Markov chain Monte Carlo," *Chance*, **13**: pp. 48-51.

Nadaraya, E. A. 1964. "On estimating regression," *Theory of Probability and its Applications*, **10**: pp. 186-190.

Nason, Guy. 1995. "Three-dimensional projection pursuit," *Applied Statistics*, **44**: pp. 411–430.

Norris, J. 1997. *Markov Chains*, Cambridge: Cambridge University Press.

Parzen, E. 1962. "On estimation of probability density function and mode," *Annals of Mathematical Statistics*, **33**: pp. 1065-1076.

Pearson, K. and A. Lee. 1903. "On the laws of inheritance in man. I. Inheritance of physical characters," *Biometrika*, **2**: pp. 357-462.

Pinder, D. A. and M. E. Witherick. 1977. "The principles, practice and pitfalls of nearest neighbor analysis," *Geography*, **57**: pp. 277–288.

Polansky, Alan M. 1999. "Upper bounds on the true coverage of bootstrap percentile type confidence intervals," *The American Statistician*, **53**: pp. 362-369.

Politis, D. N., J. P. Romano, and M. Wolf. 1999. *Subsampling*, New York: Springer-Verlag.

Port, Sidney C. 1994. *Theoretical Probability for Applications*, New York: John Wiley & Sons.

Posse, Christian. 1995a. "Projection pursuit exploratory data analysis," *Computational Statistics and Data Analysis*, **29**: pp. 669–687.

Posse, Christian. 1995b. "Tools for two-dimensional exploratory projection pursuit," *Journal of Computational and Graphical Statistics*, **4**: pp. 83–100.

Priebe, C. E. 1993. *Nonparametric maximum likelihood estimation with data-driven smoothing*, Ph.D. Dissertation, Fairfax, VA: George Mason University.

Priebe, C. E. 1994. "Adaptive mixture density estimation," *Journal of the American Statistical Association*, **89**: pp. 796-806.

Priebe, C. E., R. A. Lori, D. J. Marchette, J. L. Solka, and G. W. Rogers. 1994. "Nonparametric spatio-temporal change point analysis for early detection in mammography," *Proceedings of the Second International Workshop on Digital Mammography, SIWDM*, pp. 111-120.

Priebe, C. E. and D. J. Marchette. 2000. "Alternating kernel and mixture density estimates," *Computational Statistics and Data Analysis*, **35**: pp. 43-65.

Quenouille, M. 1949. "Approximate tests of correlation in time series," *Journal of the Royal Statistical Society, Series B*, **11**: pp. 18-44.

Quenouille, M. 1956. "Notes on bias estimation," *Biometrika*, **43**: pp. 353-360.

Rafterty, A. E. and V. E. Akman. 1986. "Bayesian analysis of a Poisson process with a change-point," *Biometrika*, **85**: pp. 85-89.

Raftery, A. E. and S. M. Lewis. 1992. "How many iterations in the Gibbs sampler?", in *Bayesian Statistics 4*, J. M. Bernardo, J. Berger, A. P. Dawid and A. F. M. Smith, eds., Oxford: Oxford University Press, pp. 763-773.

Raftery, A. E. and S. M. Lewis. 1996. "Implementing MCMC," in *Markov Chain Monte Carlo in Practice*, W. R. Gilks, S. Richardson, and D. J. Spiegelhalter, eds., London: Chapman and Hall, pp. 115-130.

Rao, C. R. 1993. *Computational Statistics*, The Netherlands: Elsevier Science Publishers.

Redner, A. R. and H. F. Walker. 1984. "Mixture densities, maximum likelihood and the EM algorithm," *SIAM Review*, **26**: pp. 195-239.

Ripley, B. D. 1976. "The second-order analysis of stationary point processes," *Journal of Applied Probability*, **13**: pp. 255-266.

Ripley, B. D. 1981. *Spatial Statistics*, New York: John Wiley & Sons.

Ripley, Brian D. 1996. *Pattern Recognition and Neural Networks*, Cambridge: Cambridge University Press.

Robert, C. P. 1995. "Convergence control techniques for Markov chain Monte Carlo algorithms," *Statistical Science*, **10**: pp. 231-253.

Robert, C. P. and G. Casella. 1999. *Monte Carlo Statistical Methods*, New York: Springer-Verlag.

Roberts, G. O. 1996. "Markov chain concepts related to sampling algorithms," in *Markov Chain Monte Carlo in Practice*, W. R. Gilks, S. Richardson, and D. J. Spiegelhalter, eds., London: Chapman and Hall, pp. 45-57.

Roberts, G. O. 2000. *Computer Intensive Methods*, Course Notes, Lancaster University, UK, www.maths.lancs.ac.uk/~robertgo/notes.ps.

Rohatgi, V. K. 1976. *An Introduction to Probability Theory and Mathematical Statistics* by New York: John Wiley & Sons.

Rohatgi, V. K. and A. K. Nd. Ehsanes Saleh. 2000. *An Introduction to Probability and Statistics*, New York: John Wiley & Sons.

Rosenblatt, M. 1956. "Remarks on some nonparametric estimates of a density function," *Annals of Mathematical Statistics*, **27**: pp. 832-837.

Ross, Sheldon. 1994. *A First Course in Probability, Fourth Edition*. New York: Macmillan College Publishing.

Ross, Sheldon. 1997. *Simulation, Second Edition*, New York: Academic Press.

Ross, Sheldon. 2000. *Introduction to Probability Models, Seventh Edition*, San Diego: Academic Press.

Rousseeuw, P. J. and A. M. Leroy. 1987. *Robust Regression and Outlier Detection*, New York: John Wiley & Sons.

Rousseeuw, P, J., I. Ruts, and J. W. Tukey. 1999. "The bagplot: A bivariate boxplot," *The American Statistician*, **53**: pp. 382-387.

Rubin, Donald B. 1987. "Comment on Tanner and Wong: The calculation of posterior distributions by data augmentation," *Journal of the American Statistical Association*, **82**: pp. 543-546.

Rubin, Donald B. 1988. "Using the SIR algorithm to simulate posterior distributions (with discussion)," in *Bayesian Statistics 3*, J. M. Bernardo, M. H. DeGroot, D. V. Lindley, and A. F. M. Smith, eds., Oxford: Oxford University Press, pp. 395-402.

Rubinstein, Reuven Y. 1981. *Simulation and the Monte Carlo Method*, New York: John Wiley & Sons.

Rutherford, E. and M. Geiger. 1910. "The probability variations in the distribution of alpha-particles," *Philosophical Magazine*, Series 6, **20**: pp. 698-704.

Safavian, S. R. and D. A. Landgrebe. 1991. "A survey of decision tree classifier methodology," *IEEE Transactions on Systems, Man and Cybernetics*, **21**: pp. 660-674.

Sasieni, Peter and Patrick Royston. 1996. "Dotplots," *Applied Statistics*, **45**: pp. 219-234.

Scott, David W. 1979. "On optimal and data-based histograms," *Biometrika*, **66**: pp. 605-610.

Scott, David W. 1985. "Frequency polygons," *Journal of the American Statistical Association*, **80**: pp. 348-354.

Scott, David W. 1992. *Multivariate Density Estimation: Theory, Practice, and Visualization*, New York: John Wiley & Sons.

Shao, J. and D. Tu. 1995. *The Jackknife and Bootstrap*, New York: Springer-Verlag.

Silverman, B. W. 1985. "Some aspects of the spline smoothing approach to nonparametric curve fitting," *Journal of the Royal Statistical Society, Series B*, **47**: pp. 1-52.

Silverman, B. W. 1986. *Density Estimation for Statistics and Data Analysis*, London: Chapman and Hall.

Simon, J. 1999. *Resampling: The New Statistics*, Arlington, VA: Resampling Stats, Inc.

Simonoff, J. S. 1996. *Smoothing Methods in Statistics*, New York: Springer-Verlag.

Snedecor, G. W. and G. C. Cochran. 1967. *Statistical Methods, Sixth Edition*, Ames: Iowa State University Press.

Snedecor, G. W. and G. C. Cochran. 1980. *Statistical Methods, Seventh Edition*, Ames: Iowa State University Press.

Solka, J., W. L. Poston, and E. J. Wegman. 1995. "A visualization technique for studying the iterative estimation of mixture densities," *Journal of Computational and Graphical Statistics*, **4**: pp. 180-198.

Solka, J. 1995. *Matching Model Information Content to Data Information*, Ph.D. Dissertation, Fairfax, VA: George Mason University.

Spath, Helmuth. 1980. *Cluster Analysis Algorithms for Data Reduction and Classification of Objects*, New York: Halsted Press.

Strang, Gilbert. 1988. *Linear Algebra and its Applications*, Third Edition, San Diego: Harcourt Brace Jovanovich.

Swayne, D. F., D. Cook, and A. Buja. 1991. "XGobi: Interactive dynamic graphics in the X window system with a link to S," *ASA Proceedings of the Section on Statistical Graphics*. pp. 1-8.

Tanner, Martin A. *Tools for Statistical Inference: Methods for the Exploration of Posterior Distributions and Likelihood Functions, Third Edition*, New York: Springer-Verlag.

Tapia, R. A. and J. R. Thompson. 1978. *Nonparametric Probability Density Estimation*, Baltimore: Johns Hopkins University Press.

Teichroew, D. 1965. "A history of distribution sampling prior to the era of the computer and its relevance to simulation," *Journal of the American Statistical Association*, **60**: pp. 27-49.

Terrell, G. R. 1990. "The maximal smoothing principle in density estimation," *Journal of the American Statistical Association*, **85**: p. 470-477.

Thisted, R. A. 1988. *Elements of Statistical Computing*, London: Chapman and Hall.

Tibshirani, R. 1988. "Variance stabilization and the bootstrap," *Biometrika*, **75**: pp. 433-444.

Tierney, L. 1994. "Markov chains for exploring posterior distributions (with discussion)," *Annals of Statistics*, **22**: pp. 1701-1762.

Tierney, L. 1996. "Introduction to general state-space Markov chain theory," in *Markov Chain Monte Carlo in Practice*, W. R. Gilks, S. Richardson, and D. J. Spiegelhalter, eds., London: Chapman and Hall, pp. 59-74.

Tinkler, K. J. 1971. "Statistical analysis of tectonic patterns in areal volcanism: the Bunyaruguru volcanic field in west Uganda," *Mathematical Geology*, **3**: pp. 335–355.

Titterington, D. M., A. F. M. Smith, and U. E. Makov. 1985. *Statistical Analysis of Finite Mixture Distributions*, New York: John Wiley & Sons.

Tripathi, R. C. and R. C. Gupta. 1988. "Another generalization of the logarithmic series and the geometric distribution," *Communications in Statistics - Theory and Methods*, **17**: pp. 1541-1547.

Tufte, E. 1983. *The Visual Display of Quantitative Information*, Cheshire, CT: Graphics Press.

Tufte, E. 1990. *Envisioning Information*, Cheshire, CT: Graphics Press.

Tufte, E. 1997. *Visual Explanations*, Cheshire, CT: Graphics Press.

Tukey, John W. 1958. "Bias and confidence in not quite large samples," *Annals of Mathematical Statistics*, **29**: pp. 614.

Tukey, John W. 1977. *Exploratory Data Analysis*, New York: Addison-Wesley.

Upton, G. and B. Fingleton. 1985. *Spatial Data Analysis by Example: Volume I: Point Pattern and Quantitative Data*, New York: John Wiley & sons.

Utts, Jessica. 1996. *Seeing Through Statistics*, New York: Duxbury Press.

van Oost, B. A., B. Veldhayzen, A. P. M. Timmermans, and J. J. Sixma. 1983. "Increased urinary $\beta$-thromoglobulin excretion in diabetes assayed with a modified RIA kit-technique," *Thrombosis and Haemostasis*, **9**: pp. 18-20.

Venables, W. N. and B. D. Ripley. 1994. *Modern Applied Statistics with S-Plus*, New York: Springer-Verlag.

Wadsworth, H. M. (ed.). 1990. *Handbook of Statistical Methods for Engineers and Scientists*, New York: McGraw-Hill.

Wainer, H. 1997. *Visual Revelations: Graphical Tales of Fate and Deception from Napoleon Bonaparte to Ross Perot*, New York: Copernicus/Springer-Verlag.

Walpole, R. E. and R. H. Myers. 1985. *Probability and Statistics for Engineers and Scientists*, New York: Macmillan Publishing Company.

Wand, M.P. and M. C. Jones. 1995. *Kernel Smoothing*, London: Chapman and Hall.

Watson, G. S. 1964. "Smooth regression analysis," *Sankhya Series A*, **26**: pp. 101-116.

Webb, Andrew. 1999. *Statistical Pattern Recognition*, Oxford: Oxford University Press.

Wegman, E. 1986. *Hyperdimensional Data Analysis Using Parallel Coordinates*, Technical Report No. 1, George Mason University Center for Computational Statistics.

Wegman, E. 1988. "Computational statistics: A new agenda for statistical theory and practice," *Journal of the Washington Academy of Sciences*, **78**: pp. 310-322.

Wegman, E. 1990. "Hyperdimensional data analysis using parallel coordinates," *Journal of the American Statistical Association*, **85**: pp. 664-675.

Wegman, E. and J. Shen. 1993. "Three-dimensional Andrews plots and the grand tour," *Proceedings of the 25th Symposium on the Interface*, pp. 284-288.

Wegman, E., D. Carr, and Q. Luo. 1993. "Visualizing multivariate data," in *Multivariate Analysis: Future Directions*, C. R. Rao, ed., The Netherlands: Elsevier Science Publishers, pp. 423-466.

Weiss, Neil. 1999. *Introductory Statistics*, New York: Addison Wesley Longman.

Wilcox, Rand R. 1997. *Introduction to Robust Estimation and Hypothesis Testing*, New York: Academic Press.

Wilk, M. and R. Gnanadesikan. 1968. "Probability plotting methods for the analysis of data," *Biometrika*, **55**: pp. 1-17.

Wilkinson, Leland. 1999. *The Grammar of Graphics*, New York: Springer-Verlag.

# *Index*